Yanpei Liu
Topological Theory of Graphs

Also of Interest

Algebraic Graph Theory
Morphisms, Monoids and Matrices
Knauer, 2011
ISBN 978-3-11-025408-2, e-ISBN 978-3-11-025509-6

Introduction to Topology
Yan, 2016
ISBN 978-3-11-037815-3, e-ISBN 978-3-11-037816-0

Discrete Algebraic Methods
Arithmetic, Cryptography, Automata and Groups
Diekert, Kufleitner, Rosenberger, Hertrampf, 2016
ISBN 978-3-11-041332-8, e-ISBN 978-3-11-041333-5

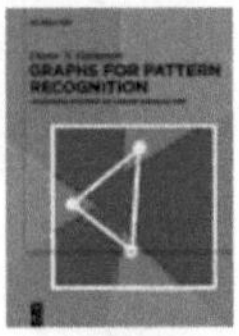

Graphs for Pattern Recognition
Infeasible Systems of Linear Inequalities
Gainanov, 2016
ISBN 978-3-11-048013-9, e-ISBN 978-3-11-048106-8

Generalized Network Design Problems
Modeling and Optimization
Pop, 2012
ISBN 978-3-11-026758-7, e-ISBN 978-3-11-026768-6

Yanpei Liu

Topological Theory of Graphs

DE GRUYTER

Author
Prof. Dr. Yanpei Liu
Department of Mathematics
School of Science
Beijing Jiaotong University
Haidian District
3 Shangyuancun
100044 Beijing
People's Republic of China
ypliu@bjtu.edu.cn

ISBN 978-3-11-047669-9
e-ISBN (PDF) 978-3-11-047949-2
e-ISBN (EPUB) 978-3-11-047922-5
Set-ISBN 978-3-11-047950-8

Library of Congress Cataloging-in-Publication Data
A CIP catalog record for this book has been applied for at the Library of Congress.

Bibliographic information published by the Deutsche Nationalbibliothek
The Deutsche Nationalbibliothek lists this publication in the Deutsche Nationalbibliografie; detailed bibliographic data are available on the Internet at http://dnb.dnb.de.

Typesetting: Integra Software Services Pvt. Ltd.
Printing and binding: CPI books GmbH, Leck
Cover image: ktsimage/iStock/thinkstock
♾ Printed on acid-free paper
Printed in Germany

www.degruyter.com

Preface to DG Edition

The De Gruyter (DG) edition is based on my previous monograph *Topological Theory on Graphs*, published by the University of Science and Technology of China (USTC) Press in 2008, with updates and improvements from two main sources.

One is that the new developments of four ways, with five pairs of theorems, enable us to get criteria, for determining the embeddability of a graph, on a surface (orientable and non-orientable) of genus arbitrarily given, as shown in Liu [229, 231, 232, 236]. Before them, only few results were known for surfaces of small genera ($\neq 0$), such as Abrams and Slilaty [1], Archdeacon [11], Archdeacon and Huneke [12], Archdeacon and Siran [14], Glover and Huneke [100], Glover et al. [101].

Notably, for the specific case of genus zero, one of the five pairs leads to the criteria for planarity, as given in the pair of Theorems 4.2.5 and 4.3.2 relating to a pair of homology and cohomology. They have a number of corollaries, including the three theorems obtained by Lefschetz [171], MacLane [261] and Whitney [392] for the planarity of a graph at a time. This causes item 4.5.10, in the Notes Section 4.5 of Chapter 4. Subsequently, item 4.5.9 is completed in the core theoretical stage of the three stages: theoretization, efficientization and intelligentization, involving with my research.

The other is the progresses in applications and usages of joint tree model described in Section 9.1.

Each Notes section in chapters is accompanied by at least one new item. I would like to mention the following:

Notes Section 1.5 in Chapter 1. Item 1.5.5 was provided for reminding readers of the universality of vector space, sketched in Section 1.4 as an abstract linear space, motivated from the background in Liu [237]. Item 1.5.6 is for accessing the efficiency of theoretical results in this book, or polynomial complexity as shown in Cook [54] and Karp [159].

Notes Section 2.6 in Chapter 2. Item 2.6.6 was presented for perspective developments available in the theory of polyhedra, shown in Liu [230, 233, 234, 238], with relevant references as complimentary for the reader.

Notes Section 3.7 in Chapter 3. Formula (3.7.1) was put into item 3.7.3 to show that the topological classification of surfaces can be done, via only the three types of transformations. Item 3.7.9 was provided to enhance readers' understanding, intrinsically from topology.

Notes Section 4.5 in Chapter 4. Further to item 4.5.10 mentioned earlier, item 4.5.11 illustrates one of the new approaches, to investigate the structure of cycle spaces in a graph, via an example.

Notes Section 5.6 in Chapter 5. Item 5.6.7 was suggested to generalize the polyhedral form, from Jordan curve axiom to the surface closed curve axiom, for recognizing whether, or not, a graph can be embedded onto a surface of given genus not zero.

Notes Section 6.5 in Chapter 6. Item 6.5.12 shows that the relationship among graphs, polyhedra, embedding and maps can be clarified via symmetries.

Notes Section 7.5 in Chapter 7. Item 7.5.6 enables us to go in a new way, for classifying knots, or links, by observing the relationship between embeddings of general networks with binary weight of edges on surfaces and knots, or links, via the correspondence between a 4-regular graph and a pair of two general graphs mutually dual.

Notes Section 8.6 in Chapter 8. Item 8.6.10 presents a theoretical framework, inspired from the pair of homology and cohomology in Sections 4.2 and 4.3, to detect a type of homology on a graph, as the dual of cohomology in Theorem 8.2.1, and to establish a new pair of criteria for the embeddability of a graph, on a surface of genus arbitrarily given, via the polyhedral theory, described in Liu [227].

Notes Section 9.5 in Chapter 9. Items 9.5.5–9.5.10 reflect a series of progresses, for determining the up-embeddability on surfaces, handle (orientable genus) and crosscap (momorientable genus) polynomials, maximum genus, genus (minimum!), average genus, etc., of graphs, or digraphs, based on the joint tree model described in Section 9.1.

Notes Section 10.5 in Chapter 10. Recent result on the planarity of a graph, by a single forbidden configuration, was mentioned in item 10.5.5, as shown in Ref. [238].

Notes Sections 11.5 and 12.5 in, respectively, in Chapters 11 and 12. Both items 11.4.9 and 12.5.8 indicate the reason why the minors are not available as a forbidden configuration, for the properties considered with the inheritness.

Notes Section 13.8 in Chapter 13. Both items 13.8.8 and 13.8.9 reflect new progresses on minimality and maximality on graphs with surfaces, based on joint tree model.

Notes Section 14.6 in Chapter 14. Item 14.6.6 shows a new approach, hopefully to recognize whether, or not, a regular matroid is graphic, or cographic, to strengthen and expand Theorems 14.4.1 and 14.5.1.

In Notes Section 15.6 in Chapter 15. Item 15.6.8 provides a theoretical framework to characterize whether, or not, two knots, or two links, are in the same class of pan-polynomial equivalence on the basis of Section 15.3.

In addition, Theorem 13.5.1 was improved and revised. Proofs of certain conclusions are completed and concise, or more accurate, such as in Lemma 3.1.2, Theorem 5.4.1, Lemma 5.5.1, Theorem 13.5.1, etc.

Last but not least, I would like to take this opportunity to express my sincere thanks to Rongxia Hao, Erling Wei and Liangxia Wan for their careful reading of the manuscript with corrections on grammar. Some of researches were partially supported by NNSFC under Grant No. 11371052.

Y. P. Liu

Beijing
September 2016

Preface to USTC Edition

The subject of this book reflects new developments established mainly by the author himself in company with a few cooperators, most of them being his former and present graduate students in the foundation, as mentioned in Liu [216, 217]. The central idea is to extract suitable parts of a topological object such as a graph which is not necessarily to be with symmetry, as linear spaces which are all with symmetry for exploiting global properties in construction of the objects. This is a way of combinatorizations and further algebraications of an object via relationship among their subspaces.

Graphs are dealt with three vector spaces over GF(2) generated by 0 (dimensional)-cells, 1 (dimensional)-cells and 2 (dimensional)-cells, with the finite field of order 2. The first two spaces were known from, e.g., Lefschetz [172] by taking 0-cells and 1-cells as, respectively, vertices and edges. Of course, a graph is only a 1-complex without two cells.

Since the 1950s, in Wu [402] and Tutte [335, 346], the chain groups generated by 0-cells and 1-cells over, respectively, GF(2) and the real field were independently used for describing a graph. And they both, after ten years, adopted non-adjacent pair of edges as a 2-cell for which the cohomology on a graph was allowed to be established.

Their results especially in Wu [402–406] enabled the author to create a number of types of planarity auxiliary graphs induced from the graph considered for the study of the efficiency of theorems in Liu [192, 193, 202, 205, 225] as one approach. Another approach can be seen in Liu [206–208, 226].

More interestingly, two decades after Liu [192], in Archdeacon and Siran [14], a theta graph (network) was used for characterizing the planarity of a given graph. The theta graph can be seen to be a type of planarity auxiliary graph (network) because planarity auxiliary graphs are subgraphs of the theta graph. However, in virtue of the order of theta network upper bounded by a exponential function of the size of given graph and that of planarity auxiliary network by a quadratic polynomial of the size of given graph, theorems deduced from a theta network are all without efficiency while those from planarity auxiliary graphs are all with efficiency.

The effects of planarity auxiliary graphs are reflected in Chapters 8, 10, 11, 12 and 13 with a number of extensions.

On the other hand, in Liu [214] a graph was dealt with a set of polyhedra via double covering the edge set by travels under certain condition so that travels were treated as 2-cells. These enable us to discover homology and another type of cohomology for showing the sufficiency of Eulerian necessary condition in this circumstance. Further, all the results for the planarity of a graph in Whitney [392] on the duality, MacLane [261, 262] on a circuit basis and Lefschetz [171] on a circuit double covering have a universal view in this way. In fact, our polyhedra are all on such surfaces, i.e., 2-dimensional compact manifolds without boundary. If a boundary is allowed on a surface, the Eulerian necessary condition is not always sufficient in general. Some people used to miss the boundary condition.

The effects of this theoretical thinking are reflected in Chapters 4, 5, 7 and 14.

Because of the clarification of the joint tree model of a polyhedron in Liu [218, 219] by the present author recently on the basis of Liu [195, 196], we are allowed to write a brief description on the theories of surfaces and polyhedra in Chapters 2 and 3 and related topics in Chapters 6, 9 and 15.

Although quotient embeddings (current graph and its dual, voltage graph) were quite active in constructing an embedding of a graph on a surface with its genus minimum in a period of decades, this book has no space for them. One reason is that some writers such as White [382], Ringel [288] and Liu [216, 217], etc., have already mentioned them. Another reason is that only graphs with higher symmetry are suitable for quotient embeddings, or for employing the covering space method, whence this book is for general graphs without such a limitation of symmetry.

In spite of refinements and simplifications for known results, this book still contains a number of new results, for example Section 5.2, the sufficiency in the proof of Theorem 5.2.1, Sections 9.4, 11.3 and 11.4, 13.1 and 13.2, 13.4 and 13.5 etc., only to name a few. Researches were partially supported by the NNSF in China under Grant No. 60373030 and No. 10571013.

Y. P. Liu

Beijing
December 2007

Contents

1 Preliminaries

Throughout, for the sake of brevity, the usual logical conventions are adopted: disjunction, conjunction, negation, implication, equivalence, universal quantification and existential quantification denoted, respectively, by the familiar symbols:$\vee$, $\wedge$, $\neg$, $\Rightarrow$, $\Leftrightarrow$, $\forall$ and $\exists$. And, *x.y* is for the section *y* in Chapter *x*.

In the context, (*i.j.k*) refers to item *k* of section *j* in Chapter *i*.

1.1 Sets and relations

A *set* is a collection of objects with some common property, which might be numbers, points, symbols, letters or whatever even sets except itself to avoid paradoxes. The objects are said to be *elements* of the set. We always denote elements by italic lower case letters and sets by upper case letter. The statement "x is (is not) an element of M" is written as $x \in M(x \notin M)$. A set is often characterized by a property. For example,

$$M = \{x \mid x \leq 4, \text{positive integer}\} = \{1, 2, 3, 4\}.$$

The *cardinality* of a set M (or the number of elements of M if finite) is denoted by $\mid M \mid$.

Let A, B be two sets. If $(\forall a)\ (a \in A \Rightarrow a \in B)$, then A is said to be a *subset* of B which is denoted by $A \subseteq B$. Further, we may define the three main operations: union, intersection and subtraction, respectively, as $A \cup B = \{x \mid (x \in A) \vee (x \in B)\}$, $A \cap B = \{x \mid (x \in A) \wedge (x \in B)\}$ and $A \setminus B = \{x \mid (x \in A) \wedge (x \notin B)\}$.

If $B \subseteq A$, then $A \setminus B = A - B$ is denoted by $\overline{B}(A)$, which is said to be the *complement* of B in A. If all the sets are considered as subsets of Ω, then the complement of A in Ω is simply denoted by $\overline{A}$. The *empty* denoted by $\emptyset$ is the set without element. For those operations on subsets of Ω mentioned above, we have the following laws:

Idempotent law $\forall A \subseteq \Omega, A \cap A = A \cup A = A.$

Commutative law $\forall A, B \subseteq \Omega, A \cup B = B \cup A; A \cap B = B \cap A.$

Associative law $\forall A, B, C \subseteq \Omega, A \cup (B \cup C) = (A \cup B) \cup C; A \cap (B \cap C) = (A \cap B) \cap C.$

Absorption law $\forall A, B \subseteq \Omega, A \cap (A \cup B) = A \cup (A \cap B) = A.$

Distributive law $\forall A, B, C \subseteq \Omega, A \cup (B \cap C) = (A \cup B) \cap (A \cup C); A \cap (B \cup C) = (A \cap B) \cup (A \cap C).$

Universal bound law $\forall A \subseteq \Omega, \emptyset \cap A = \emptyset,\ \emptyset \cup A = A; \Omega \cap A = A, \Omega \cup A = \Omega.$

Unary complement law $\forall A \subseteq \Omega, A \cap \overline{A} = \emptyset; A \cup \overline{A} = \Omega.$

The unary complement law is also called the *excluded middle law* in logic.

From the laws described earlier, we may obtain a large number of important results. Here, only a few are listed for usage in this context.

DOI 10.1515/9783110479492-001

Theorem 1.1.1. $\forall A \subseteq \Omega$,

$$\begin{cases} (\forall X \subseteq \Omega)\Big((A \cap X = A) \vee (A \cup X = X)\Big) \\ \qquad \Rightarrow A = \emptyset; \\ (\forall X \subseteq \Omega)\Big((A \cap X = X) \vee (A \cup X = A)\Big) \\ \qquad \Rightarrow A = \Omega. \end{cases} \tag{1.1.1}$$

Theorem 1.1.2. $\forall A, B \subseteq \Omega$,

$$A \cap B = A \Leftrightarrow A \cup B = B. \tag{1.1.2}$$

Theorem 1.1.3. $\forall A, B, C \subseteq \Omega$,

$$(A \cap B = A \cap C) \wedge (A \cup B = A \cup C) \Leftrightarrow B = C. \tag{1.1.3}$$

Theorem 1.1.4. $\forall A \subseteq \Omega$,

$$\overline{\overline{A}} = A. \tag{1.1.4}$$

Theorem 1.1.5. $\forall A, B \subseteq \Omega$,

$$\overline{A \cup B} = \overline{A} \cap \overline{B}; \overline{A \cap B} = \overline{A} \cup \overline{B}. \tag{1.1.5}$$

From those described above, it is seen that $\overline{\emptyset} = \Omega$ and $\overline{\Omega} = \emptyset$. Further, the symmetry (or duality) that any proposition related to $\cup, \cap, \emptyset, \Omega$ can be transformed into another by interchanging $\cup$ and $\cap$, $\emptyset$ and Ω.

For $A, B \subseteq \Omega$, an *injection* (or 1-to-1 correspondence) between A and B is a mapping $\alpha : A \to B$, such that $\forall a, b \in A,\ a \neq b \Rightarrow \alpha(a) \neq \alpha(b)$. A *surjection* between A and B is a mapping $\beta : A \to B$, such that $(\forall b \in B)(\exists a \in A)(\beta(a) = b)$. If a mapping is both an injection and a surjection, then it is called a *bijection*. Two sets are said to be *isomorphic* if there is a bijection between them. Two isomorphic sets A and B, or write $A \sim B$, are always treated as the same. Of course, for finite sets, it is trivial to justify if two sets are isomorphic by the fact: $\forall A, B \subseteq \Omega,\ A \sim B \Leftrightarrow \mid A \mid = \mid B \mid$.

For a set M, let $M \times M = \{\prec x, y \succ \mid \forall x, y \in M\}$ which is said to be the *Cartesian product* of M. Here, $\prec x, y \succ \neq \prec y, x \succ$ in general.

A *binary relation* R on M is a subset of $M \times M$. The adjective “binary” of the relation will often be omitted in the context. If the relation R holds for $x, y \in M$, then we write $\prec x, y \succ \in R$, or xRy. An *order*, denoted by $\preceq$, is a relation R which satisfies the following three laws:

Reflective law $\forall x \in M, xRx$.

Antisymmetry law $\forall x, y \in M, xRy \wedge yRx \Rightarrow x = y$.

Transitive law $\forall x, y, z \in M, xRy \wedge yRz \Rightarrow xRz$.

The set M with the order $\preceq$ is said to be a *poset* (or partial order set) denoted by $(M, \preceq)$.

Theorem 1.1.6. *In a poset $(M, \preceq)$, $\forall x_1, x_2, \ldots, x_n \in M$,*

$$x_1 \preceq x_2 \preceq \cdots \preceq x_n \preceq x_1 \Rightarrow x_1 = x_2 = \cdots = x_n. \tag{1.1.6}$$

The theorem is sometimes called the *anti-circularity law*. If a relation only satisfies Reflective law and Transitive law but not Anti-symmetry law, then it is called the *quasi-order*, which is denoted by $\bullet\prec$. A set M with $\bullet\prec$ is said to be a *quoset* denoted by $(M, \bullet\prec)$.

Theorem 1.1.7. *Any subset S of a quoset $(M, \bullet\prec)$ is itself a quoset with the restriction of the quasi-order to S.*

If a quasi-order R on M satisfies the symmetry law described below, then it is called an *equivalent relation*, or simply an *equivalence* denoted by $\sim$.

Symmetry law $\forall x, y \in M, xRy \Rightarrow yRx$.

For the equivalence $\sim$ on M, we are allowed to define the set $x(M) = \{y \mid \forall y \in M, y \sim x\}$, which is said to be the *equivalent class* for $x \in M$. The set that consists of all the equivalent classes is called the *quotient set* of $(M, \sim)$ denoted by $M/\sim$. In a quoset $(M, \bullet\prec)$, let $\sim_{\bullet\prec}$ be defined by

$$\forall x, y \in M, x \sim_{\bullet\prec} y \Leftrightarrow (x \bullet\prec y) \wedge (y \bullet\prec x). \tag{1.1.7}$$

Then, it is seen that $\sim_{\bullet\prec}$ is an equivalence on M and that $(M/\sim_{\bullet\prec}, \bullet\prec)$ is also a quoset.

Theorem 1.1.8. *A quoset $(M, \bullet\prec)$ is a poset if, and only if, $M/\sim_{\bullet\prec} = M$, or say, it satisfies the anti-circularity law.*

In a poset $(M, \preceq)$, we define the *strict inclusion* , denoted by $\prec$, of the order by the *anti-reflective law*: $\neg x \in M, x \prec x$ and the *transitive law*: $(x \prec y) \wedge (y \prec z) \Rightarrow x \prec z$ while noticing that $x \preceq y \Leftrightarrow (x \prec y) \vee (x = y)$. If an order $\preceq$ on M satisfies the alternative law described below, then it is called a *total order*, or a *linear order*.

Alternative law $\forall x, y \in M, x \npreceq y \Rightarrow y \preceq x$.

A set with a total order is said to be a *chain*. The *length* of a chain with n elements is defined to be $n - 1$. From Theorem 1.1.7 and the definitions, we may have

Theorem 1.1.9. *Any subset of a poset is a poset and any subset of a chain is a chain.*

The *converse* of a relation R on M is, by definition, the relation $R^* : \forall x, y \in M, xR^*y \Leftrightarrow yRx$. It is obvious from inspection of the three laws for order to have

Theorem 1.1.10 (Duality principle). *The converse of any order is itself an order.*

In a poset $(M, \preceq)$, there may have an element $a : \forall x \in M, a \preceq x$. Because of Anti-symmetry law, such an element, if it exists, is a unique one which is called the *least element* denoted by O. In a dual case, the *greatest element*, if it exists, is denoted by I. The elements O and I, when they exist, are called *universal bounds* of the poset.

Theorem 1.1.11. *A chain has the universal bounds if it is finite.*

In a poset $(M, \preceq)$, an element $a \in M : \forall x \in M, x \preceq a \Rightarrow x = a$ is called a *minimal element*. Dually, a *maximal element* is defined as $a \in M : \forall x \in M, a \preceq x \Rightarrow a = x$.

Theorem 1.1.12. *Any finite non-empty poset $(M, \preceq)$ has minimal and maximal elements.*

A mapping $\tau : M \to N$ from a poset $(M, \preceq)$ to a poset $(N, \preceq)$ is called *order preserving*, or *isotone* if it satisfies

$$\forall x, y \in M,\ x \preceq y \Leftrightarrow \tau(x) \preceq \tau(y). \tag{1.1.8}$$

Further, if an isotone $\tau : M \to N$ satisfies

$$\forall x, y \in M, \tau(x) \preceq \tau(y) \Rightarrow x \preceq y, \tag{1.1.9}$$

then it is called an *isomorphism*. Two posets $(M, \preceq)$ and $(N, \preceq)$ are said to be *isomorphic*, that is $(M, \preceq) \cong (N, \preceq)$, if there is an isomorphism between them. All isomorphic posets are treated as the same. However, it is not trivial as for sets to justify if two posets are isomorphic in general.

An *upper bound* of a subset X of a poset $(M, \preceq)$ is an element $a : \forall x \in X, x \preceq a$. The *least upper bound* (or l.u.b.) is an upper bound $b : a \preceq b \Rightarrow a = b$, where a is another upper bound of X. Dually, a *lower bound* and the *greatest lower bound* (g.l.b.). The *length* of a poset is the l.u.b. of the lengths of chains in the poset. A *lattice* is a poset if any two x and y of whose elements has a g.l.b. or *meet* denoted by $x \wedge y$ and an l.u.b. or *join* denoted by $x \vee y$. A lattice $L = (M, \preceq; \vee, \wedge)$ is *complete* if each of its subset X has an l.u.b. and a g.l.b.. Moreover, we have known that all finite-length lattices are complete.

Let 2^{Ω} be the set that consists of all subsets of Ω. From Section 1.1, we may see that $(2^{\Omega}, \subseteq; \cup, \cap)$ is a lattice. In fact, we have

Theorem 1.1.13. *A poset is a lattice if, and only if, it satisfies the idempotent, commutative, associative and absorption laws.*

Two lattices $(M, \preceq; \vee, \wedge)$ and $(N, \preceq; \vee, \wedge)$ are *isomorphic* if there is an isomorphism τ between $(M, \preceq)$ and $(M, \preceq)$ such that $\forall x, y \in M$,

$$(\tau(x \vee y) = \tau(x) \vee \tau(y)) \wedge (\tau(x \wedge y) = \tau(x) \wedge \tau(y)). \tag{1.1.10}$$

Of course, it is non-trivial as well to justify if two lattices are isomorphic in general.

1.2 Partitions and permutations

A *partition* of a set X is such a set of subsets of X that any two subsets are without common element and the union of all the subsets is X.

Theorem 1.2.1. *A partition $P(X)$ of a set X determines an equivalence on X such that the subsets in $P(X)$ are the equivalent classes.*

Let $P(X) = \{p_1, p_2, \dots, p_{k_1}\}$ and $Q(X) = \{q_1, q_2, \dots, q_{k_2}\}$ be two partitions of X. If for any q_j, $1 \leq j \leq k_1$, there exists a p_i, $1 \leq i \leq k_2$ such that $q_j \subset p_i$, then $Q(X)$ is called a *refinement* of $P(X)$ and $P(X)$, an *enlargement* of $Q(X)$ except only for $P(X) = Q(X)$. The partition of X with each subset of a single element, or only one subset which is X in its own right is, respectively, called the 0-*partition*, or 1-*partition* and denoted by $0(X)$, or $1(X)$.

Theorem 1.2.2. *For a set X and its partition $P(X)$, the* 0-*partition* $0(X)$ *(or* 1-*partition* $1(X)$*) can be obtained by refinements (or enlargements) for at most $O(\log |X|)$ times in the worst case.*

Proof. In the worst case, it suffices to consider $P(X) = 1(X)$(or $0(X)$) and only one more subset produced in a refinement. Because of

$$1 + 2 + 2^2 + \dots + 2^{\log |X|} = \frac{2^{1+\log |X|} - 1}{2 - 1} = O(|X|), \tag{1.2.1}$$

the times of refinements (or enlargements) needed for getting $0(X)$ (or $1(X)$) is $O(\log |X|)$. The theorem is obtained. □

For two partitions $P = \{p_1, p_2, \dots, p_s\}$ and $Q = \{q_1, q_2, \dots, q_t\}$ of a set X, the *family intersection* of P and Q is defined to be

$$P \cap Q = \bigcup_{i=1}^{s} \{p_i \cap q_1, p_i \cap q_2, \dots, p_i \cap q_t\}. \tag{1.2.2}$$

Actually, $\{p_i \cap q_1, p_i \cap q_2, \dots, p_i \cap q_t\}$ for $i = 1, 2, \dots, l$ are partitions of p_i.

Theorem 1.2.3. *The family intersection satisfies the commutative and associate laws. And further, $P \cap Q$ is a refinement of both P and Q.*

A *permutation* of a set X is a bijection of X to itself. Because elements in a set are no distinction, they are allowed to be distinguished by natural numbers as $X = \{x_1, x_2, \dots\}$, or simply $X = \{1, 2, \dots\}$. So, a permutation of set $L = \{1, 2, \dots, l\}$ can be expressed as

$$\begin{pmatrix} 1 & 2 & 3 & \dots & l \\ i_1 & i_2 & i_3 & \dots & i_l \end{pmatrix}. \tag{1.2.3}$$

If $i_j = j$ for all $1 \le j \le l$, the permutation is called the *identity*. From Theorem 1.1.4, the identity is unique.

Theorem 1.2.4. *Let π be a permutation of set $L = \{1, 2, \dots, l\}$, then for any $i \in L$ there is an integer $n \ge 0$ such that $p^n i = i$.*

Proof. By contradiction. If there is no such an integer, by the 1–to–1 property it is a contradiction to the finiteness of l. □

On the basis of this theorem, the set $X_i = \{i, \pi i, \pi^2 i, \dots, \pi^{n-1} i\}$ is called the *orbit* of i. Because any element in X_i has the same orbit as i, it can also be called an orbit of π, denoted by $\mathrm{Orb}_\pi\{i\}$, or simply $\{i\}_\pi$. Because any two orbits of a permutation are either same or disjoint, all orbits form a partition of L.

An orbit with the order in its own right is called a *cyclic permutation*, or in brief, a *cycle*. The cycle corresponding to $\mathrm{Orb}_\pi\{i\}$ is denoted by $\mathrm{Orb}_\pi(i)$, or simply $(i)_\pi$. Because of the disjointness among orbits, by considering that the composite of disjoint cycles satisfies the commutative law and the associate law, a permutation can always be expressed as a product of cycles. The *order* of a cycle is one greater than its *length*, i.e., the number of elements in the cycle. A cycle of order 1 is called a *fixed point* of the permutation. All the fixed points in a permutation are always omitted in its cyclic expression.

As an example,

$$\begin{pmatrix} 1\,2\,3\,4\,5\,6\,7 \\ 2\,5\,1\,6\,3\,4\,7 \end{pmatrix} = (1, 2, 5, 3)(4, 6)(7)$$
$$= (1, 2, 5, 3)(4, 6).$$

However, the product of two cycles with a common element is not commutative in general. For example, $P_1 = (1, 3, 2)$ and $P_2 = (1, 2, 4)$,

$$P_1P_2 = (2, 4, 3) \neq (1, 3, 4) = P_2P_1.$$

Because $C^{sk} = 1$ on the order k of a cycle C and any positive integer s, it can be seen from Theorem 1.2.4 that if permutation $\pi = C_1C_2 \cdots C_n$, where C_i, $1 \leq i \leq n$, are all the disjoint cycles of order n_i, then π has its order $[l_1, l_2, \ldots, l_n]$, the least common multiple ($\mathrm{lcm}\{l_1, l_2, \ldots, l_n\} = [l_1, l_2, \ldots, l_n]$) of $l_1, l_2, \ldots$, and l_n.

Theorem 1.2.5. *The unique inverse of a cycle* $C = (c_1, c_2, \ldots, c_n)$ *is the cycle* $C^{-1} = (c_n, c_{n-1}, \ldots, c_1)$. ■

Let $\Sigma_{|L|}$ be the set of all permutations on L. The cardinality of L is also called the *degree* of permutations. For two permutations π and σ in $\Sigma_{|L|}$, if there is a permutation $\rho \in \Sigma_{|L|}$ such that $\pi = \rho\sigma\rho^{-1}$, then π and σ are *conjugates* for ρ.

Let $\gamma = (x_1, x_2, \ldots, x_r)$ be a cycle and τ, another permutation in $\Sigma_{|L|}$. For $y \in L$, if $x = \tau^{-1}y$ is not in γ, then $\tau\gamma\tau^{-1}y = \tau x = \tau(\tau^{-1}y) = y$. Otherwise, if $x = \tau^{-1}y = x_i (1 \leq i \leq r)$, then $\tau\gamma\tau^{-1}y = \tau x_i = x_{i+1}$. This implies that

$$\begin{aligned} \tau\gamma\tau^{-1} &= \tau(x_1, x_2, \ldots, x_r)\tau^{-1} \\ &= (\tau x_1, \tau x_2, \ldots, \tau x_r). \end{aligned} \tag{1.2.4}$$

For $\pi \in \Sigma_{|L|}$, let $c(\pi)$ be the number of cycles in its cyclic partition and l_i, the number of cycles of length i, $1 \leq i \leq c(\pi)$. The *cyclic type* of permutation π is defined to be the decreased sequence of l_i, $1 \leq i \leq c(\pi)$.

Theorem 1.2.6. *Two permutations are conjugate if, and only if, they have a same cyclic type.*

Proof. The necessity is obvious because of eq. (1.2.4) for cyclic partition representation of permutations. Conversely, for any two permutations with a same cyclic type, assume with one cycle each without generality as $\pi = (x_1, x_2, \ldots, x_r)$ and $\sigma = (y_1, y_2, \ldots, y_r)$, it is seen from eq. (1.2.4) that let

$$\tau = \begin{pmatrix} x_1\ x_2 \ldots x_r \\ y_1\ y_2 \ldots y_r \end{pmatrix},$$

then $\tau\pi\tau^{-1} = \sigma$. Therefore, π and σ are conjugates. □

Two particular cases should be mentioned for conjugate pair $\{\pi, \sigma\}$ of permutations. One is for $\pi = \sigma$ and the other, $\pi = \sigma^{-1}$. The former is called *self-conjugate* and the later, *inverse conjugate*. If $\pi = (x_1, x_2, \dots, x_r)$ and $\sigma = (y_1, y_2, \dots, y_r)$, then the self-conjugate is only for $\tau = 1_r$, the identity of degree r and the inverse conjugate is for

$$\tau = (x_1, x_r)(x_2, x_{r-1}) \dots (x_{\lfloor r/2 \rfloor}, x_{\lfloor r/2 \rfloor + 1}).$$

Let $D = \langle a_1, a_2, \dots, a_d \rangle$ be a set with linear order $a_1 < a_2 < \cdots < a_d$. An ordered pair $\langle a_i, a_j \rangle$ is called an *inversion* if $1 \le j < i \le d$. Let $\mathrm{sgn}(\pi)$ denote the total number of inversions in the sequence $\prec x_1, x_2, \dots, x_d \succ$ with linear order $x_1 \prec x_2 \prec \dots \prec x_d$ for

$$\pi = \begin{pmatrix} d_1 \ d_2 \dots d_r \\ x_1 \ x_2 \dots x_r \end{pmatrix}.$$

The permutation π is said to be *even* or *odd* accordingly as $\sin(\pi)$ is even or odd. The mapping $(-1)^{\mathrm{sgn}(\pi)}$ from a permutation π to $\{1, -1\}$ is called the *parity* of π. A cycle of length 2 is called *transposition*. A transposition (x_i, x_j), assume $x_i < x_j$ and $i < j$ without loss of generality, is always an odd permutation because of odd number of inversions as $\langle x_j, x_i \rangle$ with pairs (x_j, x_k) and (x_k, x_i) for $i < k < j$.

By observing that a cycle

$$\begin{aligned}(a_1, a_2, \dots, a_l) = (a_1, a_l)(a_1, a_{l-1}) \cdots \\ (a_1, a_3)(a_1, a_2),\end{aligned} \tag{1.2.5}$$

any permutation can be represented by a composite of transpositions.

Because for $1 \le j < k < l$,

$$(a_j, a_{k+1}) = (a_k, a_{k+1})(a_j, a_k)(a_k, a_{k+1}), \tag{1.2.6}$$

the transposition representations of a permutation may have different numbers of transpositions. A transposition in form as a_i, a_{i+1}, $1 \le i < l$, is said to be *adjacent*.

Theorem 1.2.7. *Any permutation π of degree at least 2 has an adjacent transposition representation of the same congruent number of transpositions modulo 2 as $\mathrm{sgn}(\pi)$.*

Proof. First, we show the existence of such a representation. By virtue of eqs. (1.2.5) and (1.2.6), an adjacent transposition representation can be found. Then, by considering that a transposition and the two sides of eq. (1.2.6) have all an odd number of inversions, such a representation has its total number of inversions the congruent number of transpositions as $\mathrm{sgn}(\pi)$. □

Theorem 1.2.8. *For any two permutations π and σ,*

$$\mathrm{sgn}(\pi\sigma) = \mathrm{sgn}(\pi) + \mathrm{sgn}(\sigma) \pmod 2. \tag{1.2.7}$$

Proof. Since each transposition involves odd number of inversions, from Theorem 1.2.7, expression (1.2.7) holds. □

By virtue of eq. (1.2.6), we have

$$(-1)^{\mathrm{sgn}(\pi\sigma)} = (-1)^{\mathrm{sgn}(\pi)}(-1)^{\mathrm{sgn}(\sigma)}. \tag{1.2.8}$$

i.e., the parity of composite of two permutations is the product of their parities.

Theorem 1.2.9. *All transposition representations of a permutation have the same parity of the permutation.*

Proof. A direct conclusion of Theorem 1.2.8 in the case that one of π and σ is the identity. □

1.3 Graphs and networks

A *graph* denoted by $G = (V, E)$ is a set V, the *vertex set* whose elements are called *vertices*, with a binary relation $E \subseteq V * V = \{(u, v) \mid \forall u, v \in V, u \neq v\}$. Here, $(u, v) = (v, u)$. E is said to be an *edge set* whose elements are called *edges*. Occasionally, (u, u) and repetition of an element in E are allowed to be called a *loop* and a *multi-edge*, respectively. $|V|$ is the *order* of G, which is denoted by ν, and $|E|$, the *size* denoted by ϵ. Of course, only *finite* graphs, which are those of finite order, are considered without specific explanation in this book. The graph whose edge set is $V * V$ is called a *complete graph* denoted by K_ν, or simply K when without confusion. If a graph $H = (V(H), E(H))$ satisfies $V(H) \subseteq V$ and $E(H) \subseteq E$, then it is called a *subgraph* of G denoted by $H \subseteq G$. It is easily seen that all graphs are subgraphs of a complete graph and that the *empty graph* denoted by $\emptyset$ as well is a subgraph of any graph. A graph without an edge is an *isolated graph* and the graph with a single vertex, *trivial graph*.

Theorem 1.3.1. $\forall V_1 \subseteq V_2, E_1 \subseteq E_2$,

$$(V_1, E_1) = G_1 \subseteq G_2 = (V_2, E_2) \Leftrightarrow E_1 \subseteq V_1 * V_1. \tag{1.3.1}$$

Similarly to the case for sets in Section 1.1, we can define the operations: *union* and *intersection* as follows: $\forall G_1 = (V_1, E_1), G_2 = (V_1, E_2) \subseteq K$,

$$G_1 \cup G_2 = (V_1 \cup V_2, E_1 \cup E_2); \tag{1.3.2}$$

$$G_1 \cap G_2 = (V_1 \cap V_2, E_1 \cap E_2). \tag{1.3.3}$$

It is easily shown that $(2^K, \subseteq)$, 2^K is the set of all subgraphs of K, is a poset with the idempotent, commutative, associative and absorption laws for $\cup$ and $\cap$ defined earlier. Therefore, from Theorem 1.1.13, $(2^K, \subseteq; \cup, \cap)$ is a lattice.

For an edge $e = (u, v) \in E$, u and v are said to be *adjacent*, or simply write "u adj v", and e is said to be *incident with* u or v, or write "e ind u" or "e ind v". Conversely, u or v is said to be *incident to* e, or write "u ind e" or "v ind e" as well. An edge can be considered to consist of two *semi-edges*: $[u, v)$ and $(u, v]$. The *valency* of vertex v, denoted by $\rho(v)$, is the number of semi-edges incident with v. A vertex is *odd* if $\rho(v) = 1 \pmod 2$; otherwise, *even*. A vertex of valency k is said to be *k-valent* for $k \geq 0$. A 0-valent vertex is called an *isolated* vertex. An *articulate* vertex is 1-valent.

Theorem 1.3.2. *In a graph, the number of odd vertices is even.*

A subgraph H of G is called a *vertex-induced* subgraph denoted by $H = G[V(H)]$ if $E(H) = \{(u, v) \mid \forall u, v \in V(H), (u, v) \in E\}$. If a subgraph H of G satisfies that $V(H) = \{v \mid \exists e \in E(H), v \text{ ind } e\}$, then it is called an *edge-induced* subgraph denoted by $H = G[E(H)]$. We may see that $\forall H \subseteq G$,

$$H = G[V(H)] \Leftrightarrow \forall u, v \in V(H),\ \neg e = (u, v) \in E \setminus E(H)$$

and

$$H = G[E(H)] \Leftrightarrow \neg v \in V(H),\ \rho_H(v) = 0.$$

Let $2^{[G;v]}$ and $2^{[G;e]}$ be the sets of all vertex- and edge-induced subgraphs of G, respectively. It is easily shown from inspection of the three laws for partial order in Section 1.1 that both $(2^{[G;v]}, \subseteq)$ and $(2^{[G;e]}, \subseteq)$ are posets. Further, both $(2^{[G;v]}, \subseteq)$ and $(2^{[G;e]}, \subseteq)$ are lattices, although the union and the intersection of induced subgraphs are not closed on them in general.

A *trail* between two vertices u and v in G denoted by $T_{rl}(u, v)$ is a sequence of edges $e_1, e_2, \ldots, e_l$, such that $e_i = (v_i, v_{i+1}), i = 1, 2, \ldots, l$, $u = v_1, v = v_{l+1}$. Here, l is called the *length*. When $u = v$, the trail $T_{rl}(u, v)$ is called a *travel* denoted by $T_{rl}(u)$, or simply T_{rl}. If all the edges in $T_{rl}(u, v)$ are distinct, then the trail is called a *walk*, denoted by $T_r(u, v)$. When $u = v$, the walk $T_r(u, v)$ is called a *tour*, denoted by $T_r(u)$, or simply T_r. If the edge-induced subgraph $H = G[E(T_r(u, v))]$ satisfies that $(\rho_H(u) = \rho_H(v) = 1) \wedge (\rho_H(v_i) = 2, i = 1, 2, \ldots, l-1)$, then the walk is called a *path*, denoted by $P(u, v)$. When $u = v$, the path

$P(u, v)$ is a *circuit* denoted by $C(u)$, or C. Of course, walks and paths can be both seen as edge-induced subgraphs. Two vertices are said to be *connected* if there is a path between them. If all pairs of vertices in G are connected, then G is a *connected graph*. It is easy to check by the reflective, symmetry and transitive laws in Section 1.1 that the connectedness between two vertices is an equivalence on the vertex set, which is denoted by $\sim_c$.

Theorem 1.3.3. *A graph $G = (V, E)$ is connected if, and only if, $|V/\sim_c| = 1$.*

Let $\sigma = |V/\sim_c|$, which is called the *number of components* of G. For a vertex v, we define $G - v = (V \setminus \{v\}, E \setminus E_v)$, where $E_v = \{e \mid \forall e \in E, e \text{ ind } v\}$. A vertex v is called a *cut-vertex* if $\sigma(G - v) > \sigma$. Similarly, a *cut-edge* $e: \sigma(G - e) > \sigma$, $G - e = (V, E \setminus \{e\})$. A *tree* is such a graph that it is connected and is of least size. We may show that all trees of order v are of the same size, which is $v - 1$. A graph whose components are all trees is called a *forest*.

Theorem 1.3.4. *A graph of order v is a tree if, and only if, its size is $v - 1$ and all its edges are cut edges.*

A graph that has neither isolated vertex nor cut vertex is called a *block*, or a *non-separable* one. It is obvious from inspection of O1, Õ2 and O3 in Section 1.2 that the statement "two edges are on the same circuit" defines an equivalence denoted by $\sim_b$ on the edge set of a graph.

Theorem 1.3.5. *A graph without isolated vertex is non-separable if, and only if, $|E/\sim_b| = 1$.*

A subgraph H of G is said to be of *spanning* if $V(H) = V$. A spanning circuit is called a *Hamiltonian* circuit and a spanning tour on which each edge of the graph occurs, a *Eulerian* tour in the graph. If a graph has a Hamiltonian circuit, or a Eulerian tour, then it is a *Hamiltonian*, or *a Eulerian graph*, respectively.

Theorem 1.3.6. *A connected graph is Eulerian if, and only if, all the valencies of its vertices are even.*

For a graph G, if $V = A + B$ (i.e., $A \cup B$ provided $A \cap B = \emptyset$) and both $G[A]$ and $G[B]$ are isolated graphs, then G is called a *bipartite graph* denoted by $G = (A, B; E)$. If $E = \{e = (u, v) \mid \forall (u \in A)(v \in B)\}$, then the bipartite graph $(A, B; E)$ is called a *complete* one denoted by $K_{\alpha,\beta}$, where $\alpha = |A|$ and $\beta = |B|$.

Theorem 1.3.7. *A graph is bipartite if, and only if, it is without a circuit of odd length.*

If any pair of elements in a subset of V or E is not adjacent, then the subset is said to be *independent*. An independent subset of E for a graph $G = (V, E)$ is also called a *matching*. If a matching induces a spanning subgraph of G, then it is said to be *perfect*. For $a \in V$, let $N_a = \{v \mid \forall v \in V, v \text{ adj } a\}$ and for $A \subseteq V$, let

$$N(A) = \bigcup_{a \in A} N_a \backslash A.$$

Theorem 1.3.8. *A bipartite graph $G = (X, Y; E)$ has a perfect matching if, and only if, $\forall A \subseteq X$ and $\forall A \subseteq Y$, $\mid N(A) \mid \geq \mid A \mid$.*

It is known that any graph can be realized as a subset of 3-Euclidean space such that the edges are represented by sections of curves (in fact, straight segments here) any of whose pairs is without common point except only for the end points of the sections, which represent the common end of the corresponding edges. Such a representation of a graph is called an *embedding* in the space. However, not all graphs have an embedding in the plane, or 2-Euclidean space. If a graph has an embedding in the plane, then it is said to be *planar*.

A *bisection* is an operation of transforming $G = (V, E)$ into a graph $(V + \{w\}, (E \setminus \{(u, v)\}) + \{(u, w), (w, v)\})$. If a graph can be obtained from another one by a series of bisections and/or the inverses, then the two graphs are said to be *homeomorphic*.

Theorem 1.3.9. *A graph is planar if, and only if, it has no subgraph homeomorphic to K_5 or $K_{3,3}$.*

Two graphs $G_1 = (V_1, E_1)$ and $G_2 = (V_2, E_2)$ are said to be *isomorphic* if there is a bijection $\tau : V_1 \to V_2$ such that

$$\forall u, v \in V_1,\ (u, v) \in E_1 \Leftrightarrow (\tau(u), \tau(v)) \in E_2. \tag{1.3.4}$$

The bijection τ defined by eq. (1.3.4) is called an *isomorphism* between G_1 and G_2. An *automorphism* of G is an isomorphism between G and itself. It would be the most difficult problem among those are mentioned to justify if two graphs are isomorphic in general.

Similarly, a *digraph* (or a directed graph) denoted by $D = (V, A)$ is a set V, which is also called the *vertex set*, with a binary relation $A \subseteq V \times V = \{\prec u, v \succ \mid \forall u \in V, \forall v \in V\}$, which is called the *arc set*. All the above discussions have analogues in the directed case. Particularly, a poset $P = (M; \preceq)$ can be represented by a digraph $Dos = (M, Aos)$, where $\prec x, y \succ \in Aos \Leftrightarrow (x \preceq y) \wedge (\neg z, x \prec z \prec y)$, or say x is *covered* by y for $x, y \in M$. If a graph of order ν is associated with an injection (almost in any case, a bijection) from its vertex set to (onto) the integer set $(\{1, 2, \ldots, \nu\})$, then it is said to be *labelled*. The injection is called the *labelling*. The image of a vertex under the labelling is called its *label*. Of course, an isomorphism between labelled (directed) graphs has to be considered with the labels on vertices (directions on edges).

A *network* N is such a graph $G = (V, E)$ with a real function $w(e) \in \mathbf{R}, e \in E$ on E, and hence write $N = (G; w)$. Usually, a network N is denoted by the graph G itself if no confusion occurs.

Finite recursion principle On a finite set A, choose $a_0 \in A$ as the initial element at the 0th step. Assume a_i is chosen at the ith, $i \geq 0$, step with a given rule. If not all elements available from a_i are already chosen, choose one of them as a_{i+1} at the $i + 1$st step by the rule, then a chosen element will be encountered in finite steps unless all available elements of A have been chosen.

Finite restrict recursion principle On a finite set A, choose $a_0 \in A$ as the initial element at the 0th step. Assume a_i is chosen at the ith, $i \geq 0$, step with a given rule. If a_0 is not available from a_i, choose one of elements available from a_i as a_{i+1} at the $i + 1$st step by the rule, then a_0 will be encountered in finite steps unless all available elements of A are chosen.

The two principles above are very useful in finite sets, graphs and networks, even in a wide range of combinatorial optimizations.

Let $N = (G; w)$ be a network where $G = (V, E)$ and $w(e) = -w(e) \in Z_n = \{0, 1, \ldots, n-1\}$, i.e., mod n, $n \geq 1$, integer group. For example, $Z_1 = \{0\}$, $Z_2 = B = \{0, 1\}$, and so on. Suppose $x_v = -x_v \in Z_n$, $v \in V$, are variables. Let us discuss the system of equations

$$x_u + x_v = w(e) \pmod{n}, \ e = (u, v) \in E \tag{1.3.5}$$

on Z_n.

Theorem 1.3.10. *System of equations (1.3.5) has a solution on Z_n if, and only if, there is no circuit C such that*

$$\sum_{e \in C} w(e) \neq 0 \pmod{n} \tag{1.3.6}$$

on N.

Proof. Necessity. Assume C is a circuit satisfying eq. (1.3.6) on N. Because the restricted part of eq. (1.3.5) on C has no solution, the whole system of equations (1.3.5) has to have no solution either. Therefore, N has no such circuit. This is a contradiction to the assumption

Sufficiency. Let $x_0 = a \in Z_n$, start from $v_0 \in V$. Assume $v_i \in V$ and $x_i = a_i$ at step i. Choose one of $e_i = (v_i, v_{i+1}) \in E$ without used (otherwise, backward 1 step as the step i). Choose v_{i+1} with $a_{i+1} = a_i + w(e_i)$ at step $i + 1$. If a circuit such as $\{e_0, e_1, \ldots, e_l\}$, $e_j = (v_j, v_{j+1}), 0 \leq j \leq l, v_{l+1} = v_0$, occurs within a permutation of indices, then from eq. (1.3.6)

$$\begin{aligned}a_{l+1} &= a_l + w(e_l)\\ &= a_{l-1} + w(e_{l-1}) + w(e_l)\\ &\dots\\ &= a_0 + \sum_{j=0}^{l} w(e_j) = a_0.\end{aligned}$$

Because the system of equations obtained by deleting all the equations for all the edges on the circuit from eq. (1.3.5) is equivalent to the original system of equations (1.3.5). By virtue of the finite recursion principle a solution of eq. (1.3.5) can always be extracted. □

When $n = 2$, this theorem has a variety of applications. In Ref. [194], where Theorem 1.3.7 is a special case, some applications can be seen. Further, its extension on a non-Abelian group can also be done while the system of equations are not yet linear but quadratic.

A graph is said to be *even* if the valency of each vertex is even.

Theorem 1.3.11. *A graph is even if, and only if, its edges set has a cycle partition.*

Proof. Since what is obtained from an even graph by deleting all the edges on a cycle is still an even graph, based on the finite recursion principle, the theorem is done. □

Let $G = (V, E)$ be a graph where $V = \P(X)$, and $E = \{Bx | x \in X\}$ where $\P(X)$ is a partition on

$$B(X) = \bigcup_{x \in X} Bx$$

and $Bx = \{x(0), x(1)\}$ for a set X. Two graphs $G_1 = (V_1, E_1)$ and $G_2 = (V_2, E_2)$ are *isomorphic* if, and only if, there exists a bijection $\iota\colon X_1 \to X_2$ such that the diagrams

$$\begin{array}{ccc} X_1 & \xrightarrow{\ \iota\ } & X_2 \\ {\scriptstyle \sigma_1}\downarrow & & \downarrow{\scriptstyle \sigma_2} \\ X_1 & \xrightarrow[\ \iota\ ]{} & X_2 \end{array} \tag{1.3.7}$$

for $\sigma_i = B_i, \P_i$, $i = 1, 2$, are commutative. Let $\mathrm{Aut}(G)$ be the automorphism group of G.

On the other hand, a *semi-arc isomorphism* between two graphs $G_1 = (V_1, E_1)$ and $G_2 = (V_2, E_2)$ is defined to be such a bijection $\tau: B_1(X_1) \to B_2(X_2)$ that

$$\begin{array}{ccc} B_1(X_1) & \xrightarrow{\quad\tau\quad} & B_2(X_2) \\ {\scriptstyle\sigma_1}\downarrow & & \downarrow{\scriptstyle\sigma_2} \\ B_1(X_1) & \xrightarrow[\quad\tau\quad]{} & B_2(X_2) \end{array} \qquad (1.3.8)$$

for $\sigma_i = B_i, \P_i$, $i = 1, 2$, are commutative. Let $\mathrm{Aut}_{1/2}(G)$ be the semi-arc automorphism group of G.

Theorem 1.3.12. *If* Aut(G) *and* $\mathrm{Aut}_{1/2}$(G) *are, respectively, the automorphism and semi-arc automorphism groups of graph G, then*

$$\mathrm{Aut}_{1/2}(\mathrm{G}) = \mathrm{Aut}(\mathrm{G}) \times S_2^l, \qquad (1.3.9)$$

where l is the number of self-loops on G and S_2 is the symmetric group of degree 2.

Proof. Since each automorphism of G just induces two semi-arc isomorphisms of G for a self-loop, the theorem is true. □

1.4 Groups and spaces

If a *group* denoted by $\Gamma = (X, \lozenge)$ is a set X with a binary operation $\gamma{:}X \times X \to X$, it would be better to write $x\lozenge y$ for $\prec x, y \succ \gamma$ referring to "$\lozenge$" as the operation, such that the laws Γ1, Γ2 and Γ3 described below are satisfied.

Γ1 (Associative law) $\forall x, y, z \in X$, $(x\lozenge y)\lozenge z = x\lozenge(y\lozenge z)$.

Γ2 (Identity law) $(\exists 1_\Gamma$ (or simply 1) $\in X)(\forall x \in X, x\lozenge 1_\Gamma = x)$.

Γ3 (Inverse law) $(\forall x \in X)(\exists y \in X, x\lozenge y = 1_\Gamma)$.

The element 1 in Γ2 is called a *right identity* and the element y in Γ3 a *right inverse* of x. We may also define a *left identity* and a *left inverse* of an element. However, it is easily shown that they are all unique and the left one equals to the right. So we are allowed to call 1 the *identity* and x^{-1} the *inverse* of x.

The *order* of a group $\Gamma = (X, \lozenge)$ is defined to be $|\Gamma| = |X|$. We can see that $(1, \lozenge)$ is a group, which is called the *trivial group* or the *identity group*. In this book, a group $\Gamma = (X, \lozenge)$ is always written as $\Gamma = X$ without specific indication. If a group Γ satisfies the condition Γ4 below, then it is said to be *Abelian*.

Γ4 (Commutative law) $\forall x, y \in \Gamma, x\lozenge y = y\lozenge x$.

There are two commonly used ways of writing the group operation of Γ. One is the *additive* notation by writing $x\lozenge y$ as a "sum" $x + y$ with the identity 0_Γ (or 0) and the inverse $-x$ of x especially for Abelian groups. The other is the *multiplicative* notation by using a "product" $x \bullet y$ (or xy), 1_Γ (or 1) and x^{-1} as $x \lozenge y$, the identity and the inverse of x, respectively, for general groups.

Let $\Gamma = (X, \bullet)$ be a group and let $Y \subseteq X$. If $\Lambda = (Y, \bullet)$ is a group, then Λ is called a *subgroup* of Γ, denoted by $\Lambda \subseteq \Gamma$. Of course, the identity group is a subgroup of any group and a group is a subgroup of itself.

Theorem 1.4.1. $\forall Y,\ \emptyset \neq Y \subseteq X, \Lambda = (Y, \bullet) \subseteq \Gamma = (X, \bullet) \Leftrightarrow (\forall x, y \in Y)(xy^{-1} \in Y)$.

Let $\Gamma_i = (X_i, \bullet) \subseteq \Gamma = (X, \bullet), i \in I$. It is easily seen that $\cap_{i \in I}\Gamma_i = (\cap_{i \in I}X_i, \bullet) \subseteq \Gamma$, which is called the *intersection*. For an $S \subseteq X$, the intersection, denoted by $\langle S \rangle$, of all subgroups that contains S is called the subgroup *generated* by S in Γ. The subgroup $\langle \cup_{i \in I} X_i \rangle$ denoted by $\cup_{i \in I}\Gamma_i$ is called the *join* of subgroups $\Gamma_i,\ i \in I$. Let $\mathbf{\Gamma}$ be the set that consists of all subgroups of Γ. Then, it is obvious from inspection of the laws O1–O3 in Section 1.1 and Theorem 1.1.13 that $(\mathbf{\Gamma}, \subseteq; \cup, \cap)$ is a lattice, more precisely, a complete lattice because any subset of $\mathbf{\Gamma}$ has the l.u.b., which is the intersection, and the g.l.b., which is the join, in $\mathbf{\Gamma}$.

A subgroup Λ of a group Γ is said to be *normal*, or write $\Lambda \lhd \Gamma$, if it satisfies one of the following three equivalent conditions:

$$\begin{aligned} \forall x \in \Gamma, x\Lambda = \Lambda x &\Leftrightarrow \forall x \in \Gamma, x^{-1}\Lambda x = \Lambda \\ &\Leftrightarrow \forall x \in \Gamma, \forall y \in \Lambda, x^{-1}yx \in \Lambda. \end{aligned} \tag{1.4.1}$$

It is easily seen that any subgroup of an Abelian group is normal. However, in general, there exist subgroups that are not normal for a non-Abelian group. One may also see that the set of all normal subgroups of a group forms a complete lattice with the inclusion as the order and with the intersection and the join as the two operations.

Because it can be shown that the relation, denoted by $\sim_N$:

$$x \sim_N y \Leftrightarrow \exists h \in N, x = hy, \tag{1.4.2}$$

provides an equivalence on the set X of the group $\Gamma(X, \bullet)$ for $N \lhd \Gamma$. We are allowed to define the *quotient* (or *factor*) group of N in Γ to be

$$\Gamma/N = (X/{\sim_N}, \bullet), \tag{1.4.3}$$

where $(Nx)(Ny) = N(xy)$. The order of Γ/N is called the *index* of N in Γ.

Let Γ and Λ be two groups. A function $\alpha : \Gamma \to \Lambda$ is called a *homomorphism* from Γ to Λ if

$$\forall x, y \in \Gamma,\ \alpha(xy) = \alpha(x)\alpha(y). \tag{1.4.4}$$

Because $o : \Gamma \to 1_\Lambda$ is a homomorphism, which is called *zero homomorphism*, the set $\mathrm{Hom}\,(\Gamma, \Lambda)$ of all homomorphisms from Γ to Λ is always non-empty. A homomorphism from Γ to Γ itself is said to be an *endomorphism* of Γ. The identity function $\iota : \Gamma \to \Gamma$ is an endomorphism of Γ.

For a homomorphism α from Γ to Λ, let

$$\begin{cases} \mathrm{Im}\,\alpha = \alpha(\Gamma) = \{\alpha(x) \mid \forall x \in X\}; \\ \mathrm{Ker}\,\alpha = \{x \mid \forall x \in X, \alpha(x) = 1_\Lambda\}, \end{cases} \tag{1.4.5}$$

which is said to be the *image*, the *kernel* of α, respectively. It is easy to check by Theorem 1.4.1 that $\mathrm{Im}\,\alpha \subseteq \Lambda$ and $\mathrm{Ker}\,\alpha \lhd \Gamma$. If a homomorphism α from Γ to Λ satisfies $\mathrm{Ker}\,\alpha = 1_\Gamma$, then α is called a *monomorphism*. If a homomorphism α from Γ to Λ has $\mathrm{Im}\,\alpha = \Lambda$, then α is called an *epimorphism*. A homomorphism that is both a monomorphism and an epimorphism is said to be an *isomorphism*. Two groups Γ and Λ are said to be *isomorphic*, or written as $\Gamma \cong \Lambda$, when there is an isomorphism between them. An isomorphism from Γ to Γ itself is called an *automorphism* of Γ. It can be easily shown from inspection of the laws $\Gamma1$–$\Gamma3$ that the set of all automorphisms of Γ is a group, which is called the *automorphism group* of Γ, denoted by $\mathrm{Aut}\,\Gamma$.

Theorem 1.4.2 (First isomorphism law). $\forall \alpha \in \mathrm{Hom}(\Gamma, \Lambda)$,

$$\Gamma/\mathrm{Ker}\,\alpha \cong \mathrm{Im}\,\alpha.$$

Based on Theorem 1.4.2, we are allowed to call $\Gamma/\mathrm{Ker}\,\alpha$ the *coimage* of α. If $N \lhd \Gamma$, then the mapping $\phi : x \mapsto Nx$ is an epimorphism from Γ to Γ/N with $\mathrm{Ker}\,\phi = N$. We call ϕ the *canonical homomorphism*.

For two groups $\Lambda = (Y, \bullet) \subseteq (X, \bullet) = \Gamma$, let $\Gamma\Lambda = (XY, \bullet)$, where $XY = \{xy \mid \forall x \in X, \forall y \in Y\}$. One may see that $\forall \Lambda \subseteq \Gamma, N \lhd \Gamma \Rightarrow \Lambda \cap N \lhd \Lambda$.

Theorem 1.4.3 (Second isomorphism law). $\forall \Lambda \subseteq \Gamma, \forall N \lhd \Gamma$,

$$\Lambda/N \cap \Lambda \cong N\Lambda/N.$$

Let N and Q be two normal subgroups of a group Γ and let $N \subseteq Q$. Then, it is known that $Q/N \lhd \Gamma/N$.

Theorem 1.4.4 (Third isomorphism law). $\forall N, Q \lhd \Gamma$,

$$N \subseteq Q \Rightarrow (\Gamma/N)/(Q/N) \cong \Gamma/Q.$$

Let Φ be a group, S a non-empty set and $\sigma : S \to \Phi$, a function. Then, Φ, or precisely (Φ, σ), is said to be *free* on S if for each function $\alpha : S \to \Gamma$, there is a unique homomorphism $\beta : \Phi \to \Gamma$ such that $\alpha = \beta\sigma$. A group which is free on some set is called a *free group*. From the definition it can be derived that σ is injective and that Im σ generates Φ. In fact, it can be shown that for any non-empty set S there exists a group Φ and a function $\sigma : S \to \Phi$ such that Φ is free on S and $\Phi = \langle \mathrm{Im}\ \sigma \rangle$.

Theorem 1.4.5. *If Φ_1 is free on S_1 and Φ_2 is free on S_2, then $\Phi_1 \cong \Phi_2 \Leftrightarrow | S_1 | = | S_2 |$.*

This theorem allows us to define the *rank* of a free group as the cardinality of any set on which it is free. Further, we have known that any group is an image of a free group. Such an image is called a presentation of the group. More precisely, a *free presentation* of a group Γ is an epimorphism $\pi : \Phi \to \Gamma$, Φ is a free group. From Theorem 1.4.2, we have $\Phi/\mathrm{Ker}\ \pi \cong \Gamma$. The elements of Ker$\pi$ are called *relators* of the presentation. Therefore, any group can be characterized by generators and relaters. Although a presentation of a group is known, to justify if two groups are isomorphic in general is still not easy because a group may have different kinds of presentations.

A *space* (or precisely a *vector space* or *linear space*) over $\boldsymbol{F}$ denoted by $(\mathscr{X}, \boldsymbol{F}; +, \bullet)$ (or simply write $\mathscr{X}$) is an Abelian group $(\mathscr{X}, +)$, or $\mathscr{X}$ as well, associated with a field $(\boldsymbol{F}, +, \bullet)$, or simply $\boldsymbol{F}$, and two binary operations: "+", called the *sum* and "•", the *scalar product*, satisfying the following four axioms: Vects.1–4. The sum is with the same symbol as the addition on the group $\mathscr{X}$ and the addition on the field $\boldsymbol{F}$. The scalar product $a \bullet A$, or simply aA, is defined for $a \in \boldsymbol{F}$ and $A \in \mathscr{X}$ and is with the same symbol as the multiplication on $\boldsymbol{F}$. Members of $\mathscr{X}$ are called *vectors*, and those of $\boldsymbol{F}$, *scalars*.

Vect.1 $\forall a \in \boldsymbol{F}, \forall A, B \in \mathscr{X}, a(A + B) = aA + aB$.

Vect.2 $\forall a, b \in \boldsymbol{F}, \forall A \in \mathscr{X}, (a + b)A = aA + bA$.

Vect.3 $\forall a, b \in \boldsymbol{F}, \forall A \in \mathscr{X}, (ab)A = a(bA)$.

Vect.4 $\forall A \in \mathscr{X}, 1A = A$.

It seems that the only notational distinction we have to make between vectors and scalars is to denote the zero elements of $\mathscr{X}$ and $\boldsymbol{F}$ by $0_{\mathscr{X}}$ and $0_{\boldsymbol{F}}$, respectively. However, since it is easily shown, from the axioms Vects.1–4, that $\forall A \in \mathscr{X}, 0_{\boldsymbol{F}}A = 0_{\mathscr{X}}$ and that $\forall a \in \boldsymbol{F}$, $a0_{\mathscr{X}} = 0_{\mathscr{X}}$, the distinction will almost always be dropped and $0_{\boldsymbol{F}}$, $0_{\mathscr{X}}$ be written simply 0.

A subset $\mathscr{Y} \subseteq \mathscr{X}$ of a space $\mathscr{X}$ over $\boldsymbol{F}$ is said to be a *subspace* , denoted by $\mathscr{Y} \subseteq_{\mathrm{vect}} \mathscr{X}$ (or simply $\mathscr{Y} \subseteq \mathscr{X}$ without confusion), of $\mathscr{X}$ if $\mathscr{Y}$ is a space over $\boldsymbol{F}$ in its own right, but with respect to the same operations as $\mathscr{X}$. The zero vector 0 belongs to any space and itself is a space called the *zero space* or *trivial space* denoted by 0 as well. Any non-zero vector of order 2 with 0 here forms a subspace, which is denoted by $\mathscr{J}$.

Theorem 1.4.6. $\forall \mathscr{Y} \subseteq \mathscr{X}, \mathscr{Y} \subseteq_{\mathrm{vect}} \mathscr{X} \Leftrightarrow$

$$(\forall A, B \in \mathscr{Y}, A + B \in \mathscr{Y}) \wedge (\forall a \in \boldsymbol{F},\ \forall A \in \mathscr{Y}, aA \in \mathscr{Y}).$$

Proof. The necessity is straight forward. Conversely, because $\mathscr{Y} \subseteq \mathscr{X}$, from the last statement, Vects.2–4 hold and from the first statement, Vect.1 holds for $\mathscr{Y}$. The sufficiency is obtained. □

Apparently, for spaces we are also allowed to introduce the two operations: $\cap$, the intersection, and $\cup$, the join, as described before for groups and find that $(2^{\mathscr{X}}, \subseteq; \cup, \cap)$ forms a lattice, of course, a complete one.

In what follows, we are only concerned with the field $\boldsymbol{F} = \mathrm{GF}(2)$, the finite field of two elements for spaces. In this case, the space is called a *binary space*. For any $A \in \mathscr{X}$, we always have $A + A = 0$, the zero vector. That is of characteristic 2. Suppose $\mathscr{X} = 2^X$ is the free Abelian group $\langle x | \, \forall x \in X \rangle$ generated by all the elements of X. Then, a vector is also a subset of X. We always employ the same symbol to denote a vector of $\mathscr{X}$ and a subset in X. Let $A \in \mathscr{X}$, then

$$A = \sum_{x \in X} A_x x = \sum_{x \in A} x, \tag{1.4.6}$$

where A_x is said to be the *coefficient*, or *component* of A on x. Of course, $A_x = 1$, if $x \in A$; 0, otherwise.

On the space $\mathscr{X}$, we define an *inner product* denoted by (A, B) for $A, B \in \mathscr{X}$ as

$$(A, B) = \sum_{x \in X} A_x B_x. \tag{1.4.7}$$

By this notation, we have the relation:

$$A_x = (A, x), \forall x \in X. \tag{1.4.8}$$

If for $A, B \in \mathscr{X}$, $(A, B) = 0$, then A and B are said to be *orthogonal* denoted by $A \perp B$ or $B \perp A$ from the symmetry: $(A, B) = (B, A)$. Here, one may see

$$\forall A, B \in \mathscr{X}, (A, B) = 0 \Leftrightarrow |A \cap B| = 0 \pmod 2. \tag{1.4.9}$$

If $(A, A) = 0$, then vector A is said to be *even*. Let $\mathscr{A}(\mathscr{X})$ be the set of all even vectors in $\mathscr{X}$. It can be seen from inspection of axioms Vects.1–4 that $\mathscr{A}(\mathscr{X})$ is a subspace of $\mathscr{X}$ and is called the *alternating* (or *symplectic*) space on X.

Further, we may also see that for $A \in \mathscr{X}$ given,

$$A = 0 \Leftrightarrow \forall B \in \mathscr{X}, (A, B) = 0. \tag{1.4.10}$$

Or, in other words, the inner product is *non-degenerate*.

If a vector A satisfies that $\forall B \in \mathscr{B}, (A, B) = 0$, then it is said to be *orthogonal* to $\mathscr{B}$ and denoted by $A \perp \mathscr{B}$.

Let $\mathscr{A}$ and $\mathscr{B}$ be two subspaces of $\mathscr{X}$. If

$$\mathscr{A} = \{A \mid \forall A \in \mathscr{X}, A \perp \mathscr{B}\}, \quad (1.4.11)$$

then $\mathscr{A}$ is said to be the *orthogonal space* of $\mathscr{B}$ in $\mathscr{X}$ and is denoted by $\mathscr{A} = \mathscr{B}^{\perp}$. Moreover, from the symmetry of the inner product, we have

$$(\mathscr{B}^{\perp})^{\perp} = \mathscr{B}. \quad (1.4.12)$$

In Chapters 4 and 8, we shall see a number of spaces related to graphs. Almost all results for them can be extended to general spaces over GF(2), the finite field of two elements.

1.5 Notes

1.5.1 This book is in principle designed to be self-contained in the background presented in this chapter. One might still like to read more materials related to topology. References can be chosen such as Alexandroff [5], Giblin [97], Greenberg [107], Massey [269], Stillwell [321], Agoston [2], or Lefschetz [173], Williams [398].

1.5.2 Permutations are established from partitions on a set. Such an idea enables us to observe embeddings, or super maps of a graph as permutations, from the graph as a partition. A description in a certain detail can be seen in Liu [218, 219, 224, 234]. Most books on basic algebra involve permutations such as Jacobson [157], Gilbert [98], particularly Dixon and Mortimer [73].

1.5.3 A graph turns out a partition of the ground set from a set by a binary group sticking on from Liu [218]. Although a great number of books on graphs have appeared in literature as Bellman et al. [24], Berge [25], Biggs [31], Bondy and Murty [35], Capobianco and Molluzzo [39], Chan [41], Chen [43], Fiorini and Wilson [86], Golumbic [105], Harary [122], Iri [156], Kaufman [162], Kuo [165], Lovasz [252], Tutte [347, 350], Zykov [427], et al. Only a few, more or less, directly related to this book are listed as Ore [273], Tutte [349], Ringel [286], White [382], Lefschetz [172], Wu [404] and Liu [216, 217], particularly more popular book: Gross and Tucker [108].

1.5.4 Those mentioned in Section 1.4 are all extracted from Liu [216]. One might also like to read more about general groups such as MacLane and Birkhoff [263], and Robinson [292].

1.5.5 One might see that the vector space can be generalized to an abstract linear space from, for an example, Theorem 1.2.1 in Liu [237]. To read more about linear space (advanced) is referred to Roman [293].

1.5.6 By considering theoretical efficiency, basic knowledge of data structure, algorithm and complexity should be known. The reader is suggested to read from Aho et al. [3], Pralts [277], Garey and Johnson [95] if necessary.

2 Polyhedra

2.1 Polygon double covers

A *polygon*, denoted by $(a, b, c, \ldots)$, is a finite set of letters in a cyclic order. In general, such a polygon can be represented by the infinite face of a connected plane graph conformed with convex polygons and articulate edges, or the inner face of a regular polygon. Hence, the letters in a polygon are allowed with repetition of each letter at most twice (with the same power or different powers: 1 always omitted and −1) in the first case. For a letter a, a^{-1} is called the *inverse* of a. The inverse satisfies the following two rules:

Inverse rule 1 For a letter a, $(a^{-1})^{-1} = a$.

Inverse rule 2 For two letters a and b, $(ab)^{-1} = b^{-1}a^{-1}$, or $(a, b)^{-1} = (b^{-1}, a^{-1})$.

Two polygons A_1 and A_2 are dealt with the same if one becomes the other by one of the following alternatives:

No.diff.gon1 For $a \in A_1$, A_2 is different from A_1 only in interchanging the positions of the two occurrences of a, if any.

No.diff.gon2 For $a, b \in A_1$, A_2 is different from A_1 only in interchange between a and b.

Let polygon $A = (a_1, a_2, \ldots, a_l)$, then polygons $(a_2, a_3, \ldots, a_1), \ldots, (a_l, a_1, \ldots, a_{l-1})$ are, respectively, called *cyclic left shift* of A in $1, 2, \ldots, l-1$ bits.

No.diff.gon3 A_2 is any of all the cyclic left shifts of A_1.

The polygon $(a_l, \ldots, a_2, a_1)$ is called a *reversion*, denoted by $(a_1, a_2, \ldots, a_l)^{\mathrm{rv}}$, of polygon $(a_1, a_2, \ldots, a_l)$.

No.diff.gon4 $A_2 = (A_1)^{\mathrm{rv}}$.

The polygon $(a_1^{-1}, a_2^{-1}, \ldots, a_l^{-1})$ is called a *conversion*, denoted by $(a_1, a_2, \ldots, a_l)^{\mathrm{cv}}$, of polygon $(a_1, a_2, \ldots, a_l)$.

No.diff.gon5 $A_2 = (A_1)^{\mathrm{cv}}$.

An *inversion* of polygon $A = (a_1, a_2, \ldots, a_l)$ is defined to be $A^{\mathrm{iv}} = (a_l^{-1}, \ldots, a_2^{-1}, a_1^{-1})$.

Proposition 2.1.1. *For any polygon A,*

$$A^{iv} = (A^{rv})^{cv} = (A^{cv})^{rv}. \tag{2.1.1}$$

Proof. Easy to check by the definitions. □

On the basis of this proposition, it is from the inverse rule 2 seen that $A^{\mathrm{iv}} = A^{-1}$.

DOI 10.1515/9783110479492-002

If a set of polygons has each letter occuring exactly twice, then it is called a *double cover* on the set of all letters in the polygons.

A *polyhedron* P is a set $\mathscr{C} = \{C_i|1 \le i \le k\}$, $k \ge 1$, of polygons which forms a double cover on a set A of letters, where C_i is called a *face* of P such that no proper subset of $\mathscr{C}$ is a double cover of a subset of A.

This is the combinatorial representation of Heffter's in Heffter [133] (1891, and more than half a century later, Edmonds' in Edmonds [83] as dual case) for a polyhedron.

Let $P = \{C_i|1 \le i \le k\}$ be a polyhedron and $X = X_P$, the set of all letters in P. An element (or letter) in X is called an *edge* of P. The property that the two occurrences of a letter with the same or different directions in a polyhedron is called the *status* of the edge. By sticking the group B of two elements on X, each edge consists of two *semi-edges* as $\{x^+, x^-\}$, or written as $\{+x, -x\}$, $\{x, x^{-1}\}$, or $\{x, \bar{x}\}$ for certain convenience. Each semi-edge $+x$, or $-x$, is compounded with its copy marked by a prime, i.e. $+x'$ or $-x'$ (x', or $x^{-1'}$) respectively. Then, an edge is further considered as

$$\{+x, +x', -x, -x'\} \text{ or simply, } \{x, x', x^{-1}, x^{-1'}\}$$

and hence $\{+x, +x'\}$ or $\{-x, -x'\}$ as well is now a semi-edge. The set

$$\mathscr{X}(P) = \sum_{x \in P}(\{x^+, x^{-1}\} + \{x+, x^{-1}\}') \tag{2.1.2}$$

is called a *ground set* of P. An element of the ground set is also called a *quarter* (of an edge).

Attention 2.1.1.

(1) For $x \in X$ and $x \in \mathscr{X}_P$, x has different meanings. The former is a letter and the latter, a quarter of an edge.

(2) For $x \in \mathscr{X}_P$, both $'$ and $^{-1}$ are seen as permutations on the ground set, i.e.,

$$' = \prod_{x \in X+X^{-1}} (x, x') \quad \text{and} \quad ^{-1} = \prod_{x \in X+X'} (x, x^{-1}), \tag{2.1.3}$$

where $X' = \{x' | \forall x \in X\}$ and $X^{-1} = \{x^{-1} | \forall x \in X\}$ for $X \subseteq \mathscr{X}_P$.

(3) For $x, y \in \mathscr{X}_P$, $(xy)' = y'x'$, $(xy)^{-1} = y^{-1}x^{-1}$, and $x'^{-1} = x^{-1'}$.

A face A in polyhedron P is seen in companion with A^{-1} on its ground set.

Proposition 2.1.2. *Let P be a polyhedron with its face set $\mathscr{A}$. Then, P is determined by the permutation π_P on its ground set as*

$$\pi_P = \prod_{A \in \mathscr{A}} (A)(A^{-1}), \tag{2.1.4}$$

in which two occurrences of a letter with the same power are distinguished by one with a prime.

Proof. By observing that all cycles appearing in eq. (2.1.4) form a partition, in view of Section 1.2 the conclusion is seen. □

Let $\sigma = {}'$ and $\delta = {}^{-1}$ be the permutations shown in eq. (2.1.3) on the ground set $\mathscr{X}_P$, i.e. for $x \in \mathscr{X}_P$,

$$\sigma(x) = \begin{cases} y', & \text{when } x = y; \\ y, & \text{when } x = y' \end{cases} \tag{2.1.5}$$

and for $x \in \mathscr{X}_P$,

$$\delta(x) = \begin{cases} y^{-1}, & \text{when } x = y; \\ y, & \text{when } x = y^{-1} \end{cases} \tag{2.1.6}$$

Then, $\pi_P^* = \pi_P \sigma \delta$ is a permutation on $\mathscr{X}_P$ as well.

Lemma 2.1.1. *On $\mathscr{X}_P$, $\delta \pi_P = \pi_P^{-1} \delta$.*

Proof. By virtue of $\pi_P \delta x = \pi_P x^{-1} = (\pi_P^{-1} x)^{-1} = \delta(\pi_P^{-1} x) = (\delta \pi_P) x$, by the arbitrariness of $x \in \mathscr{X}_P$ the lemma is obtained. □

Lemma 2.1.2. *On $\mathscr{X}_P$, $\sigma \pi_P^* = \pi_P^{*-1} \sigma$.*

Proof. By considering that

$$\begin{aligned} \sigma \pi_P^* &= \sigma \pi_P \sigma \delta = \sigma(\pi_P \delta)\sigma \text{ (by Lemma 2.1.1)} \\ &= \sigma(\delta \pi_P^{-1})\sigma = (\sigma^{-1} \delta^{-1} \pi_P^{-1})\sigma = \pi_P^{*-1} \sigma, \end{aligned}$$

the lemma is done. □

Lemma 2.1.3. *For $x \in \mathscr{X}_P$, two orbits $(x)_{\pi_P^*}$ and $(x')_{\pi_P^*}$ are disjoint and conjugate.*

Proof. By virtue of Lemma 2.1.2, the two orbits have the same type. From Theorem 1.2.6, they are conjugate. □

Theorem 2.1.1. *Permutation π_P^* on $\mathscr{X}_P$ determines a polyhedron.*

Proof. On the basis of Lemma 2.1.3, each pair of the conjugate orbits determine a polygon when the prime is omitted. Then, the set of all such polygons form a polyhedron. □

The polyhedron P^* obtained by omitting the power -1 and then replacing the prime by -1 from the permutation π_P^* shown in this theorem is called a *dual* of P. A face of the dual P^* is defined to be a *vertex* of P.

For a polyhedron P determined by permutation π_P on the ground set $\mathscr{X}_P$, the transposition

$$(x^{-1}, \pi_P x) = (\delta x, \pi_P x)$$

is called an *angle*. Two semi-edges incident with the same angle is said to be *V-adjacent*. Then, an equivalence called *V-adjacence* by appending the transitive law on the V-adjacent relation is obtained on the set of all semi-edges.

Theorem 2.1.2. *A set of semi-edges of a polyhedron forms a vertex if, and only if, it is an equivalent class under V-equivalence.*

Proof. In fact, a conjugate pair of cycles on π_P^* determines a equivalent class under V-equivalence. This is the theorem. □

Example 2.1.1. Only one polygon $(ae^{-1}b^{-1}cdefdb^{-1}afc^{-1})$ forms a polyhedron named by P. The permutation that determines P is

$$\pi_P = (ae^{-1}b^{-1}cde'fdb'^{-1}a'f'c'^{-1})$$
$$(a^{-1}c'f'^{-1}a'^{-1}b'd^{-1}f^{-1}e'^{-1}d^{-1}c^{-1}be).$$

Then,

$$\pi_P^* = (ab'c)(a'c'b)(df^{-1}c'^{-1})(d'c^{-1}f'^{-1})$$
$$(a^{-1}f'e'^{-1})(a'^{-1}e^{-1}f)(b^{-1}d'^{-1}e')(b'^{-1}e'd^{-1}).$$

By omitting the power −1 and then replacing the prime by −1 on π_P^*, we have

$$P^* = (ab^{-1}c)(dfc^{-1})(af^{-1}e^{-1})(bd^{-1}e^{-1}).$$

Theorem 2.1.3. *For two polyhedra P and Q, P is a dual of Q if, and only if, Q is a dual of P. Or in other words,* $P^{**} = P$.

Proof. By observing that

$$\pi_P^{**} = (\pi_P\sigma\delta)\delta\sigma = \pi_P(\sigma\delta\delta\sigma) = \pi_P(\sigma\sigma) = P,$$

the theorem is done from Theorem 2.1.1. □

2.2 Supports and skeletons

A *support* of polyhedron $P = \{C_i | 1 \le i \le k\}$ is the network formed by graph $U = (V_U, E_U)$ with a weight w on E_U where $V_U = \{C_i | 1 \le i \le k\}$, $(C_i, C_j) \in E_U$ if, and only if, C_i and C_j, $1 \le i, j \le k$, have a common letter, and

$$w(e) = \begin{cases} 0, & \text{when two powers are different;} \\ 1, & \text{otherwise} \end{cases} \tag{2.2.1}$$

for $e \in E_U$.

The support of P^* is called a *skeleton* of P, the under-graph of P with edge weights.

Example 2.2.1. The polyhedron P and its dual P^* in Example 2.1.1 have their supports, as shown in Figure 2.2.1

A polyhedron is *orientable* if there is an orientation of each cycle, clockwise or anti-clockwise, such that the two occurrences of each letter with different powers; *non-orientable*, otherwise.

Attention 2.2.1. For a polygon in clockwise, its form in anti-clockwise is the inversion. From No.Diff.gon4–5 and Proposition 2.1.1, the two forms are with no difference.

For a network $N = (G; w)$ $(G = (V, E), w(e) \in \mathrm{GF}(2), e \in E)$, the equation system about $x_v \in V$

$$x_u + x_v = w(e) \ (\mathrm{mod}\ 2) \tag{2.2.2}$$

for all $(u, v) \in E$ is called an *associate equation* on N.

Lemma 2.2.1. *If a polyhedron* $P = \{C_i | 1 \leq i \leq k\}$ *is orientable, then the associate equation (2.2.2) on the support has a solution.*

Proof. First, let P be in form as the two occurrences of each letter with different powers. Since the weights of all edges are the constant 0, eq. (2.2.2) has a solution of $x_i = 0$ for all $C_i \in V_P$, $1 \leq i \leq k$.

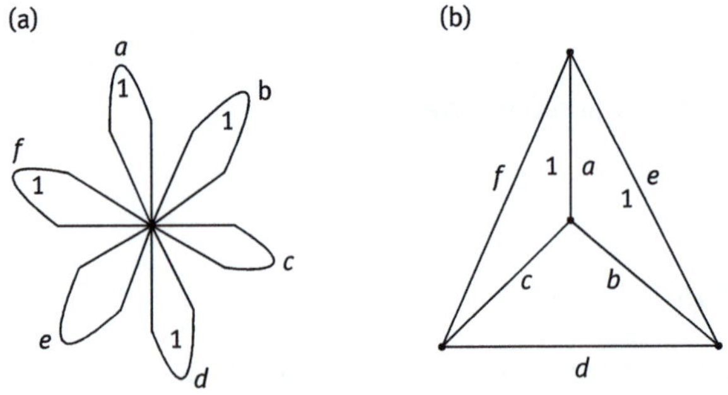

Figure 2.2.1: Support and skeleton: (a) a support of P and (b) a skeleton of P

Then, by considering that the consistency of eq. (2.2.2) is not changed from switching the orientation of a cycle between clockwise and anti-clockwise while interchanging the weights between 0 and 1 of all the edges in the cycle on the support, the conclusion is obtained. □

The set of all edges with weight 1 in a network is called the 1-*set* of the network.

Lemma 2.2.2. *If eq. (2.2.2) on a network has a solution, then the 1-set on the network forms a cocycle.*

Proof. Assume eq. (2.2.2) has a solution. The vertices of the network can be divided into two classes. It is seen that the cocycle consisted of all the edges between the two classes is the 1-set of the network. This is the lemma. □

Lemma 2.2.3. *If the* 1-*set on a network forms a cocycle, then the network has no odd-weight circuit.*

Proof. Since any circuit meets even number of edges with a cocycle, all circuits are with even weight. This means there is no odd-weight circuit. □

Lemma 2.2.4. *If a network has no odd-weight circuit, then the network has no odd-weight fundamental circuit for a given spanning tree.*

Proof. On account of that, a fundamental circuit is a circuit in its own right, the lemma follows. □

Lemma 2.2.5. *If the support of a polyhedron has no odd-weight fundamental circuit for a spanning tree, then what obtained by contracting all edges of weight* 0 *on the support is a bipartite graph.*

Proof. By considering that the contraction of an edge with weight 0 does not change the parity of circuits, the final graph with all edges of weight 1 has no odd-length circuit. From Theorem 1.3.7, it is bipartite. □

Lemma 2.2.6. *If the graph obtained by contracting all edges of weight* 0 *on the support of a polyhedron P is a bipartite graph, then P is orientable.*

Proof. Because of bipartiteness, vertices of the support are partitioned into two classes by the equivalence that two vertices are joined by even-weight path. By switching the orientation of all vertices in one of the two classes and those in the other class unchanged, a support of the polyhedron P without weight 1 edge is found. This implies that P is orientable. □

Theorem 2.2.1. *A polyhedron $P = \{C_i|1 \le i \le k\}$ is orientable if, and only if, one of the following statements is satisfied:*

(1) *What is obtained by contracting all edges of weight 0 on its support is a bipartite graph.*
(2) *No odd-weight fundamental circuit is on the support.*
(3) *No odd-weight circuit is on the support.*
(4) *The 1-set on the support forms a cocycle.*
(5) *The associate equation (2.2.2) on the support is consistent.*

Proof. Let $\tau = (123456)$ be the permutation on $\{1, 2, 3, 4, 5, 6\}$. For (i), the necessity is from Lemmas $2.2.i - \tau^4(i)$ and the sufficiency is from Lemma $2.2.\tau^5(i)$ when $i = 1, 2, 3, 4$ or 5. □

On the basis of statement (2) in the theorem, all the statements including the orientability of a polyhedron are polynomially recognizable.

2.3 Orientable polyhedra

For a polyhedra P, let $J = \{\pi_P, \delta, \sigma\}$, then the group generated by J is denoted by Ψ_P.

Lemma 2.3.1. *For a polyhedra P, Ψ_P has exactly one orbit on its ground set $\mathscr{X}_P$ if, and only if, the support of P is connected.*

Proof. First, let $y = \psi x$, $x, y \in \mathscr{X}_P$, $\psi \in \Psi_P$. Because ψ can be written in form as

$$\prod_{j=1}^{s} (\pi_P)^{i_j} \delta^{m_j} \sigma^{n_j},$$

where $i_j \ge 0$, $1 \le j \le s$ are all integers and m_j, n_j, $1 \le j \le s$ are all binary numbers, i.e. 0 or 1, there is a walk of length $s-1$ from the vertex at x to the vertex at y in the support. This is the necessity of the lemma.

Then, let $L = (a_1 a_2 a_3 \dots a_s)$ be a walk with $a_i \doteq \{x_i, \sigma x_i, \delta x_i, \sigma\delta x_i\}$, $i = 1, 2, 3, \dots, s$ such that $(\pi_P)^{l_i} \sigma\delta x_i = x_{i+1}$, $i = 1, 2, 3, \dots, s-1$ in the support of P. For $x = x_1$ and $y = \delta x_s$, we have $y = \psi x$, $\psi \in \Psi_P$, where

$$\psi = \delta(\pi_P)^{l_{s-1}} \sigma\delta \dots (\pi_P)^{l_2} \sigma\delta(\pi_P)^{l_1} \sigma\delta.$$

From the arbitrariness of x and y, Ψ_P has exactly one orbit on $\mathscr{P}_P$. This is the sufficiency of the lemma. □

Lemma 2.3.2. *The support of any polyhedron P is always connected.*

Proof. By contradiction. Suppose the support of some polyhedron P is not connected. Without loss of generality, let A and B be the two connected components. Because the edge set of A is a proper subset of the edge set of P, P has a proper subset which forms a polyhedron, a contradiction to that P is a polyhedron. □

Lemma 2.3.3. *For a polyhedron P, Ψ_P has exactly one orbit on its ground set $\mathscr{X}_P$.*

Proof. A direct result of Lemmas 2.3.1 and 2.3.2. □

Lemma 2.3.4. *For a polyhedron P, the group generated by $\{\pi_P, \sigma\delta\}$ has at most two orbits on $\mathscr{X}_P$.*

Proof. Let $\Psi_{\{\pi_P,\sigma\delta\}}$ be the group generated by $\{\pi_P, \sigma\delta\}$. Because $\pi_P, \pi_P^* \in \Psi_{\{\pi_P,\sigma\delta\}}$ and $\pi_P(\sigma\delta x) = (\pi_P\sigma\delta)x$, we have that x and $\sigma\delta x$ are in the same orbit of $\Psi_{\{\pi_P,\sigma\delta\}}$ for any edge $\{x, \sigma x, \delta x, \sigma\delta x\}$. Thus, each orbit of $\Psi_{\{\pi_P,\sigma\delta\}}$ contains at least $|(\mathscr{X})_P|/2$ elements. This implies that $\Psi_{\{\pi_P,\sigma\delta\}}$ has at most two orbits. □

Lemma 2.3.5. *If a polyhedron P is orientable, then the group $\Psi_{\{\pi_P,\sigma\delta\}}$ has exactly two orbits on $\mathscr{X}_P$.*

Proof. Because P is orientable, $\{x\}_{\pi_P\sigma\delta}$ and $\{\sigma x\}_{\pi_P\sigma\delta}$ are different orbits on $\mathscr{X}_P$. From Lemma 2.3.4, the lemma is true. □

Theorem 2.3.1. *A polyhedron P is orientable if, and only if, the group generated by $\{\pi_P, \sigma\delta\}$ has exactly two orbits on $\mathscr{X}_P$.*

Proof. The necessity is from Lemma 2.3.5. Conversely, assume the group $\Psi_{\{\pi_P,\sigma\delta\}}$ has two orbits. Because x and σx should be in different orbits on $\mathscr{X}_P$ if any, the set $\{(y)_{\pi_P^*} | y \in \{x\}_{\Psi_{\{\pi_P,\sigma\delta\}}}\}$ ($\{(y)_{\pi_P^*} | y \in \{\sigma x\}_{\Psi_{\{\pi_P,\sigma\delta\}}}\}$ as well) forms P where $\pi_P^* = \pi_P\sigma\delta$. By virtue of each edge occuring twice in P with different powers if $\sigma\delta x$ (δx) is seen as x^{-1} ($(\sigma x)^{-1}$), P is orientable. This is the sufficiency. □

Theorem 2.3.2. *A polyhedron P is orientable if, and only if, its dual is orientable.*

Proof. Because of $\Psi_{\{\pi_P,\sigma\delta\}} = \Psi_{\{\pi_P\sigma\delta,\sigma\delta\}}$, Theorem 2.3.1 leads to this theorem. □

On the basis of this theorem, the orientability of a polyhedron can be determined by its skeleton instead of its support as in Theorem 2.2.1.

Theorem 2.3.3. *A polyhedron P is orientable if, and only if, one of the following statements is satisfied:*

(1) *What is obtained by contracting all edges of weight* 0 *on its skeleton is a bipartite graph.*
(2) *No odd-weight fundamental circuit is on the skeleton.*
(3) *No odd-weight circuit is on the skeleton.*
(4) *The* 1*-set on the skeleton forms a cocycle.*
(5) *The associate equation* (2.2.2) *on the skeleton is consistent.*

Proof. Because the skeleton of a polyhedron P is the support of its dual P^*, Theorems 2.3.2 and 2.2.1 lead to the proof of the theorem. □

The following five corollaries are all deduced from Theorems 2.2.1 and 2.3.3.

Corollary 2.3.1. *For a polyhedron P, the graph obtained by contracting all edges of weight* 0 *on its support is bipartite if, and only if, that is on its skeleton.*

Corollary 2.3.2. *For a polyhedron P, no odd-weight fundamental circuit is on its support if, and only if, that is on its skeleton.*

Corollary 2.3.3. *For a polyhedron P, no odd-weight circuit is on its support if, and only if, so is on its skeleton.*

Corollary 2.3.4. *For a polyhedron P, its support has the set of all* 1*-edges a cocycle if, and only if, so does its skeleton.*

Corollary 2.3.5. *For a polyhedron P, the associate equation* (2.2.2) *on its support is consistent if, and only if, so is that on its skeleton.*

2.4 Non-orientable polyhedra

On the basis of Theorem 2.2.1 (or Theorem 2.3.3), we are allowed to design an efficient algorithm for determining if a polyhedron is orientable, or non-orientable.

A *binary network* (i.e. a network with binary weights on edges) is said to be *balanced* if it has one of the properties (1–5) as shown in Theorem 2.3.3 (or Theorem 2.1.1).

Let $P = \{C_i|1 \le i \le s\}$ be a polyhedron with S_P and T_P, respectively, its support and skeleton.

Theorem 2.4.1. *The support S_P and skeleton T_P of a polyhedron P have the same balance (the property of balance).*

Proof. Because of Theorems 2.2.1 and 2.3.3, the theorem is soon found. □

This theorem enables us to discuss one of support and skeleton, in what follows, only skeleton is employed for determining a polyhedron without loss of generality.

For the skeleton T_P of a polyhedron $P = \{\mathscr{C}\}$, assume S is a cocycle. The vertex set V of P is partitioned (or, in other words, separated) into two parts: A and $B = V - A$ by S. The operation of interchanging the weights 1 and 0 on S and then replacing each $C^* \in \mathscr{C}_A^*$ (or for B as well) by C^{iv} is called a *switch* for S on P, where $\mathscr{C}_A$ is the restriction of $\mathscr{C}^*$ on A.

Theorem 2.4.2. *Let P' be obtained by a switch on a polyhedron P, then P' is still a polyhedron of no difference with P.*

Proof. Because of No.diff.gon4 and No.diff.gon5 with Proposition 2.1.1, the theorem follows. □

Corollary 2.4.1. *A switch does not change the balance of the skeleton.*

Theorem 2.4.3. *A polyhedron of no difference has a skeleton which has no cocycle of all edges with weight* 1.

Proof. If a skeleton has a cocycle of all edges with weight 1, then by switching the cocycle what is obtained has one less cocycle of all edges with weight 1. From Theorem 2.4.2, the resultant polyhedron is of no difference from the original one. By the finite recursion principle, the theorem is found. □

Lemma 2.4.1. *For the dual P^* of a polyhedron P, Ψ_{P^*} has exactly one orbit on $\mathscr{X}_{P^*}$ if, and only if, so does Ψ_P on $\mathscr{X}_P$.*

Proof. Because $\Psi_P = \Psi_{P^*}$, the lemma is done. □

Theorem 2.4.4. *Two polyhedra are of no difference if, and only if, so are their duals.*

Proof. On the basis of Theorems 2.2.1 and 2.3.3, from Theorem 2.4.2, this theorem is true. □

Lemma 2.4.2. *For any polyhedron P, its skeleton is always connected.*

Proof. Because of $\Psi_P = \Psi_{P^*}$, from Lemma 2.3.1 and the duality between support and skeleton the lemma is obtained. □

Although the non-orientability can be determined by orientability, the following theorem provides a way to determine the non-orientability directly.

Theorem 2.4.5. *A polyhedron P is non-orientable if, and only if, there exists an element $x \in \mathscr{X}_P$ such that both x and σx are in the same orbit of the group generated by $\{\pi_P, \sigma\delta\}$ on $\mathscr{X}_P$.*

Proof. From Lemmas 2.3.4 and 2.3.5, by considering that x and σx are all in different orbits of $\Psi_{\{\pi p,\sigma\delta\}}$ when two orbits $\Psi_{\{\pi p,\sigma\delta\}}$ has, the theorem is found. □

This theorem enables us to design an algorithm in time linear for the size to justify if a polyhedron is non-orientable.

2.5 Classic polyhedra

On the basis of Section 2.4, a classification can be done. That is the purpose of this section.

Theorem 2.5.1. *The orientability of a polyhedron does not change under switches.*

Proof. From the definition of orientability, the theorem is true. □

Lemma 2.5.1. *For a polyhedron P, there is another polyhedron P' whose skeleton has no cocycle of all edges with weight* 1 *such that P' is of no difference from P.*

Proof. By the similar procedure in the proof of Theorem 2.4.3, the theorem is proved. □

Such a polyhedron P' in Lemma 2.5.1 is said to be *classic*. From Lemma 2.5.1, it is seen that only classic polyhedra are enough to distinguish all polyhedra by no difference.

Let T be a minimal set of edges having an edge in common with every cocycle in the skeleton of a polyhedron. In fact, because of Lemma 2.4.2, it can be seen that T is a spanning tree.

Lemma 2.5.2. *A polyhedron P is classic if, and only if, whose skeleton has a spanning tree of all edges with weight* 0.

Proof. If a spanning tree T has an edge e with weight 1, then by doing a switch for the fundamental cocycle e is in, T has 1 less edges with weight 1. From Theorem 2.4.2, by the finite recursion principle, such a spanning tree T without an edge of weight 1 is obtained. This is the lemma. □

This lemma enables us to find a classic polyhedron among all those without difference efficiently.

Theorem 2.5.2. *A classic polyhedron is orientable if, and only if, all edges as letters have their two occurrences with different powers. A classic polyhedron is non-orientable if, and only if, the set of letters each of which has its two occurrences with the same power does not contain a cocycle.*

Proof. The first statement is deduced from Theorem 2.2.1(3). The second statement is, by contradiction, done from Theorems 2.3.3(3) and 2.5.1. □

By virtue of Lemma 2.5.2, this theorem enables us to classify all classic polyhedra with the same underlying graph of their skeletons based on a spanning tree.

Theorem 2.5.3. *Two orientable classic polyhedra with the same underlying graph of their skeletons are different if, and only if, they have a pair of different corresponding faces.*

Proof. On the basis of Theorems 2.5.2 and 2.4.4, the theorem can be soon found. □

Corollary 2.5.1. *Two orientable classic polyhedra with the same underlying graph of their skeletons are different if, and only if, their duals have a pair of corresponding faces different.*

Proof. A direct result of Theorems 2.5.2 and 2.4.4. □

Theorem 2.5.4. *Two non-orientable classic polyhedra with the same underlying graph of their supports with all pairs of corresponding faces with no difference are different if, and only if, for a spanning tree they have the sets of their edges with two occurrences of the same power different subsets of the set of cotree edges.*

Proof. By Theorem 2.4.1, Lemma 2.5.2 and Theorem 2.5.2, the theorem is proved. □

Corollary 2.5.2. *Two non-orientable classic polyhedra with the same underlying graph of their skeletons with all pairs of corresponding vertices with no difference are different if, and only if, for a spanning tree they have the sets of their edges with two occurrences of the same-power different subsets of the set of cotree edges.*

Proof. A direct result of Theorems 2.4.1 and 2.5.4. □

The cycles in P are called *faces* and those in P^* are *vertices*. Cycles in δ are edges. Let $\nu(P)$, $\epsilon(P)$ and $\phi(P)$ be, respectively, the number of vertices, edges and faces on P, then $\mathrm{Eul}(P) = \nu(P) - \epsilon(P) + \phi(P)$ is the *Eulerian characteristic* of P.

Theorem 2.5.5. *P^* is non-orientable if, and only if, so is P. And for any polyhedron P,* $\mathrm{Eul}(P) = \mathrm{Eul}(P^*)$.

Proof. The first statement is from Theorems 2.3.3 and 2.4.1. The second statement is from the facts that $\nu(P) = \phi(P^*)$ and $\epsilon(P) = \epsilon(P^*)$. □

2.6 Notes

2.6.1 Polyhedron is an abstraction of a polytope, which is a geometric concept in a Euclidean space, or general vector space. One might like to read more references such as Heffter [133], Edmonds [83], Ringel [287, 288], White [382], Liu [195, 196, 214, 216, 217, 224], etc.

2.6.2 Polygon is, in fact, a cyclic sequence of letters with each letter appearing at most twice as an abstraction of a polygon in geometry. Further, in next chapter, a surface can be dealt as a polyhegon (an abbreviation of polyhedral polygon), which is a polygon as well as a polyhedron, seen in Liu [200].

2.6.3 Switch on a polyhedron can be seen as a topological transformation from one to another shown in Liu [221, 222, 224].

2.6.4 For a spanning tree T on a graph G, all classic (general as well) polyhedra whose skeletons have all the graph G can be topologically classified as each class has exactly one classic polyhedron whose skeleton is with all tree edges of weight 0, or, in other words, each letter corresponding to a tree edge is with its two occurrences of different powers. The earliest references of this initial idea can be seen in Liu [195, 196].

2.6.5 In Theorems 2.2.1 and 2.3.3, because of (2), the recognitions of statements (1–5), as well as the orientability and the non-orientability, are all polynomially realizable by efficient algorithms, from the computing complexity. Such a type of results has no mention in topological books and articles before. The initiation of this idea is from Liu [192, 193].

2.6.6 For more new developments on the theory of polyhedra, one might like to read Liu [227, 233, 234], and others related such as Stahl [318] and Tutte [349].

3 Surfaces

3.1 Polyhegons

A polygon is called *polyhedral* if its edges are pairwise marked by occurrences of letters such that each pair of occurrences are with the same letter in same power or different powers: 1 (always omitted!) and −1. Naturally, any polyhedral polygon has even number of edges. For the sake of brevity, a polyhedral polygon is always abbreviated as a *polyhegon*. The number of occurrences of letters on a polyhegon is called its *size*.

The following items are used for clarifying the no distinction among polyhegons:

NoDist.1 Two letters interchanged, or a letter in a polyhegon is replaced by a new letter out of the polygon.

NoDist.2 For a polyhegon $P = (a, b, c, \dots, s, t)$,

$$\begin{aligned} P &= (b, c, \dots, s, t, a) = (c, \dots, s, t, a, b) = \dots \\ &= (t, a, b, c, \dots, s). \end{aligned}$$

NoDist.3 A segment in a polyhegon P moved to any place in P whenever the segment can be seen as a polyhegon itself.

For two polyhegons $X = (x_1, x_2, \dots, x_{2l})$ and $Y = (y_1, y_2, \dots, y_{2l})$, if $Y = (x_{2l}, \dots, x_2, x_1)$ then Y is called the *reversion* of X, or written as $Y = X^{rv}$. Of course, if a polyhegon X is the reversion of Y, then Y is also a reversion of X from $(X^{rv})^{rv} = X$.

Theorem 3.1.1. *If two polyhegons are the reversion of one to another, then they are without distinction.*

Proof. Since a polyhegon is a polyhedron itself, the conclusion is from No.Diff.gon3 in Section 2.1. □

For two polyhegons $X = (x_1, x_2, \dots, x_{2l})$ and $Y = (y_1, y_2, \dots, y_{2l})$, if $Y = (x_1^{-1}, x_2^{-1}, \dots, x_{2l}^{-1})$ then Y is called the *conversion* of X, or written as $Y = X^{cv}$. Of course, if a polyhegon X is the conversion of Y, then Y is also a conversion of X from $(X^{cv})^{cv} = X$.

Theorem 3.1.2. *If two polyhegons are the conversion of one to another, then they are without distinction.*

Proof. From NoDist.1, via interchanging each letter by its inverse the conclusion is soon obtained. □

DOI 10.1515/9783110479492-003

For two polyhegons $X = (x_1, x_2, \ldots, x_{2l})$ and $Y = (y_1, y_2, \ldots, y_{2l})$, if $Y = (x_{2l}^{-1}, \ldots, x_2^{-1}, x_1^{-1})$ then Y is called the *inversion* of X, denoted by $Y = X^{-1}$. Of course, if a polyhegon X is the inversion of Y, then Y is also the inversion of X.

Corollary 3.1.1. *If two polyhegons are the inversion of one to another, then they are without distinction.*

Proof. Because of $(X^{rv})^{cv} = (X^{cv})^{rv}$ and then $X^{-1} = (X^{rv})^{cv}$, a direct result of Theorems 3.1.1 and 3.1.2. □

A *topological surface* can be also defined to be a set of polyhegons in the sense of elementary equivalence. A polyhegon is therefore dealt with a *combinatorial surface.*

Theorem 3.1.3. *A polyhegon is non-orientable if, and only if, it has a letter of its two occurrences with the same power.*

Proof. Because of the support which is a bouquet without cocycle, by Theorem 2.2.1 the theorem is done. □

Example 3.1.1. All distinct polyhegons of size 6 are listed as

$$P_1 = (1, 1^{-1}, 2, 2^{-1}, 3, 3^{-1}),\ P_2 = (1, 2^{-1}, 2, 1^{-1}, 3, 3^{-1}),$$
$$P_3 = (1, 2, 1^{-1}, 2^{-1}, 3, 3^{-1}),\ P_4 = (1, 2, 3^{-1}, 1^{-1}, 3, 2^{-1}),$$
$$P_5 = (1, 2, 3, 3^{-1}, 1^{-1}, 2^{-1}),$$

where numbers are used as letters for some convenience and a comma is inserted between two numbers (letters) to avoid confusion.

Among all $5! = 120$ polyhegons as cycles of length 6, only 5 are distinct as shown above.

More about distinct polyhegons of small size can be seen in Appendix II in Liu [219].

The following three types of operations on polyhegons are established for carrying out the classification of surfaces.

Transform 1. $(AaBa^{-1}) \Leftrightarrow (AB)$ whenever (A) or (B) is a polyhegon.

Two cases should be clarified on Transform 1. One is that A and/or B is allowed to be empty. The other is that the empty is seen as a polyhegon in its own right.

Transform 2. For A, B, C and D in a polyhegon,

$$(AxByCx^{-1}Dy^{-1}) \Leftrightarrow (A^{rv}B^{rv}C^{rv}D^{rv}xyx^{-1}y^{-1}).$$

By virtue of that $A^{rv}B^{rv}C^{rv}D^{rv}$ is a polyhegon itself, since

$$(A^{rv}B^{rv}C^{rv}D^{rv}) = (ADCB)^{rv},$$

we have $(A^{rv}B^{rv}C^{rv}D^{rv}xyx^{-1}y^{-1})$

$$\begin{aligned} &= ((ADCB)^{rv}xyx^{-1}y^{-1}) \text{ (by Theorem 3.1.1)} \\ &= (ADCBxyx^{-1}y^{-1}). \end{aligned}$$

That is

$$(AxByCx^{-1}Dy^{-1}) \Leftrightarrow (ADCBxyx^{-1}y^{-1}).$$

In particular, if EA is used instead of A, then

$$(AxByCx^{-1}Dy^{-1}E) \Leftrightarrow (ADCBExyx^{-1}y^{-1}). \qquad (3.1.1)$$

Transform 3. $(AxBx) \Leftrightarrow (AB^{-1}xx)$.

By virtue of that AB^{-1} is a polyhegon itself, it is also seen that

$$(AB^{-1}xx) = (B^{-1}Axx).$$

In particular, if CA is used instead of A, then

$$(AxB^{-1}xC) \Leftrightarrow (AB^{-1}Cxx). \qquad (3.1.2)$$

Lemma 3.1.1. $(Axyx^{-1}y^{-1}zz) \Leftrightarrow (Axxyyzz)$.

Proof. Because of A itself being a polyhegon, we are allowed only to discuss $(xyx^{-1}y^{-1}zz)$. By Transform 3 for z, $(xyx^{-1}y^{-1}zz)$

$$\begin{aligned} &\Leftrightarrow (xyzyxz), \text{ by. eq (3.1.1) for } y, \\ &\Leftrightarrow (xz^{-1}xzyy), \text{ by. eq (3.1.1) for } x, \\ &\Leftrightarrow (zzyyxx), \text{ by NoDist.1 and NoDist.2}, \\ &= (xxyyzz). \end{aligned}$$

Then, in a similar manner, by considering the polyhegon, the lemma is soon derived. □

In Transforms 2–3 and Lemma 3.1.1, all capital letters A, B, C and D have no restriction except only for making sense as a polyhegon.

Lemma 3.1.2. *Let P be an orientable polyhegon not in form as $(AxByCx^{-1}Dy^{-1}E)$, then $P \Leftrightarrow (aa^{-1})$.*

Proof. Assume $P = (XxYx^{-1}Z)$ for some x without loss of generality by P not in form as $(AxByCx^{-1}Dy^{-1}E)$. If such $Y = \emptyset$, then P is replaced by $P' = (X, Z)$, which has its size 2 less than that of P with the same condition. Otherwise, because of $Y \neq \emptyset$ and P not in form as $(AxByCx^{-1}Dy^{-1}E)$, there exists another y in Y such that $Y = \langle Y_1yB_1y^{-1}Y_2\rangle$. Let $X_1 = XxY_1$ and $X_2 = Y_2x^{-1}Z$, then P is replaced by $P' = (X_1yB_1y^{-1}X_2)$ in which B_1 has its size 2 less than that of Y under the same condition that P has. By Transform 1 and the finite recursion principle, the lemma is derived. □

Theorem 3.1.4. *Let P be an orientable polyhegon, then*

$$P \Leftrightarrow \begin{cases} (aa^{-1}), & \text{not in form as } (AxByCx^{-1}Dy^{-1}E); \\ \left(\prod_{i=1}^{p} a_ib_ia_i^{-1}b_i^{-1}\right), & \text{otherwise} \end{cases}$$

where $p \geq 1$ is an integer.

Proof. When P is not in form as $(AxByCx^{-1}Dy^{-1}E)$, Lemma 3.1.2 leads to the conclusion. Otherwise, by eq. (3.1.1) and the finite recursion principle, the theorem is done. □

Theorem 3.1.5. *For a non-orientable polyhegon Q,*

$$Q \Leftrightarrow \left(\prod_{i=1}^{q} a_ia_i\right),$$

where $q \geq 1$ is an integer.

Proof. Because of the non-orientability, $Q = (AxBxC)$ for some x in Q. By eq. (3.1.2) and the finite recursion principle, there is an integer $k \geq 1$ such that

$$Q \Leftrightarrow \left(A\prod_{i=1}^{k} a_ia_i\right),$$

where A is not yet non-orientable. If $A = \emptyset$, then $k = q$ and hence the theorem is done. Otherwise, since (A) is orientable, by Theorem 3.1.4, there exists an integer $s \geq 1$ such that

$$(A) \Leftrightarrow \left(\prod_{i=1}^{s} a_ib_ia_i^{-1}b_i^{-1}\right).$$

Then, by Lemma 3.1.1 and the finite recursion principle, the theorem is done for $q = 2s + k$. □

The last two theorems show the topological classification of surfaces under the transformations given in Transforms 1–3.

3.2 Surface closed curve axiom

According to Section 3.1, a *surface* is seen as a polyhegon (i.e., polyhedral polygon) with edges pairwise identified by the same letter in the same direction or in different directions according as its powers are the same or not. This is corresponding to a compact 2-manyfold without boundary in classic topology. It has turned out that a surface can be seen as the result of deforming a polyhegon which is topologically equivalent to a disc on the plane, more precisely, identifying edges pairwise with directions coincide

Very simple and fundamental surfaces are first observed for use in generalization.

On the plane, or the sphere when infinity is considered as a point shown in Figure 3.2.1(a), the Jordan closed curve axiom tells us that any closed curve divides the plane into two open domains with the curve as their common boundary. This shows that the closed curve has two sides because no curve segment connects one point near the boundary in one domain to that in another domain of the curve without intersecting.

A general surface is not necessary to have such a fact in any case, e.g., each closed curve represented by a dotted line in the projective plane or the Klein bottle shown, respectively, in Figure 3.2.1(b) or Figure 3.2.2(b) is not with two sides, but with only one side.

However, a general surface satisfies the following fact as an axiom. A curve here can always be seen as the limit of a polygon when their edges tend to 0 in length.

Surface closed curve axiom A closed curve C on a surface has one of the two possibilities: one side and two sides. Any curve between two points in different connected components of the complement of a two-sided closed curve C has a point in common with C. The complement of a one-sided closed curve C has exactly one connected component.

A curve with two sides is called a *doubled sided curve*; otherwise, a *single-sided curve*. As shown in Figure 3.2.1(a), any closed curve on a sphere is a doubled sided curve. However, it is different from the sphere for the projective plane shown in Figure 3.2.1(b), or the Klein bottle shown in Figure 3.2.2(a), in that both a single (shown by a dotted line) and a double-sided closed curve are allowed to occur.

How do we justify whether a closed curve on a surface is of single side or not?

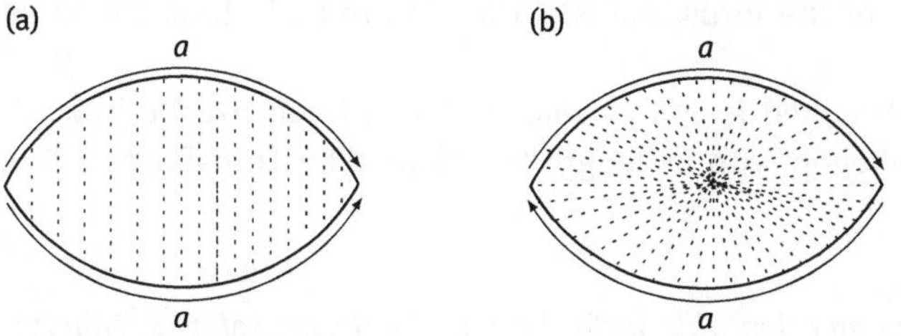

Figure 3.2.1: Sphere and projective plane: (a) the plane and (b) the projective plane.

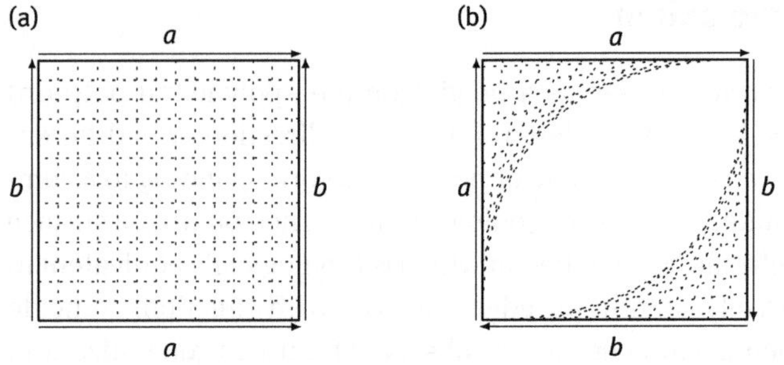

Figure 3.2.2: Torus (a) and Klein bottle (b).

In order to answer this question precisely, the concept of contractibility of a curve has to be clarified. If a closed curve on a surface can be continuously contracted along one side into a point, then it is said to be *contractible*, or *homotopic* to 0.

It is seen that a single-sided closed curve is never homotopic to 0 and a double-sided curve has the possibility of being homotopic to 0. For example, in Figure 3.2.2(a), i.e. in the torus, each of the dotted lines is double sided but not contractible. In Figure 3.2.2(b), i.e. in the Klein bottle, all the closed curves represented by dotted lines are of single side, and hence, none of them is contractible.

A surface is *orientable* if, and only if, no closed curve on the surface is with a single side. For example, in Figure 3.2.1, the sphere is orientable and the projective plane is non-orientable. In Figure 3.2.2, the torus is orientable and the Klein bottle is non-orientable.

From Theorem 3.1.4, all orientable surfaces of genus not 0 can be composed only by the torus. From Theorem 3.1.5, all non-orientable surfaces, only by the projective plane.

The maximum number of closed curves cutting along without destroying the continuity on a surface is called the *pregenus* of the surface.

In view of surface closed curve axiom, there is no such closed curve on the sphere because the complement of any closed curve has two connected components. Thus, the pregenus of sphere is 0. On the projective plane, only one such curve is available (each of dotted lines is such a closed curve in Figure 3.2.1(a)) and hence the pregenus of projective plane is 1.

Similarly, the pregenera of the torus and Klein bottle are both 2, as shown in Figure 3.2.2.

A polyhegon of size $2l$ is called *length irreducible* (or in brief, *irreducible*) if it never becomes another polyhegon with its length less than $2l$ by Transforms 1–3 in Section 3.1.

Lemma 3.2.1. *The pregenus of an orientable surface is half the the size of its irreducible polyhegon.*

Proof. Because of a handle with pregenus 2, the lemma holds. □

Lemma 3.2.2. *The size of an irreducible orientable polyhegon has 4 as a factor.*

Proof. Because of each pregenus contributing two occurrences to the polyhegon obtained by cutting such a closed curve, Lemma 3.2.1 leads to this lemma. □

Theorem 3.2.1. *The pregenus of an orientable surface is a non-negative even number.*

Proof. By virtue of Lemmas 3.2.1 and 3.2.2, the theorem follows. □

Lemma 3.2.3. *The pregenus of a non-orientable surface is half the size of its irreducible polyhegon.*

Proof. Because of each crosscap with pregenus 1, the lemma is obtained soon. □

Lemma 3.2.4. *The size of a non-orientable polyhegon has 2 as a factor.*

Proof. Because of each pregenus contributing one occurrence to the polyhegon obtained by cutting such a closed curve, Lemma 3.2.3 leads to this lemma. □

Theorem 3.2.2. *The pregenus of a non-orientable surface is a positive integer.*

Proof. A direct result of Lemmas 3.2.3 and 3.2.4. □

Based on this theorem, the genus of an orientable surface can be defined to be half its pregenus, called an *orientable genus*. The genus of a non-orientable surface, called *non-orientable genus*, is its pregenus itself.

The sphere is written as aa^{-1}, where a^{-1} is in the opposite direction of a on the boundary of the polyhegon. Thus, the projective plane, torus and Klein bottle are, respectively, aa, $aba^{-1}b^{-1}$ and $aabb$. In general,

$$\begin{aligned} O_p &= \prod_{i=1}^{p} a_i b_i a_i^{-1} b_i^{-1} \\ &= \left(\prod_{i=1}^{p-1} a_i b_i a_i^{-1} b_i^{-1} \right) a_p b_p a_p^{-1} b_p^{-1} \end{aligned} \tag{3.2.1}$$

and

$$Q_q = \prod_{i=1}^{q} a_i a_i = \left(\prod_{i=1}^{q-1} a_i a_i \right) a_q a_q \tag{3.2.2}$$

denote, respectively, a surface of orientable genus $p \geq 1$ and a surface of non-orientable genus $q \geq 1$. Of course, O_0, Q_1, O_1 and Q_2 are, respectively, the sphere, projective plane, torus and Klein bottle.

From Theorem 3.1.3, all O_p, $p \geq 0$, are orientable and all Q_q, $q \geq 1$, are non-orientable.

Forms shown in eqs. (3.2.1) and (3.2.2) are said to be *standard*. Here, $O_p(p \geq 0)$ is called the orientable standard surface of genus p and $Q_q(q \geq 1)$, the non-orientable standard surface of genus q.

3.3 Topological transformations

A surface is seen as a set of polyhedra. Surfaces are then classified by a type of topological equivalence. Let **P** be the set of all polyhedra.

For $P = \{(A_i)|i \geq 1\} \in \mathbf{P}$, the following three operations including their inverses are called *elementary transformation*. It is seen that they are all topological between two polyhedra on a surface.

Operation 0: For $(A) = (Xaa^{-1}Y) \in \mathbf{P}$, $(A) \Leftrightarrow (XY)$ where at least one of X and Y is not empty as shown in Figure 3.3.1.

Operation 1: For $(A) = (XabYab) \in \mathbf{P}$(or $(XabYb^{-1}a^{-1})$), $(A) \Leftrightarrow (XaYa)$(or $XaYa^{-1}$) as shown in Figure 3.3.2.

Operation 2: For $(A) = (Xa)$, $(B) = (a^{-1}Y) \in \mathbf{P}$, $(XY) \Leftrightarrow (\{(A), (B)\})$ where at least one of X and Y is not empty and not in the case that both (X) and (Y) are polyhedra as shown in Figure 3.3.3.

If a polyhedron P can be obtained by elementary transformation into another polyhedron Q, then they are called *elementary equivalence*, denoted by $P \sim_{el} Q$. In topology, the elementary equivalence is topological in 2-dimensional sense.

Theorem 3.3.1. *For $P \in \mathbf{P}$, if $P = (XY)$ with both (X) and (Y) as polyhedra, then for any $a \notin A \cup B$, $P \sim_{el} (XaYa^{-1})$ as shown in Figure 3.3.4 (as Operation 3).*

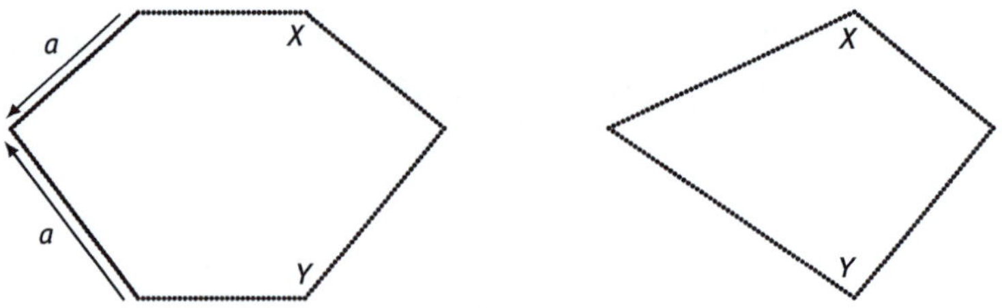

Figure 3.3.1: Operation 0: $Aaa^{-1} \Longleftrightarrow A$.

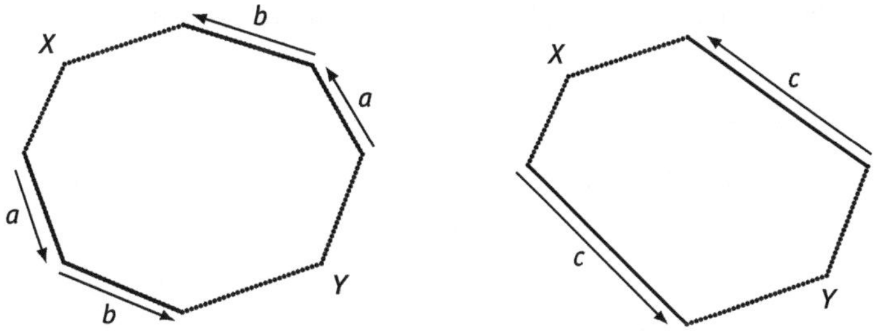

Figure 3.3.2: Operation 1: $AabBab \Longleftrightarrow AcBc$.

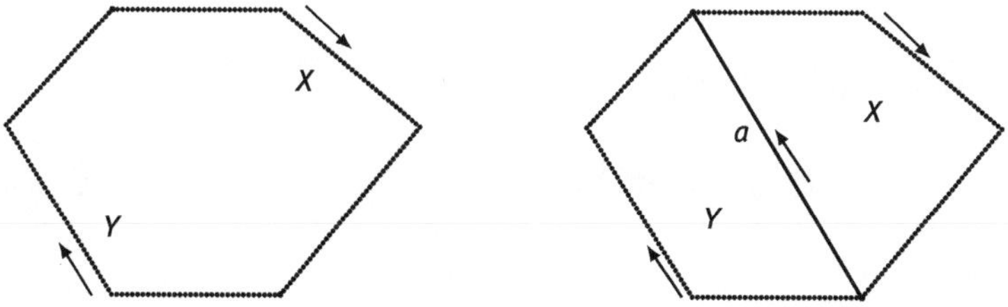

Figure 3.3.3: Operation 2: $(XY) \Longleftrightarrow \{(Xa), (a^{-1}Y)\}$.

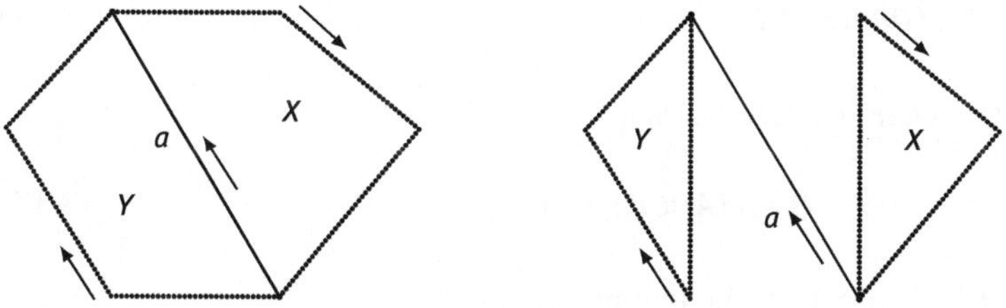

Figure 3.3.4: Operation 3: $\{(A), (B)\} \Longleftrightarrow (XaYa^{-1})$.

Proof. From Operation 0,

$$\begin{aligned} P &\sim_{el} (Xaa^{-1}Y), \text{ by NoDist.3,} \\ &= ((X)(aa^{-1}Y)), \text{ by NoDist.1–2,} \\ &= ((X)(aYa^{-1})) = (XaYa^{-1}). \end{aligned}$$ □

Theorem 3.3.2. *For $P \in \mathbf{P}$, there exists a polyhedron $Q = (X) \in \mathbf{P}$ where X is a linear order such that $P \sim_{el} Q$.*

Proof. Let $P = (\{(A_i)|1 \leq i \leq k\})$. If $k = 1$, P is in the form as Q itself. If $k \geq 2$, by employing Operation 2 step by step to reduce the number of cycles one by one if any, from no proper subpolyhedron the form Q can be found. □

From Theorems 3.3.2, for classifying **P** it suffices only to discuss polygons as Q.

Lemma 3.3.1. *Let* $Q = (AxByCx^{-1}Dy^{-1})$, *then*

$$Q \sim_{el} (ADxyBx^{-1}Cy^{-1}). \tag{3.3.1}$$

Proof. It is seen that

$$\begin{aligned} Q &\sim_{el} ((Axz)(z^{-1}ByCx^{-1}Dy^{-1})) \text{ (by Operation 2)} \\ &\sim_{el} (zADy^{-1}z^{-1}ByC) \text{ (by Operation 2)} \\ &= (ADxyBx^{-1}Cy^{-1}). \end{aligned}$$

□

Lemma 3.3.2. *Let* $Q = (AxByCx^{-1}Dy^{-1})$, *then*

$$Q \sim_{el} (BAxyx^{-1}DCy^{-1}). \tag{3.3.2}$$

Proof. It is seen that

$$\begin{aligned} Q &\sim_{el} ((x^{-1}Dy^{-1}Axz)(ByCz^{-1})) \text{ (by Operation 2)} \\ &\sim_{el} (BAxzx^{-1}DCz^{-1}) \text{ (by Operation 2)} \\ &= (BAxyx^{-1}DCy^{-1}). \end{aligned}$$

□

Theorem 3.3.3. *Let* $Q = (AxByCx^{-1}Dy^{-1})$, *then*

$$Q \sim_{el} (ADCBxyx^{-1}y^{-1}). \tag{3.3.3}$$

Proof. From Lemmas 3.3.2 and 3.3.1, the lemma is soon done. □

According to Theorem 3.3.3, if A is replaced by EA in polyhedron $(ADCB)$, then the relation is soon derived as

Relation 1 : For A, B, C, D and E in a polyhegon,

$$(AxByCx^{-1}Dy^{-1}E) \sim_{el} (ADCBExyx^{-1}y^{-1}).$$

Theorem 3.3.4. *Let* $Q = (AxBx) \in \mathbf{P}$, *then* $Q \sim_{el} (AB^{-1}xx)$.

Proof. It is seen that

$$\begin{aligned} Q &\sim_{el} ((Axz)(z^{-1}Bx)) \\ &= ((zAx)(x^{-1}B^{-1}z)) \text{ (by Operation 2)} \\ &\sim_{el} (zAB^{-1}z)\text{(by Operation 2)} \\ &= (AB^{-1}xx). \end{aligned}$$

□

According to Theorem 3.3.4, if A is replaced by CA in polyhedron (AB^{-1}), then the relation is soon derived as

Relation 2 : $(AxBxC) \sim_{el} (AB^{-1}Cxx)$.

Theorem 3.3.5. *Let* $Q = (Axyx^{-1}y^{-1}zz) \in \mathbf{P}$, *then* $Q \sim_{el} (Axyzyxz)$.

Proof. It is seen that

$$\begin{aligned} Q &\sim_{el} ((zAxyt)(t^{-1}x^{-1}y^{-1}z)) \text{ (by Operation 2)} \\ &\sim_{el} (Axytyxt) \text{ (by Operation 2)} \\ &= (Axyzyxz). \end{aligned}$$

□

According to Theorem 3.3.5, then by Relation 2 twice for x and y, the relation is soon derived as

Relation 3 :$(Axyx^{-1}y^{-1}zz) \sim_{el} (Axxyyzz)$.

Theorem 3.3.6. *If* $Q \in \mathbf{P}$ *orientable not in form as*

$$(AxByCx^{-1}Dy^{-1}E),$$

then $Q \sim_{el} (xx^{-1})(= O_0)$.

Proof. Because Q is not in the above form, Q has to be in the form of $(Axx^{-1}B)$. If both A and B are empty, thenm $Q \sim_{el} (xx^{-1})$; otherwise, $Q \sim_{el} (AB)$. Because (AB) still satisfies the given condition, by the finite recursion principle, (xx^{-1}) can be found. □

Theorem 3.3.7. *For any A, B and C meaningful on a polyhegon,* $(ABxCx^{-1}) \sim_{el} (BAxCx^{-1})$.

Proof. It is seen that $(ABxCx^{-1})$

$$\begin{aligned} &\sim_{el} ((Bxy)(y^{-1}Cx^{-1}A)) \text{ (by Operation 2)} \\ &= ((yBx)(x^{-1}Ay^{-1}C)) \text{ (by NoDist.2 in Section 3.1)} \\ &\sim_{el} (yBAy^{-1}C) \text{ (by Operation 2)} \\ &= (BAy^{-1}Cy) \text{ (by NoDist.2 in Section 3.1).} \end{aligned}$$

Then, by substituting x for y^{-1}, the theorem is soon obtained from NoDist.1 in Section 3.1. □

Theorem 3.3.8. *For any A, B, C and D meaningful on a polyhegon* $(AxBx^{-1}CyDy^{-1}) \sim_{el} (ACxDx^{-1}yBy^{-1})$.

Proof. It is seen that $(AxBx^{-1}CyDy^{-1})$

$$\begin{aligned}&\sim_{el} ((Axz)(x^{-1}CyDy^{-1})) \text{ (by Operation 2)}\\&= ((zAx)(x^{-1}CyDy^{-1}z^{-1}B)) \text{ (by NoDist.2 in Section 3.1)}\\&\sim_{el} (zACyDy^{-1}z^{-1}B) \text{ (by Operation 2)}\\&= (ACyDy^{-1}z^{-1}Bz) \text{ (by NoDist.2 in Section 3.1).}\end{aligned}$$

Then, by substituting x for z^{-1}, the theorem is soon obtained from NoDist.1 in Section 3.1. □

Theorem 3.3.9. *For any A, B, C and D meaningful on a polyhegon,*

$$(AxBx^{-1}CyDy^{-1}) \sim_{el} (ACyxBx^{-1}yDy^{-1}).$$

Proof. It is seen that $(AxBx^{-1}CyDy^{-1})$

$$\begin{aligned}&\sim_{el} ((x^{-1}Cyz)(z^{-1}Dy^{-1}AxB)) \text{ (by Operation 2)}\\&= ((Cyzx^{-1})(xBz^{-1}Dy^{-1}A)) \text{ (by NoDist.2 in Section 3.1)}\\&\sim_{el} (CyzBz^{-1}Dy^{-1}A) \text{ (by Operation 2)}\\&= (ACyzBz^{-1}Dy^{-1}) \text{ (by NoDist.2 in Section 3.1).}\end{aligned}$$

Then, by substituting x for z, the theorem is soon obtained from NoDiff.1 in Section 3.1. □

By observing Theorems 3.3.7–3.3.9, suitably commutative principles can be naturally extracted for certain usages.

3.4 Complete invariants

On the basis of the last section, it suffices to determine the invariants of polyhegons under the elementary equivalence for all polyhedra on surfaces.

Theorem 3.4.1. *Every class of polyhedra under elementary equivalence is with the same orientability.*

Proof. By virtue of Theorem 2.2.1, it is easily checked that all the operations 0–2 on polyhedra in Section 3.3 do not change the orientability. Therefore, the theorem is true. □

This theorem tells us that the orientability on polyhedra is an invariant under the elementary equivalence.

Theorem 3.4.2. *An orientable polyhegon is of genus 0 if, and only if, it is not in form of* $(AxByCx^{-1}Dy^{-1})$.

Proof. If an orientable polyhegon is in the form of $(AxByCx^{-1}Dy^{-1})$, then by Relation 1 and that $(ADCBE)$ is also orientable, the polyhegon has its genus at least 1, never 0. This is the necessity.

Conversely, Theorem 3.3.6 leads to the sufficiency. □

Lemma 3.4.1. *A polyhegon not of genus 0 has to be in one of the following two distinct forms:* $(AxByCx^{-1}Dy^{-1}E)$ *and* $AxBx$.

Proof. Since a polyhegon P is either orientable or non-orientable, two cases are discussed.

Case 1. When P is orientable. Because of Theorem 3.4.2, P is in the form of $(AxByCx^{-1}Dy^{-1}E)$.

Case 2. When P is non-orientable. On account of Theorem 3.1.3, P is in the form of $AxBx$.

Therefore, the lemma is true. □

Theorem 3.4.3. *For a polyhegon P in the form of*

$$(AxByCx^{-1}Dy^{-1}E)$$

is orientable if, and only if, polyhegon $(ADCBE)$ *is orientable. If P is orientable, then its genus is* $p(\geq 1)$ *if, and only if,* $(ADCBE)$ *is with genus* $p-1$.

Proof. From Relation 1 in Section 3.3, $P \sim_{el} (ADCBExyx^{-1}y^{-1})$. Because of $(ADCBE)$ as a polyhegon as well, the first statement is true.

Further, because of $(ADCBE)$ orientable, the genus p_1 of

$$(ADCBE)$$

is non-negative. Since $p = p_1 + 1$, we have $p_1 = p - 1$. This is the second statement. □

Theorem 3.4.4. *For a polyhegon P in the form of*

$$(AxByCx^{-1}Dy^{-1}E)$$

is non-orientable if, and only if, the polyhegon (ADCBE) is non-orientable. If P is non-orientable, then its genus is $q(\geq 3)$ if, and only if, (ADCBE) has genus $q-2$.

Proof. By employing Relation 1 in Section 3.3, $P \sim_{el} (ADCBExyx^{-1}y^{-1})$. Because of $(ADCBE)$ as a polyhegon as well and $xyx^{-1}y^{-1}$ orientable, the first statement is found.

Further, because of $(ADCBE)$ non-orientable, in view of Relation 3 in Section 3.3, the genus q_1 of $(ADCBE)$ is at least 3. By considering $q = q_1 + 2$, $q_1 = q - 2$. This is the second statement. □

Theorem 3.4.5. *Given a polyhegon P in the form of (AxBxC). If polyhegon $(AB^{-1}C)$ is non-orientable, then P has non-orientable genus $q(\geq 2)$ if, and only if, polyhegon $(AB^{-1}C)$ is non-orientable genus $q-1$. If polyhegon $(AB^{-1}C)$ is orientable, then P has non-orientable genus $q \geq 1 (q = 1 (\mod 2))$ if, and only if, polyhegon $(AB^{-1}C)$ is of orientable genus $(q-1)/2$.*

Proof. From Theorem 3.1.3, P is non-orientable. From Relation 2 in Section 3.3, $P \sim_{el} (AB^{-1}Cxx)$.

Because $(AB^{-1}C)$ is a polyhegon as well as non-orientable, assume the non-orientable genus of $(AB^{-1}C)$ is q_1. By virtue of Relation 3 in Section 3.3, $q = q_1 + 1$, i.e. $q_1 = q - 1$. This is the first conclusion of the theorem.

Further, if $(AB^{-1}C)$ is orientable, assume the orientable genus of $(AB^{-1}C)$ is p_1. By Relation 3, $q = 2p_1 + 1$, i.e. $p_1 = (q-1)/2$. This is the second conclusion of the theorem. □

On the basis of Lemma 3.1.2 and Theorems 3.1.4 and 3.1.5, surfaces in topology are in fact the classes of polyhedra under the elementary equivalence. Surfaces O_0, $O_p(p \geq 1)$ in eq. (3.2.1) are, respectively, orientable *standard surfaces* of genus 0, $p(p \geq 1)$. Surfaces $Q_q(q \geq 1)$ in eq. (3.2.2) are non-orientable standard surfaces of genus q.

3.5 Graphs on surfaces

A topological representation of a graph on a surface as a point set without interior point in common between two edges as curve segments is called an *embedding*. If an intersection at an interior point between two curve segments is permitted, then what is obtained is an *immersion*. However, embeddings in this book are all with a condition (nothing in common sense) that its complement on the surface has all connected components contractible. So, an embedding of a graph G on a surface can be combinatorially seen as a polyhedron whose skeleton has its underlying graph G.

Two ways have been employed for constructing an orientable embedding of a graph on a surface as shown, for example, in Ringel [288].

One goes from the following procedure.

Procedure 3.5.1. First, choose a rotation of semi-edges at each vertex.

Second, immerse the graph on the plane in correspondence with the rotation clockwise at each vertex. This can always be done because of the Jordan curve axiom on the plane.

Third, travel along edges on one side never crosses to another side till meeting the starting position to get a face.

Final, halt as each edge passes through twice.

The other is based on partitioning the vertices into two classes marked by hollow dots and thick dots with their rotations, respectively, clockwise and anticlockwise in the immersion.

The only difference is, in travelling, that crossing from one side to another whenever the two ends with different marks.

The former is shown in Figure 3.5.1(a) and the latter in Figure 3.5.1(b).

Lemma 3.5.1. *The curve segment with its two ends v and u has to have a point in common with a closed curve C on a surface S if C is contractible and the contractible connected component of S – C does not contain both u and v.*

Proof. Because of $S - C$ with a component contractible, C is two sides, on account of the surface closed curve axiom, the lemma is soon found. □

Theorem 3.5.1. *What is obtained by one of the two ways above is both an orientable embedding of the graph.*

Proof. Since the procedure guarantees always on one side of a curve, from the finite recursion principle, each closed curve obtained by travelling forms a face. Therefore,

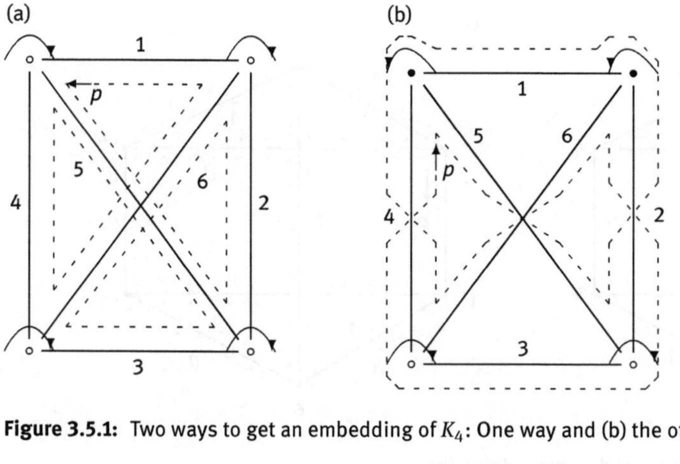

Figure 3.5.1: Two ways to get an embedding of K_4: One way and (b) the other way.

from the surface closed curve axiom, the result is an embedding of the graph. It is orientable because of each pair of occurrences with different directions whenever each starting side is in the second occurrence of the edge, the travelling direction is chosen to be different from that of the first occurrence on the basis of G connected and each face boundary of two sides. □

In order to establish a procedure more general for finding any embedding of a graph, orientable and non-orientable, the second menthod mentioned above reminds us to design each edge of the graph as weight 0 or 1 while a rotation is done at each vertex. Such a weighted graph is called a *joint network* of the graph shown in Figure 3.5.2(a).

Procedure 3.5.2. First, immerse the graph on the plane according to the rotation at each vertex. Mark each edge of weight 1 as a sign of cross, called a *twist*.

Second, travel along edges on one side never crossing to another side except only for those with the sign crossing to the other side till meeting the starting position to get a face.

Finally, halt as each edge passing through twice as shown in Figure 3.5.2(b).

Example 3.5.1. Given a graph underlined by the joint network shown in Figure 3.5.2(a). The embedding obtained by following Procedure 3.5.1 is $P = \{A, B\}$ where

$$A = (abcdbigfhgec^{-1}ih^{-1})$$

and

$$B = (a^{-1}f^{-1}ed).$$

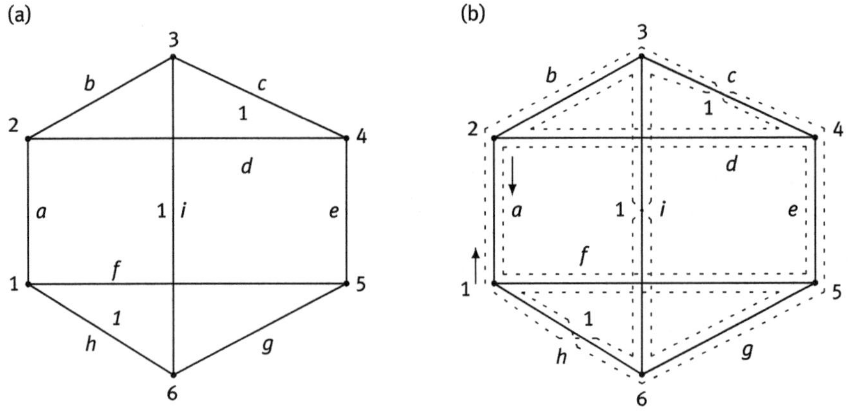

Figure 3.5.2: A joint network (a) and an embedding (b).

By employing the prime to distinguish the two occurrences of each letter, $A = (abcdb'igfhg'ec'^{-1}i'h'^{-1})$, or

$$(a^{-1}h'i'^{-1}c'e^{-1}g'^{-1}h^{-1}f^{-1}g^{-1}i^{-1}b'^{-1}d^{-1}c^{-1}b^{-1}),$$

and

$$B = (a'^{-1}f'^{-1}e'd'), \text{ or, } (a'd'^{-1}e'^{-1}f').$$

The vertex set of P is $\{1, 2, 3, 4, 5, 6\}$ where $1 = (af'^{-1}h)$, or $(a'h'f^{-1})$, $2 = (a'^{-1}bd^{-1})$, or $(a^{-1}d'^{-1}b')$, $3 = (b'^{-1}ci')$, or $(b^{-1}ic')$, $4 = (c'^{-1}de'^{-1})$, or $(c^{-1}e^{-1}d')$, $5 = (ef'g^{-1})$, or $(e'g'^{-1}f)$ and $6 = (g'i^{-1}h'^{-1})$, or $(gh^{-1}i'^{-1})$. By missing the power -1 and substituting -1 for the prime, $1 = (af^{-1}h)$, $2 = (a^{-1}bd)$, $3 = (b^{-1}ci^{-1})$, $4 = (ced^{-1})$, $5 = (e^{-1}g^{-1}f)$ and $6 = (ghi^{-1})$. It is seen that the joint network with the rotation at each vertex is just the skeleton of P.

Theorem 3.5.2. *What is obtained by Procedure 3.5.2 is a general embedding of graph G with the joint network having the same balance of its skeleton.*

Proof. In fact, as shown in Example 3.5.1 in general, Procedure 3.5.2 produces an embedding (orientable or non-orientable) whose skeleton is the joint network with the given rotation at each vertex. From Theorem 2.4.1, the theorem is done. □

By virtue of the theory described in Section 2.5, all embeddings of a graph on surfaces can be classified by no difference, or topological equivalence on the basis of a spanning tree chosen in any way.

Given a graph G, let T be a spanning tree of G, $\bar{T}$, the cotree of T. The family of all subsets of the set of all edges in $\bar{T}$ is denoted by $\mathscr{D}$. Its cardinality is the Betti number β independent of the choice of T on G. For the vertex set V of G, let $\mathscr{R}$ be the set of all rotation systems $R = \{\gamma(v)|\forall v \in V\}$ where $\gamma(v)$ is the rotation at vertex v.

An embedding of a graph is said to be *classic* if the corresponding polyhedron is classic. According to the theory mentioned in Sections 2.5 and 3.4, only classic embeddings are necessary to be discussed for the topological classification of embeddings.

Lemma 3.5.2. *In each class of embeddings of a graph G, there is exact one which is classic for a spanning tree T.*

Proof. By the procedure in the proof of Lemma 2.5.2, any class has an embedding which is classic for the tree. From Theorems 2.5.3 and 2.5.4, only one such embedding is done. □

Theorem 3.5.3. *The bi-tuple (R, D) for $R \in \mathscr{R}$ and $D \in \mathscr{D}$ is a complete invariant in a class within no distinction (or topological equivalence in other words) of classic embeddings of a graph for a spanning tree.*

Proof. A direct result of Theorems 2.5.3 and 2.5.4. □

For a given (R, D), $R \in \mathscr{R}$ and $D \in \mathscr{D}$, for spanning tree T of G, let $\mathscr{E}_T$ be the set of all classic embeddings in the class determined by (R, D).

Theorem 3.5.4. *For any spanning tree T' other than T, there is exactly one embedding which is classic for T' in $\mathscr{E}_T$.*

Proof. From Lemma 2.5.2 and Theorem 2.5.3, the theorem is derived. □

This theorem tells us that the classification presented in Theorem 3.6.3 does not depend on the choice of tree T.

3.6 Up-embeddability

The following procedure can be used for finding all embeddings of a graph on surfaces without repetition in the sense of no difference.

Procedure 3.6.1. First, given a rotation (a cyclic order of all semi-edges at a vertex) at each vertex of G.

Second, find a tree (spanning, of course) T on G and distinguish all the edges not on T by letters.

Third, replace each edge not of T by two articulate edges with the same letter with different powers, 1 and −1, or the same power.

The tree obtained from T in companion with pairs of articulate edges corresponding to cotree edges each in the above procedure is said to be *extended.*

Theorem 3.6.1. *What is obtained by Procedure 3.6.1 is an embedding of G on the surface determined by the polyhegon in coincidence with the boundary of the extended tree.*

Proof. By considering that an edge is twisted whenever its two occurrences with the same power, from Procedure 3.6.1, the theorem is deduced. □

In view of topology, the conclusion of this theorem is straightforward to some extent.

From this procedure, G is transformed into $\tilde{G}$ without changing the rotation at each vertex except for new vertices that are all articulate. Because $\tilde{G}$ is a tree, according to the rotation, all lettered articulate edges of $\tilde{G}$ form a polyhegon with β pairs of edges, and hence a surface in correspondence with a choice of indices on each pair of the same letter. For convenience, $\tilde{G}$ with a choice of indices of each pair in the same letter is called a *joint tree* of G.

Because of Theorem 3.5.4, a joint tree can be used as a model for constructing an embedding of graph on surfaces without repetition in the sense of not topological equivalence.

Theorem 3.6.2. *A graph $G = (V, E)$ is always embedded into a surface of orientable genus at most $\lfloor \beta/2 \rfloor$, or of nonorientable genus at most β, where β is the Betti number of G.*

Proof. It is seen that any joint tree of G is anembedding of G on the surface determined by its associate polyhegon. From eq. (3.2.1) for the orientable case, the surface has its genus at most $\lfloor 2\beta/4 \rfloor = \lfloor \beta/2 \rfloor$. From eq. (3.2.2) for the non-orientable case, the surface has its genus at most $2\beta/2 = \beta$. □

In Figure 3.6.1, graph G and one of its joint tree $\tilde{G}$ are shown. Here, the spanning tree T is represented by edges without letter. a, b and c are edges not on T. Because the polyhegon is

$$abcacb \sim_{top} c^{-1}b^{-1}cbaa \text{ (Relation 2 in Section 3.3)}$$
$$\sim_{top} aabbcc \text{ (Relation 3 in Section 3.3),}$$

the joint tree is, in fact, an embedding of G on a non-orientable surface of genus 3.

It is seen that any graph with a given rotation can always be immersed in the plane in agreement with the rotation and each edge has two sides. Embeddings of a graph on surfaces are distinguished by the rotation of semi-edges at each vertex and the choice of indices of the two semi-edges on each edge of the graph whenever the edges are labelled by letters. Different indices of the two semi-edges of an edge stand for one side of the edge to the other on a face boundary in an embedding.

Theorem 3.6.3. *A tree can only be embedded on the sphere. Any graph G except tree can be embedded on a non-orientable surface. Any graph G can always be embedded*

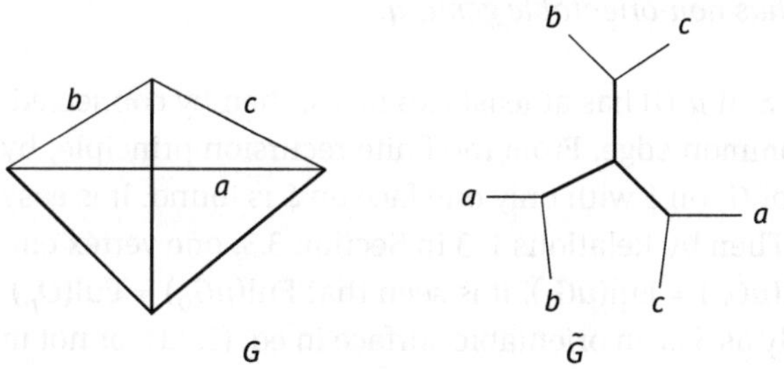

Figure 3.6.1: Graph and its joint tree.

on an orientable surface. Let $n_O(G)$ be the number of distinct embeddings on orientable surfaces, then the number of embeddings on all surfaces is

$$2^{\beta(G)}n_O(G),\ n_O(G) = \prod_{i\geq 2}((i-1)!)^{n_i}, \tag{3.6.1}$$

where $\beta(G)$ is the Betti number and n_i is the number of vertices of degree i in G.

Proof. On a surface of genus not 0, only a graph with at least a circuit is possible to have an embedding. Because a tree has no circuit, it can only be embedded on the sphere. Because a graph not a tree has at least one circuit, from Theorems 3.5.1 and 3.5.3 the second and the third statements are true. Since distinct planar embeddings of a joint tree of G with the indices of each letter different correspond to distinct embeddings of G on orientable surfaces and the number of distinct planar embeddings of joint trees is

$$n_O(G) = \prod_{i\geq 2}((i-1)!)^{n_i}.$$

Further, since the indices of letters on the $\beta(G)$ edges has $2^{\beta(G)}$ of choices for a given orientable embedding and among them only one choice corresponds to an orientable embedding, the fourth statement is true. □

For an embedding $\mu(G)$ of G on a surface, let $\nu(\mu G)$, $\epsilon(\mu G)$ and $\phi(\mu G)$ are, respectively, its vertex number, or *order*, edge number, or *size* and face number, or *coorder*.

Theorem 3.6.4. *For a surface S, all embeddings $\mu(G)$ of a graph G have $Eul(\mu G) = \nu(\mu G) - \epsilon(\mu G) + \phi(\mu G)$ the same, only dependent on S. Further, $Eul(\mu G) =$*

$$\begin{cases} 2-2p,\ p\geq 0, \\ \quad \textit{when S has orientable genus } p; \\ 2-q,\ q\geq 1, \\ \quad \textit{when S has non-orientable genus } q. \end{cases} \tag{3.6.2}$$

Proof. For an embedding $\mu(G)$ on S, if $\mu(G)$ has at least two faces, then by connectedness $\mu(G)$ has two faces with a common edge. From the finite recursion principle, by Operation 2 an embedding $\mu(G_1)$ of G_1 on S with only one face on S is found. It is easy to check that $\mathrm{Eul}(\mu G) = \mathrm{Eul}(\mu G_1)$. Then by Relations 1–3 in Section 3.3, one vertex embedding $\mu(G_0)$ is obtained for $\mathrm{Eul}(\mu G_0) = \mathrm{Eul}(\mu G_1)$. It is seen that $\mathrm{Eul}(\mu G_0) = \mathrm{Eul}(O_p)$, $p \geq 0$; or $\mathrm{Eul}(Q_q)$, $q \geq 1$ accordingly as S is an orientable surface in eq. (3.2.1), or not in eq. (3.2.2). From the arbitrariness of G, the first statement is proved.

By calculating the order, size and coorder of O_p, $p \geq 0$; or Q_q, $q \geq 1$, eq. (3.6.2) is soon obtained. This is the second statement. □

According to this theorem, for an embedding $\mu(G)$ of graph G, $\mathrm{Eul}(\mu G)$ is called its *Euler characteristic* of $\mu(G)$, or of the surface $\mu(G)$ is on. Further, $g(\mu G)$ is the genus of the surface $\mu(G)$ is on.

If a graph G is with the minimum length of circuits σ, then from Theorem 3.6.4 the genus $\gamma(G)$ of an orientable surface G can be embedded on satisfies the inequality

$$1 - \frac{\nu(G) - \epsilon(1 - \frac{2}{\sigma})}{2} \le \gamma(G) \le \left\lfloor \frac{\beta}{2} \right\rfloor, \tag{3.6.3}$$

and the genus $\tilde{\gamma}(G)$ of a non-orientable surface G can be embedded on satisfies the inequality

$$2 - \left(\nu(G) - \epsilon\left(1 - \frac{2}{\sigma}\right)\right) \le \tilde{\gamma}(G) \le \beta. \tag{3.6.4}$$

If a graph has an embedding with its genus attaining the lower (upper) bound in eq. (3.6.3) or (3.6.4), then it is called *down(up)-embeddable*. In fact, a graph is up-embeddable on non-orientable, or orientable surfaces accordingly as it has an embedding with only one face, or at most two faces.

Theorem 3.6.5. *All graphs but trees are up-embeddable on non-orientable surfaces.*

Further, if a graph has an embedding of non-orientable genus l and an embedding of non-orientable genus k, $l < k$, then for any i, $l < i < k$, it has an embedding of non-orientable genus i.

Proof. For an arbitrary embedding of a graph G on a non-orientable surface, let T be the spanning tree of G in its corresponding joint tree. From the non-orientability, the associate $2\beta(G)$-gon P has at least one letter with same index of its two occurrences. If $P = Q_q$, $q = \beta(G)$, then the embedding is an up-embedding in its own right. Otherwise, by Relation 2, or Relation 3 in Section 3.3 if necessary, whenever $s^{-1}s$ or $stst$ occurs, it is, respectively, replaced by ss or $sts^{-1}t$. By virtue of no letter missed in the procedure, from the finite recursion principle, $P' = Q_q$, $q = \beta(G)$, is obtained. This is the first statement.

From the arbitrariness of starting embedding in the procedure of proving the first statement by only using Relation 2 instead of Relation 3 ($AststB \sim_{top} Ass^{-1}Btt$ by Relation 2), because the genus of surface is increased one by one, the second statement is true. □

The second statement of this theorem is also called *interpolation theorem*. The orientable form of interpolation theorem was first given by Duke [82]. The maximum(minimum) of the genera of surfaces (orientable or non-orientable), a graph can be embedded on, is called the *maximum genus* (*minimum genus*) of the graph.

Theorem 3.6.4 shows that graphs but trees are all having that their maximum genus on non-orientable surfaces is the Betti number, with the interpolation theorem. The proof might be the simplest. However, for the orientable case, it is far from simple. Many results on up-embeddability have been obtained since 1979 (see [40, 48, 49, 75–77, 126, 127, 147–150, 177, 178, 195–197, 248–251, 255, 258–260] etc., in this aspect). On the determination of minimum genus of a graph, only a few of graphs with certain symmetry are done (see chapter 12 in Refs. [216, 217]).

3.7 Notes

3.7.1 On the basis of this chapter, one might see that a polyhegon can be chosen as a model of a topological surface, because of Transforms 1–3 in Section 3.1 all in topological equivalence as shown in Section 3.3. This idea is initially seen in Liu [216, 217].

3.7.2 The surface closed curve axiom can be seen as a type of generalization of Jordan closed curve axiom in the plane, or the sphere, because the latter is a degenerate case of the former when no noncontractible closed curve occurs.

3.7.3 The classification of polyhegons, as well as of surfaces in Section 3.1 is self-contained as the simplest model of surfaces used in literature. Shortly, the classification is determined by

$$\begin{cases} ABxCx^{-1} \sim_{ed} BAxCx^{-1}; \\ AxByCx^{-1}Dy^{-1} \sim_{ed} ADxyBx^{-1}Cy^{-1}; \\ AxBx \sim_{ed} AB^{-1}xx. \end{cases} \tag{3.7.1}$$

3.7.4 Although the embedding model described in Section 3.5 has been widely used in references, certain points encounter with confusion to some extent, for constructing non-orientable embedding, and distinguishing non-orientable embeddings. This section shows much clarification.

3.7.5 The joint tree model mentioned in Section 3.6 was initiated from the early articles of Liu [196, 197] and established in Liu's [218, 219]. It has been applied for extracting genus distributions of a number of types of graphs undirected as in Zhao andLiu [423, 425], Li and Liu [183], Zhu and Liu [426], Wan and Liu [356, 357, 359], Chen et al. [51], Shao and Liu [311, 314], Zeng and Liu [420, 421], and directed as in Hao and Liu [118, 119], Hao et al. [121].

3.7.6 Because of complication, up-embeddability in orientable case is kept for Chapter 13. In Section 13.1, up-embeddability is a special case of determining the maximum genus of a graph.

3.7.7 Although Theorem 3.6.3 is a direct result of Theorem 3.5.3, the proof of Theorem 3.6.3 presented there is rather elementary without use of Theorem 3.5.3.

3.7.8 All embeddings considered in this book are dealt with polyhedra, whose complement on the surface has all its connected components contractible, as in Liu [216, 217]. In early references as in White [382] and Ringel [288], such embeddings were called *cellular*.

3.7.9 In fact, this chapter presents a number of intrinsic ideas for proving very basic facts in topology on surfaces. One might like to read more references as in Agoston [2], Gross and Tucker [108], Lefschetz [173], Massey [269], Ringel [288], White [382], particularly in Liu [227].

4 Homology on Polyhedra

4.1 Double cover by travels

Let $G = (V, E)$ be a graph of *order* $v = |V|$ and *size* $\epsilon = |E|$. A set F of travels, $|F| = \varphi$, is also pre-assumed with the property that each edge in E appears in the elements of F exactly twice, or called a *double cover*. Of course, F always means an underlain polyhedron of G in the context without specification in its own right when G is connected. In this case, the members of F are said to be *faces*.

By no means any double cover by travels on G is an underlain polyhedron of G although all double covers on a graph are polyhedra. Such a polyhedron determined by double cover on G is an embedding of G on a surface if, and only if, its underlying graph is G.

Attention 4.1.1. In this chapter, only orientable polyhedra are considered enough for our purpose.

Example 4.1.1. Both $(abcddcba)$ and $\{(abcd), (dcba)\}$ are double covers by travels on a 4-circuit. They are, respectively, polyhedra $(abcdd^{-1}c^{-1}b^{-1}a^{-1})$ and $\{(abcd), (d^{-1}c^{-1}b^{-1}a^{-1})\}$. However, only the latter has as its underlying graph the 4-circuit while the underlying graph of the former is a path of length 4.

Let $\mathcal{G}_0 = 2^V$, $\mathcal{G}_1 = 2^E$ and $\mathcal{G}_2 = 2^F$ be the spaces called *V-space*, *E-space* and *F-space* on G generated by $\langle v| \forall v \in V\rangle$, $\langle e| \forall e \in E\rangle$ and $\langle f| \forall f \in F\rangle$ over GF(2), respectively. Here, we only concentrate on investigating constructions of the E-space. If G is not connected, suppose $G = G_1 + G_2 + \cdots + G_\omega$. Then, $\mathcal{G}_i(G) = \mathcal{G}_i(G_1) + \mathcal{G}_i(G_2) + \cdots + \mathcal{G}_i(G_\omega)$ such that $\mathcal{G}_i(G_l) \cap \mathcal{G}_i(G_s) = 0$, the trivial space, $1 \le l < s \le \omega$ for $i = 0, 1, 2$. We are allowed to discuss the spaces of a connected graph only without specific requirement.

If the *boundary* $\partial_1 e$ of an edge $e \in E$ is defined as the sum of its two incident vertices in $\mathcal{G}_0$, then ∂_1 defined on the basis of $\mathcal{G}_1$ can be extended to a linear transformation $\partial_1 : \mathcal{G}_1 \to \mathcal{G}_0$ called the 1-*boundary mapping* as

$$\forall A \in \mathcal{G}_1, \partial_1 A = \sum_{e \in A} \partial_1 e. \tag{4.1.1}$$

Because ∂_1 is a homomorphism from $\mathcal{G}_1$ to $\mathcal{G}_0$, i.e.

$$\partial_1 \in \mathrm{Hom}(\mathcal{G}_1, \mathcal{G}_0),$$

it is easily shown that $\mathrm{Ker}\, \partial_1$ is a subspace of $\mathcal{G}_1$ which is called the 1-*cycle space* of G. Similarly, $\mathrm{Im}\, \partial_1$ is a subspace of $\mathcal{G}_0$, which is called the 0-*boundary space* of G.

DOI 10.1515/9783110479492-004

If a subset A of E in $G = (V, E)$ is a union of edge–disjoint circuits, then A is said to be a *cycle* of G. Let $\mathscr{C}$ be the set of all cycles of G.

Theorem 4.1.1. $\mathrm{Ker}\partial_1 = \mathscr{C}$.

Proof. Because for $A \in \mathrm{Ker}\,\partial_1$, $\partial_1 A = \sum_{e\in A} \partial_1 e = 0$, all vertices have their valencies even in $G[A]$. Thus, each component of $G[A]$ is a tour from Theorem 1.3.6. However, each tour is an edge-disjoint union of circuits. Hence, $A \in \mathscr{C}$. Conversely, for $A \in \mathscr{C}$, because all valencies of vertices are even in A, from eq. (1.4.9), $A \in \mathrm{Ker}\,\partial_1$. □

According to Theorem 4.1.1, $\mathscr{C}$ is called the *cycle space* of G. Similarly, the *coboundary* $\delta_0 v$ of a vertex $v \in V$ is defined as the sum of all the edges incident with v, i.e.

$$\forall e \in E, (e, \delta_0 v) = (\partial_1 e, v). \tag{4.1.2}$$

Then, δ_0 defined on the basis of $\mathscr{G}_0$ is extended to a linear transformation $\delta_0 : \mathscr{G}_0 \to \mathscr{G}_1$ which is called the 0-*coboundary mapping*, i.e.

$$\forall A \in \mathscr{G}_0,\ \delta_0 A = \sum_{v\in A} \delta_0 v. \tag{4.1.3}$$

From eq. (4.1.2) and the bilinearity of the inner product with respect to ∂_1 and δ_0, the following more general relation can be deduced:

$$\forall B \in \mathscr{G}_0, \forall A \in \mathscr{G}_1, (A, \delta_0 B) = (\partial_1 A, B). \tag{4.1.4}$$

From $\delta_0 \in \mathrm{Hom}(\mathscr{G}_0, \mathscr{G}_1)$, it can be shown by Theorem 1.4.6 that $\mathrm{Im}\,\delta_0$ is a subspace of $\mathscr{G}_1$ which is called the 1-*coboundary space* of G. Similarly, $\mathrm{Ker}\,\delta_0$ is a subspace of $\mathscr{G}_0$ which is called the 0-*cocycle space* of G.

Lemma 4.1.1. $\forall A \in \mathrm{Ker}\,\partial_1, \forall B \in \mathrm{Im}\,\delta_0,\ (A, B) = 0$.

Proof. From $B \in \mathrm{Im}\,\delta_0$, $\exists X \in \mathscr{G}_0, B = \delta_0 X$. From eq. (4.1.4), $(A, B) = (A, \delta_0 X) = (\partial_1 A, X)$. Moreover, from $A \in \mathrm{Ker}\,\partial_1$, $\partial_1 A = 0$. The lemma is found. □

According to Lemma 4.1.1 and Theorem 1.3.7, it is known that $\mathrm{Im}\,\delta_0 = \mathscr{C}^\perp$. Let $\mathscr{C}^*$ be the set of all cocycles of G.

Theorem 4.1.2. $\mathscr{C}^\perp = \mathscr{C}^*$.

Proof. Because $\mathrm{Im}\,\delta_0 = \mathscr{C}^\perp$ and for $A \in \mathscr{G}_0, \delta_0 A \in \mathscr{C}^\perp$ and $\delta_0 A = E_A = \{(u, v)|\forall u \in A, \forall v \notin A\}$. Therefore, $\delta_0 A \in \mathscr{C}^*$. Conversely, for any $X \in \mathscr{C}^*$, suppose $X = E_A, A \in \mathscr{G}_0$, then $X = \delta_0 A \in \mathscr{C}^\perp$. The theorem follows. □

Based on Theorem 4.1.2, $\mathscr{C}^\perp = \operatorname{Im}\delta_0$, the 1-coboundary space, is also called the *cocycle space* of G. Because both $\mathscr{C} = \operatorname{Ker}\partial_1$ and $\mathscr{C}^\perp$ are spaces, $0 \in \mathscr{C} \cap \mathscr{C}^\perp$. Generally, it is allowed to have non-zero vectors in $\mathscr{C}\cap\mathscr{C}^\perp$. Moreover, the intersection of two spaces is always a subspace of them. Thus, $\mathscr{B} = \mathscr{C}\cap\mathscr{C}^\perp$ is called the *bicycle space* of G. Elements of $\mathscr{B}$ are said to be *bicycles*.

Theorem 4.1.3. $\dim\mathscr{C} + \dim\mathscr{C}^\perp = \epsilon(= \dim\mathscr{G}_1)$.

Proof. Suppose the dimension of $\mathscr{C}$ is $r, 0 \le r \le \epsilon$, and suppose $\{e_1, \ldots, e_r\}$ extends an ordered basis $\{C_1, C_2, \ldots, C_r\}$ of $\mathscr{C}$ while the basis of $\mathscr{G}_1$ is ordered as $\{e_1, e_2, \ldots, e_\epsilon\}$ such that $e_i \in C_i, e_i \notin C_j, j \neq i, 1 \le i, j \le r$. From (1.4.10), $\mathscr{G}_1{}^\perp = 0$. Then, $\{e_{r+1}, \ldots, e_\epsilon\}$ can be extended to find $C_i^* \in \mathscr{C}^\perp, r+1 \le i \le \epsilon$ such that $e_i \in C_i^*, e_i \notin C_j^*, j \neq i, r+1 \le i, j \le \epsilon$ hence a basis $\{C_{r+1}^*, \ldots, C_\epsilon^*\}$ of $\mathscr{C}^\perp$. Therefore, the theorem holds. □

Although the sum of the dimensions of $\mathscr{C}$ and $\mathscr{C}^\perp$ is the dimension of $\mathscr{G}_1$ from the theorem, it does not imply that any vector of $\mathscr{G}_1$ can be expressed as a sum of two vectors, one of which is in $\mathscr{C}$ and the other in $\mathscr{C}^\perp$ in every case. In fact, if $B \in \mathscr{B}, B \neq 0$, then every edge in B is never such a sum.

Theorem 4.1.4. $\mathscr{B}^\perp = \mathscr{C} + \mathscr{C}^\perp$.

Proof. Since

$$\begin{aligned} A \in (\mathscr{C} + \mathscr{C}^\perp)^\perp &\Leftrightarrow (A\perp\mathscr{C}) \wedge (A\perp C^\perp) \\ &\Leftrightarrow (A \in \mathscr{C}^\perp) \wedge (A \in \mathscr{C}) \\ &\Leftrightarrow A \in \mathscr{B}, \end{aligned}$$

the theorem is obtained by eq. (1.4.12). □

4.2 Homology

Let $G = (V, E)$ be a graph, connected without specific explanation, and $\mathscr{C}$, the cycle space of G. If a subset J of E satisfies that $\forall C \in \mathscr{C}, C \cap J \neq \emptyset$, then J is called a *joint set* of $\mathscr{C}$. If a joint set has the property that each of the complements of its proper subsets in E contains a non-empty cycle, then it is called a *minimal joint set* of $\mathscr{C}$, denoted by $R(\mathscr{C})$, or simply R.

Lemma 4.2.1. *Let $\mathscr{R}$ be the set of all minimal joint sets of $\mathscr{C}$ and $\mathscr{T}^*$, the set of all cotrees on G. Then $\mathscr{R} = \mathscr{T}^*$.*

Proof. Suppose $R \in \mathscr{R}$ and $\bar{R} = E\backslash R$. If there is a non-zero cocycle in R then from Lemma 4.1.1, the orthogonality of the cocycle to $\mathscr{C}$ and the minimality of R, the cocycle consists of only a single cut-edge. However, any cut-edge is never a member of R by the minimality. Hence, R has no non-zero cocycle. From the connectedness of G, $G[\bar{R}]$

is connected. Further, from R being a joint set of $\mathscr{C}$, $\bar{R}$ is without a non-zero cycle, hence without a circuit. Therefore, $R \in \mathscr{T}^*$ by Tr.1 in Liu [217] (Section 2.1). Conversely, it is easily seen that a cotree is a minimal joint set of $\mathscr{C}$ from Lemma 2.1.3 in Liu [217] (Section 2.1). □

Let T be a tree on a graph $G = (V, E)$. Suppose the edges in E are ordered as $\{e_1, e_2, \ldots, e_\epsilon\}$ such that $\{e_1, e_2, \ldots, e_{\epsilon-\nu+1}\}$ forms the cotree T^*, and hence $\{e_{\epsilon-\nu+2}, \ldots, e_\epsilon\}$, the tree T.

Lemma 4.2.2. *Let $\{C(i) | \forall i(1 \le i \le \epsilon - \nu + 1), C(i) = C(e_i)\}$ be the set of all fundamental circuits for T. Then, it is a linearly independent set in $\mathscr{C}$.*

Proof. $\forall \alpha_1, \alpha_2, \ldots, \alpha_{\epsilon-\nu+1} \in \mathrm{GF}(2)$,

$$\sum_{1\le i\le\epsilon-\nu+1} \alpha_i C(i) = 0 \Leftrightarrow \sum_{1\le i\le\epsilon-\nu+1} \alpha_i e_i = 0$$
$$\Leftrightarrow \forall i(1 \le i \le \epsilon - \nu + 1), \alpha_i = 0.$$

The lemma is obtained. □

Lemma 4.2.3. $\forall C \in \mathscr{C}, \exists \alpha_i(1 \le i \le \epsilon - \nu + 1) \in \mathrm{GF}(2)$,

$$C = \sum_{1\le i\le\epsilon-\nu+1} \alpha_i C(i).$$

Proof. Suppose $\{e_{i_1}, \ldots, e_{i_s}\} = \{e_i | \forall i(1 \le i \le \epsilon - \nu + 1), (C, e_i) = 1\}$. Then

$$C = \sum_{j=1}^{s} C(i_j). \tag{4.2.1}$$

Otherwise, if $\tilde{C} = \sum_{j=1}^{s} C(i_j)$ but $\tilde{C} \ne C$, then $0 \ne C + \tilde{C} \in \mathscr{C}$. However, $C + \tilde{C} = (C \cup \tilde{C}) \backslash (C \cap \tilde{C}) \subseteq T$. This is a contradiction to Tr.1 in Liu [217] (1995, Section 2.1). □

Theorem 4.2.1. *All the fundamental circuits for a tree T in G form a basis of the cycle space $\mathscr{C}$.*

Proof. From Lemmas 4.2.2 and 4.2.3, the theorem is directly derived. □

Because the number of fundamental circuits is the cyclomatic number $\beta(G) = \epsilon - \nu + 1$ (or Betti number), the dimension of the cycles space is

$$\dim \mathscr{C} = \beta(G), \tag{4.2.2}$$

which is also called the *corank* (or *null-rank*) of G.

Based on the cycle space $\mathscr{C}$, a binary relation denoted by $\sim_{\text{cycl}}$ on $\mathscr{G}_1$ can be defined as $\forall A, B \in \mathscr{G}_1$,

$$A \sim_{\text{cycl}} B \Leftrightarrow \exists C \in \mathscr{C}, A + B = C. \tag{4.2.3}$$

It is easily seen that $\sim_{\text{cycl}}$ is an equivalence on $\mathscr{G}_1$ from inspection of the reflective law, symmetry law and transitive law in Section 1.1. Further, the quotient group of $\mathscr{C}$ in $\mathscr{G}_1$,

$$\mathscr{G}_1/\mathscr{C} = \mathscr{G}_1/\sim_{\text{cycl}} \tag{4.2.4}$$

can be seen as a space on $\mathscr{G}_1$ as well as by Theorem 4.1.1. Thus, it is called the *quotient space* of $\mathscr{C}$ in $\mathscr{G}_1$.

Theorem 4.2.2. $\mathscr{G}_1 = \mathscr{C} + \mathscr{G}_1/\mathscr{C}$.

Proof. For each $e \in E$, if e is on a cycle C, it is obvious that $\exists B = C \setminus \{e\} \in \mathscr{G}_1/\mathscr{C}, e = C+B$, from eq. (4.2.3). Otherwise, $e = e + 0, 0 \in \mathscr{C}$, whence $e \in \mathscr{G}_1/\mathscr{C}$. For $A \in \mathscr{G}_1$, suppose $A = \sum_{e\in A} e$ and $e = D(\mathscr{C}; e) + B(\mathscr{G}_1/\mathscr{C}; e), D(\mathscr{C}, e) \in \mathscr{C}, B(\mathscr{G}_1/\mathscr{C}; e) \in \mathscr{G}_1/\mathscr{C}$, then by the linearity

$$\begin{aligned} A &= \sum_{e\in A} \left(D(\mathscr{C}; e) + B(\mathscr{G}_1/\mathscr{C}; e)\right) \\ &= \sum_{e\in A} D(\mathscr{C}; e) + \sum_{e\in A} B(\mathscr{G}_1/\mathscr{C}; e) \\ &= D(\mathscr{C}; A) + B(\mathscr{G}_1/\mathscr{C}; A). \end{aligned} \tag{4.2.5}$$

Moreover,

$$D(\mathscr{C}; A) = \sum_{e\in A} D(\mathscr{C}; e) \in \mathscr{C}$$

and

$$B(\mathscr{G}_1/\mathscr{C}; A) = \sum_{e\in A} B(\mathscr{G}_1/\mathscr{C}; e) \in \mathscr{G}_1/\mathscr{C}.$$

The theorem is obtained. □

By a similar discussion, one may see that $\mathscr{C}/\mathscr{B}, \mathscr{B} = C \cap C^{\perp}$ is the quotient space of the bicycle space $\mathscr{B}$ in $\mathscr{C}$ and that $\mathscr{G}_1/\mathscr{B}$ is the quotient space of $\mathscr{B}$ in $\mathscr{G}_1$. Further, the following two formulae can be found similarly to Theorem 4.2.2 as well:

$$\mathscr{C} = \mathscr{B} + \mathscr{C}/\mathscr{B}; \tag{4.2.6}$$

$$\mathscr{G}_1 = \mathscr{B} + \mathscr{G}_1/\mathscr{B}. \tag{4.2.7}$$

Theorem 4.2.3. $\mathscr{G}_1/\mathscr{C} \cong (\mathscr{G}_1/\mathscr{B})/(\mathscr{C}/\mathscr{B})$.

Proof. From Theorem 1.4.4, for $\mathcal{G}_1/\mathcal{C}, \mathcal{G}_1/\mathcal{B}$, and $\mathcal{C}/\mathcal{B}$ as groups, the relation in the theorem holds. Moreover, it is easily checked that the isomorphism for groups can be naturally extended to that for the corresponding spaces. Therefore, the theorem is found. □

Define the 0-*boundary mapping* $\partial_0 : \mathcal{G}_0 \to \mathcal{J}$ by that $\forall A \in \mathcal{G}_0, \partial_0 A = 0$, if $A \in \mathcal{A}(\mathcal{G}_0)$, the alternating space of $\mathcal{G}_0$; 1, otherwise. Then, $\mathrm{Ker}\,\partial_0 = \mathcal{A}(\mathcal{G}_0)$. Here, $\mathrm{Ker}\,\partial_0$ is called 0-*cycle space* of G, and $\mathrm{Im}\,\partial_0$ is called the (−1)-*boundary space* of G, which is of course the space $\mathcal{J}$.

Lemma 4.2.4. $\mathrm{Im}\,\partial_1 = \mathrm{Ker}\,\partial_0$ *if, and only if, G is connected.*

Proof. Because of Theorem 1.3.2, for any $A \in \mathcal{G}_1, \partial_1 A \in \mathcal{A}(\mathcal{G}_0)$. Then, $\mathrm{Im}\,\partial_1 \subseteq \mathcal{A}(\mathcal{G}_0) = \mathrm{Ker}\,\partial_0$. Further, it is easily shown that $\{v + u \mid \forall u \in \mathcal{G}_0, v \in \mathcal{G}_0\}$ contains a basis of $\mathcal{A}(\mathcal{G}_0)$. However, $u + v \in \mathrm{Im}\,\partial_1$ if, and only if, there is a path $p(u, v)$ in G. That implies the necessity. Conversely, because for any two vertices u and v, there is a path P and $\partial_1 P = u + v$, from $\mathrm{Im}\,\partial_1$ being a space, $\mathrm{Im}\,\partial_1 = \mathcal{A}(\mathcal{G}_0) = \mathrm{Ker}\,\partial_0$. This is the sufficiency. □

Lemma 4.2.5. $\mathcal{G}_1/\mathcal{C} \cong \mathrm{Ker}\,\partial_0$ *if, and only if, G is connected.*

Proof. From Theorem 1.4.2 (First isomorphism law),

$$\mathcal{G}_1/\mathcal{C} = \mathcal{G}_1/\mathrm{Ker}\,\partial_1 \cong \mathrm{Im}\,\partial_1.$$

Then, from Lemma 4.2.4, the lemma follows. □

Because $\mathrm{Im}\,\partial_1$ is a subspace (normal of course) of $\mathrm{Ker}\,\partial_0$, then

$$\mathcal{H}_0 = \mathcal{H}_0(G; \mathrm{GF}(2)) = \mathrm{Ker}\,\partial_0/\mathrm{Im}\,\partial_1$$

is the quotient space of $\mathrm{Im}\,\partial_1$ in $\mathrm{Ker}\,\partial_0$, which is called the 0-*homology space* of G.

Theorem 4.2.4. $\mathcal{H}_0 = 0$ *if, and only if, G is connected.*

Proof. From Lemma 4.2.4, the theorem is obtained. □

Similarly, we define the boundary $\partial_2 f$ of $f \in F$ as the sum of all edges incident to f. Of course, ∂_2 defined on a basis of $\mathcal{G}_2$ can be extended to a linear transformation $\partial_2 : \mathcal{G}_2 \to \mathcal{G}_1$ called the 2-*boundary mapping*, i.e.

$$\forall A \in \mathcal{G}_2,\ \partial_2 A = \sum_{f \in A} \partial_2 f. \tag{4.2.8}$$

It can also be shown that $\partial_2 \in \text{Hom}(\mathscr{G}_2, \mathscr{G}_1)$ in the case of spaces. Hence, $\text{Im}\,\partial_2$ is a subspace of $\mathscr{G}_1$, which is called the 1-*boundary space* of G.

Lemma 4.2.6. $\partial_1\partial_2 = 0$.

Proof. Because for any $f \in F, \partial_2 f$ is a cycle in $\mathscr{G}_1$, from $\text{Im}\,\partial_2$ being a space, for any $A, \partial_2 A$ is also a cycle by eq. (4.2.8). Thus, $\text{Im}\,\partial_2 \subseteq \mathscr{C}$. However, from Theorem 4.1.1, $\mathscr{C} = \text{Ker}\,\partial_1$. Therefore, $\partial_1\partial_2 A = 0$ and then the lemma. □

For a graph $G = (V, E)$ with a set F of travels given, another graph denoted by $G^* = (V^*, E^*)$ with F^* can be constructed as

$$\begin{cases} V^* = F; \; F^* = V; \\ e^* = (f_1, f_2) \in E^* \\ \quad \Leftrightarrow \exists e \in E, (\partial_2 f_1, e) = (\partial_2 f_2, e) = 1. \end{cases} \tag{4.2.9}$$

In this case, e and e^* are said to be a corresponding pair of edges. Notice that G^* may have loops when e is contained in the boundary of only one face $f \in F$ even without loop for G. Here, G^* is called a *dual* of G. Clearly, the V-space, $\mathscr{G}_0^* = 2^{V^*}$ of G^* is isomorphic to the F-space $\mathscr{G}_2$ of G and the 2-boundary mapping ∂_2 on $\mathscr{G}_2$ corresponds to the 0-coboundary mapping δ_0^* on $\mathscr{G}_0^*$. Moreover, $\mathscr{G}_1^* \cong \mathscr{G}_1$ and $\mathscr{G}_2^* \cong \mathscr{G}_0$ where $f^* \in F^*$ in G^* is determined by the corresponding vertex $v \in V$, in G. Further, it is easy to see that $(G^*)^* = G$.

Lemma 4.2.7. $\text{Im}\,\partial_2 \cong \mathscr{A}(\mathscr{G}_2)$ *if, and only if, G is connected.*

Proof. Because it can be shown that $\text{Ker}\,\partial_2 = \{0, F\} \cong \mathscr{J}$ if, and only if, G is connected. From Theorem 1.4.2 (First isomorphism law), $\mathscr{G}_2/\text{Ker}\,\partial_2 \cong \text{Im}\,\partial_2$. Moreover, we have $\mathscr{G}_2/\text{Ker}\,\partial_2 \cong \mathscr{A}(\mathscr{G}_2)$. Then, the lemma follows. □

Lemma 4.2.8. *A connected graph G is planar if, and only if,* $\mathscr{C} \cong \mathscr{A}(\mathscr{G}_2)$.

Proof. Suppose G is planar. From the Euler characteristic being two in this case, $\varphi = |F| = \epsilon - \nu + 2$. Then, $\text{Im}\,\partial_2$ contains a basis of $\mathscr{C}$. Hence, $\text{Im}\,\partial_2 = \mathscr{C}$. From Lemma 4.2.7, $\mathscr{C} \cong \mathscr{A}(\mathscr{G}_2)$. Conversely, if $\mathscr{C} \cong \mathscr{A}(\mathscr{G}_2)$, then from Lemma 4.2.7, $\text{Im}\,\partial_2 = \mathscr{C}$. Since $\dim(\text{Im}\,\partial_2) = \epsilon - \nu + 1$, we can find $\varphi = |F| = \epsilon - \nu + 2$. Thus, the polyhedron $\Sigma = \Sigma(G; F)$ determined by F has $\text{Eul}(\Sigma(G; F)) = 2$, i.e. $G \cong G(\Sigma) \subset P_0$. Therefore, G is planar. □

According to Lemma 4.2.6, Im ∂_2 is a subspace of $\mathscr{C}$. It is allowed to define a 1-*homology space* of G as the quotient subspace Ker ∂_1/Im $\partial_2 = \mathscr{C}/\text{Im}\partial_2$ denoted by $\mathscr{H}_1(\Sigma; \text{GF}(2))$, or simply $\mathscr{H}_1$ when Σ and GF(2) are apparently known.

Theorem 4.2.5. *A connected graph G is planar if, and only if, there is a set F of travels on G such that $\mathscr{H}_1 = 0$.*

Proof. A direct result of Lemmas 4.2.7 and 4.2.8. □

If the polyhedron Σ is on a surface $S \in S$, then the dual G^* defined by (4.2.9) is said to be a *S-dual* of G. Specifically, if S is the sphere, then G^* is called the *planar dual* of G. From what discussed above and in Section 1.6, the following corollaries are straightforward.

Corollary 4.2.1. *A connected graph G is planar if, and only if, there is a set F of travels in G such that the Eulerian characteristic*

$$\text{Eul}(\Sigma(G; F)) = \nu - \epsilon + \varphi = 2.$$

In fact, this is the MacLane's planarity characterization in MacLane [261, 262]. The following corollary is the Whitney's planarity characterization of graphs in Whitney [391, 392].

Corollary 4.2.2. *A connected graph G is planar if, and only if, G has a planar dual.*

More generally, we can find the corollaries below for characterizing the embeddability of a graph on a surface. The first of them is due to Lefschetz [171, 173].

Corollary 4.2.3. *A connected graph G is embeddable on a surface S if, and only if, there exists a set F of travels such that*

$$\text{Eul}(\Sigma(G; F)) = \text{Eul}(S).$$

Corollary 4.2.4. *A connected graph G is embeddable on a surface S if, and only if, G has an S-dual.*

4.3 Cohomology

For the cocycle space $\mathscr{C}^\perp$ of a graph $G = (V, E)$, if a subset J^* of E satisfies that for any $C^* \in \mathscr{C}^\perp$, $C^* \cap J^* \neq \emptyset$, then J^* is called a *joint set* of $\mathscr{C}^\perp$. If the complement of any proper subset of a joint set J^* of $\mathscr{C}^\perp$ in E contains a non-empty cocycle, then J^* is called a *minimal joint set* of $\mathscr{C}^\perp$, denoted by $R^*(\mathscr{C}^\perp)$, or simply R^*. Easy to check that any tree on G is a minimal joint set of $\mathscr{C}^\perp$.

Lemma 4.3.1. *Let $\mathscr{R}^*$ be the set of all minimal joint set of $\mathscr{C}^\perp$ and $\mathscr{T}$, the set of all trees on G (of course, connected here). Then $\mathscr{R}^* = \mathscr{T}$.*

Proof. First, we prove that for any $R^* \in \mathscr{R}^*$, R^* is without circuit. Otherwise, if $C \subseteq R^*$ is a circuit. Then from the minimality, for $e \in C$, $E\backslash(R^* - e)$ has a non-empty cocycle C^*. Because R^* is a joint set of $\mathscr{C}^\perp$, we have $e \in C^*$. However, from Lemma 4.1.1, the orthogonality, there is an edge $e' \in C - e$, hence $e' \notin E\backslash(R^* - e)$ such that $e' \in C^*$. This is a contradiction to $C^* \subseteq E\backslash(R^* - e)$. Then, we prove that for any $R^* \in \mathscr{R}^*$, $G[R^*]$ is connected. Otherwise, suppose $G[R^*] = G_1 + G_2$. Thus, $E_{V(G_1)} \in \mathscr{C}^\perp$. However, $E_{V(G_1)} \subseteq E\backslash R^*$, a contradiction to that R^* is a joint set of $\mathscr{C}^\perp$. □

Based on the lemma, similarly to the Lemmas 4.2.2 and 4.2.3, it can be shown that the set of all the fundamental cocircuits for a tree form a basis of the cocycle space $\mathscr{C}^\perp$. Suppose the basis of $\mathscr{G}_1$ is ordered as $\{e_1, e_2, \dots, e_{\nu-1}, e_\nu, \dots, e_\epsilon\}$ such that $\{e_1, e_2, \dots, e_{\nu-1}\}$ forms a tree T on G. Let $C^*(i) = C^*_{e_i}$, $1 \le i \le \nu - 1$ be the fundamental cocycle generated by e_i with the cotree T^* corresponding to T. Then, we have

$$\forall C^* \in \mathscr{C}^\perp, C^* = \sum_{(C^*, e_i)=1 1 \le i \le \nu-1} C^*(i). \tag{4.3.1}$$

Therefore, from Theorems 4.1.3 and 4.2.1, $\{C^*(i) | i = 1, \dots, \nu - 1\}$ is a basis of $\mathscr{C}^\perp$. The dimension $\nu - 1$ of $\mathscr{C}^\perp$ is also called the *rank* of G.

Although a minimal join set of $\mathscr{C}^\perp$, or a tree on G determines a basis of $\mathscr{C}^\perp$ that is the set of fundamental cocircuits with the tree, it does not mean that any basis of $\mathscr{C}^\perp$ consists of fundamental cocircuits. Therefore, the number of trees on G provides a lower bound of the number of basis. In fact, it is easily shown that E_v for any $\nu - 1$ vertices v in V form a basis of $\mathscr{C}^\perp$. However, they are not a set of fundamental cocircuits in general.

Theorem 4.3.1. $\mathscr{C}^\perp = \mathrm{Im}\, \delta_0$, *the* 1-*coboundary space of* G.

Proof. Because $\delta_0 v = E_v$, and a set of E_v for $\nu - 1$ vertices $v \in V$ forms a basis of $\mathscr{C}^\perp$, from the definition of δ_0, the theorem is soon obtained. □

For a cocycle space $\mathscr{C}^\perp$ given, similarly to the case of the cycle spaces discussed in Section 4.2, we can also see that $\mathscr{G}_1/\mathscr{C}^\perp$ is a space and have the following formulae:

$$\mathscr{G}_1 = \mathscr{C}^\perp + \mathscr{G}_1/\mathscr{C}^\perp; \tag{4.3.2}$$

$$\mathscr{C}^\perp = \mathscr{B} + \mathscr{C}^\perp/\mathscr{B}; \tag{4.3.3}$$

$$\mathscr{G}_1/\mathscr{C}^\perp \cong (\mathscr{G}_1/\mathscr{B})/(\mathscr{C}^\perp/\mathscr{B}). \tag{4.3.4}$$

If the 1-coboundary $\delta_1 e$ of $e \in \mathscr{G}_1$ is defined to be the sum of its incident travels in F. Then, the 1-coboundary $\delta_1 A$ of $A \in \mathscr{G}_1$ is determined by the linearity of the space as

$$\delta_1 A = \sum_{e \in A} \delta_1 e. \tag{4.3.5}$$

That implies $\delta_1 : \mathscr{G}_1 \to \mathscr{G}_2$ is a linear mapping, which is called the 1-*coboundary mapping* of G. Moreover, it is easily seen that $\delta_1 \in \mathrm{Hom}\,(\mathscr{G}_1, \mathscr{G}_2)$. The subspace $\mathrm{Ker}\,\delta_1$ of $\mathscr{G}_1$ is called the 1-*cocycle space* of G and the subspace $\mathrm{Im}\,\delta_1$, the 2-*coboundary space* of G.

Lemma 4.3.2. $\delta_1 \delta_0 = 0$.

Proof. In fact, whenever noticing the following diagram commutative

$$\begin{array}{ccccc} \mathscr{G}_0 & \xrightarrow{\delta_0} & \mathscr{G}_1 & \xrightarrow{\delta_1} & \mathscr{G}_2 \\ \tau\downarrow & & \tau\downarrow & & \tau\downarrow \\ \mathscr{G}_2^* & \xrightarrow{\partial_2^*} & \mathscr{G}_1^* & \xrightarrow{\partial_1^*} & \mathscr{G}_0^* \end{array}, \tag{4.3.6}$$

where

$$\mathscr{G}_i = \mathscr{G}_i(\Sigma, \mathrm{GF}(2)),\ \mathscr{G}_i^* = \mathscr{G}_i(\Sigma^*, \mathrm{GF}(2)),$$

for $i = 0, 1, 2$, $\Sigma^* = \Sigma(G^*; F^*)$ is a dual of $\Sigma = \Sigma(G; F)$ and τ is the mapping that determines the dual. From Lemma 4.2.6, it is known $\partial_1^* \partial_2^* = 0$. The lemma is deduced by the duality. □

The lemma tells us that $\mathrm{Im}\,\delta_0 \subseteq \mathrm{Ker}\,\delta_1$, i.e. the 1-coboundary space of G is a subspace of the 1-cocycle space. Generally speaking, the 1-cocycle space of G is not the cocycle space $\mathscr{C}^\perp$, the 1-coboundary space.

Lemma 4.3.3. $\mathrm{Im}\,\delta_0 = \mathrm{Ker}\,\delta_1$ *if, and only if, G with F is planar.*

Proof. According to the diagram shown in eq. (4.3.6), $\mathrm{Im}\,\delta_0 \cong \mathrm{Im}\,\partial_2^*$ and $\mathrm{Ker}\,\delta_1 \cong \mathrm{Ker}\,\partial_1^*$. From Theorem 4.2.5 in the dual case, the lemma follows. □

Let $\overline{\mathscr{H}_1}(\Sigma; \mathrm{GF}(2)) = \mathrm{Ker}\,\delta_1/\mathrm{Im}\,\delta_0$, or $\overline{\mathscr{H}_1}$ directly when G is obviously known. We call $\overline{\mathscr{H}_1}$ a 1-*cohomology space* of G.

Theorem 4.3.2. *A graph $G = (V, E)$ is planar if, and only if, there is a set F of travels on G such that $\overline{\mathscr{H}_1} = 0$.*

Proof. A direct result of Lemma 4.3.3. □

Now, one might like to find what the 0-cohomology and 2-cohomology spaces of a graph G with F look like by introducing (−1)- coboundary and 2-coboundary mappings: δ_{-1} and δ_2, respectively. Because no much difficulty will appear in doing them similarly to the discussion in Section 4.2, they are only left to the reader.

In what follows, we investigate the constructive property of the cycle space $\mathscr{C}$ and the cocycle space $\mathscr{C}^\perp$. Suppose T be a tree and T^*, the corresponding cotree of G. Let $C^*(e)$, $e \in T$, be the fundamental cocircuit and $C(e)$, $e \in T^*$, the fundamental circuit. If we define

$$f(e) = \begin{cases} C^*(e), \text{ when } e \in T, \text{ i.e.} (e, T) = 1; \\ \sum\limits_{a \in C(e)+e} C^*(a), \text{ otherwise.} \end{cases} \quad (4.3.7)$$

Then, it is extended to a function $f : \mathscr{G}_1 \to \mathscr{C}^\perp$ by the linearity. Of course, $f \in \mathrm{Hom}(\mathscr{G}_1, \mathscr{C}^\perp)$. If f is restricted on the subspace 2^T of $\mathscr{G}_1$, then $f_T \in \mathrm{Hom}(2^T, \mathscr{C}^\perp)$. However, it is easily shown that $\mathrm{Ker} f_T = 0$. From Theorem 1.4.2 (First isomorphism law), $2^T \cong \mathrm{Im} f_T$.

Theorem 4.3.3. $2^T \cong \mathscr{C}^\perp$.

Proof. Because $\{f_T(e) | \forall e \in E, (e, T) = 1\}$ is a basis of $\mathscr{C}^\perp$ from what discussed above, we have $\mathrm{Im} f_T = \mathscr{C}^\perp$. The theorem is soon obtained. □

Similarly, if we define

$$h(e) = \begin{cases} C(e), \text{ when } e \in T^*, \text{ i.e.} (e, T^*) = 1; \\ \sum\limits_{a \in C^*(e)+e} C(a), \text{ otherwise.} \end{cases} \quad (4.3.8)$$

and extend it to a homomorphism $h \in \mathrm{Hom}\,(\mathscr{G}_1, \mathscr{C})$, then the restrict function of h on T^*, $h_{T^*} \in \mathrm{Hom}\,(2^{T^*}, \mathscr{C})$. It is also easily shown that $\mathrm{Ker}\, h_{T^*} = 0$. Therefore, from Theorem 1.4.2, $2^{T^*} \cong \mathrm{Im}\, h_{T^*}$. By the similar reason to the case of $\mathscr{C}^\perp$, it is known that $\mathrm{Im}\, h_{T^*} = \mathscr{C}$. Therefore, h_{T^*} is an isomorphism between 2^{T^*} and $\mathscr{C}$.

Theorem 4.3.4. $A \in \mathscr{C}^\perp \Leftrightarrow \sum_{e \in A} h(e) = 0$.

Proof. First, we prove that for a fundamental cocircuit $C^*(e)$, $e \in T$, the theorem holds. Because

$$\sum_{a \in C^*(e)} h(a) = \sum_{a \in C^*(e)+e} C(a) + \sum_{a \in C^*(e)+e} C(a) = 0,$$

from the uniqueness of the expression, the statement is true. Then, in general case, because any cocycle can be uniquely expressed as the sum of fundamental cocircuits by eq. (4.3.1), from the linearity of h, the theorem is obtained. □

Dually, we can obtain

$$A \in \mathscr{C} \Leftrightarrow \sum_{e \in A} f(e) = 0. \tag{4.3.9}$$

If a graph has no bicycle the condition of which we shall discuss, from Theorem 4.1.4, we know that any $A \in \mathscr{G}_1$ can be uniquely expressed as

$$A = \gamma(A) + \omega(A), \tag{4.3.10}$$

where $\gamma(A) \in \mathscr{C}$ and $\omega(A) \in \mathscr{C}^{\perp}$. In particular, any $e \in E$ has the following form:

$$e = \gamma(e) + \omega(e), \tag{4.3.11}$$

where $\gamma(e) \in \mathscr{C}$ and $\omega(e) \in \mathscr{C}^{\perp}$, which are called a *principal cycle* and *cocycle*, respectively. From the uniqueness, we have

$$\gamma(A) = \sum_{e \in A} \gamma(e); \ \omega(A) = \sum_{e \in A} \omega(e). \tag{4.3.12}$$

In fact, by eq. (4.3.12), $\gamma \in \mathrm{Hom}(\mathscr{G}_1, \mathscr{C})$ and $\omega \in \mathrm{Hom}(\mathscr{G}_1, \mathscr{C}^{\perp})$. It is easily seen that $\mathrm{Im}\,\gamma = \mathscr{C}$, $\mathrm{Ker}\,\gamma = \mathscr{C}^{\perp}$, $\mathrm{Im}\,\omega = \mathscr{C}^{\perp}$ and $\mathrm{Ker}\,\omega = \mathscr{C}$ and that $\gamma^2 = \gamma$ and $\omega^2 = \omega$.

Theorem 4.3.5. *If $\mathscr{B} = 0$, then*

$$A \in \mathscr{C}^{\perp} \Leftrightarrow \sum_{e \in A} \gamma(e) = 0; \tag{4.3.13}$$

$$A \in \mathscr{C} \Leftrightarrow \sum_{e \in A} \omega(e) = 0. \tag{4.3.14}$$

Proof. From eq. (4.3.10), $A \in \mathscr{C}^{\perp} \Leftrightarrow \gamma(A) = 0$ and $A \in \mathscr{C} \Leftrightarrow \omega(A) = 0$. This is the theorem. □

Further, we can determine a basis of $\mathscr{C}$ and a basis of $\mathscr{C}^{\perp}$ by the principal cycles and the principal cocycles, respectively, when G has no bicycle.

Theorem 4.3.6. *If G has no bicycle, then for a tree T and its corresponding cotree T^* on G, the set of cycles $\{\gamma(e) | \forall e \in T^*\}$ is a basis of the cycle space $\mathscr{C}$, and dually $\{\omega(e) | \forall e \in T\}$ is a basis of the cocycle space $\mathscr{C}^{\perp}$.*

Proof. It is only necessary to prove one of the two statements. We prove the latter. Because $\operatorname{Im}\omega = \mathscr{C}^\perp$, $\{\omega(e)|\forall e \in E\}$ generates $C^\perp$. Suppose $C(a)$ is the fundamental circuit for $a \in T^*$. From $\operatorname{Ker}\omega = \mathscr{C}$, $\omega(C(a)) = 0$. Since $C(a) = a + T \cap C(a)$,

$$\omega(a) = \sum_{e \in T \cap C(a)} \omega(e).$$

Hence, $\{\omega(e)|\forall e \in T\}$ also generates $\mathscr{C}^\perp$. From $|\{\omega(e)|\forall e \in T\}| = \nu - 1$, the dimension of $\mathscr{C}^\perp$, the theorem follows. □

Theorem 4.3.7. *In a graph $G = (V, E)$ without bicycles, we have $\forall e \in E$,*

$$f(e) = \omega(T \bigcap f^2(e)). \tag{4.3.15}$$

Proof. Because for any $A \in \mathscr{C}^\perp$, from the definition of f, we see $A = f(T \cap A)$. By $f^2(e) \in \mathscr{C}^\perp$, we have

$$f^2(e) = f(T \cap f^2(e)).$$

However, it implies that $f(e)$ and $T \cap f^2(e)$ have the same image under f. Since $\operatorname{Ker} f = \mathscr{C}$, we have

$$T \cap f^2(e) = C + f(e)$$

where $C \in \mathscr{C}$. Moreover, $f(e) \in \mathscr{C}^\perp$. From the decomposition form (4.3.10), we have $f(e) = \omega(T \cap f^2(e))$. □

By the dual discussion, we may also find that for any $e \in E$ of a graph without bicycles, we have

$$h(e) = \gamma(T^* \bigcap h^2(e)). \tag{4.3.16}$$

Further, it is easily shown that the union of $\{T \cap f^2(e)|\ \forall e \in T\}$ and $\{T^* \cap h^2(e)|\ \forall e \in T^*\}$ forms a basis of $\mathscr{G}_1$.

4.4 Bicycles

If a graph $G = (V, E)$ has a non-empty bicycle, or say $\mathscr{G}_1$ has a non-zero vector, which is a bicycle, then from Theorem 4.1.4, any edge on a bicycle cannot be expressed as a sum of a principal cycle and a principal cocycle because it is not orthogonal to the bicycle space $\mathscr{B}$. However, any edge e not belonging any bicycle has such a decomposition:

$$e = \gamma(e) + \omega(e). \tag{4.4.1}$$

Because for any $B \in \mathscr{B}$; $\gamma(e) + B$ and $\omega(e) + B$ are also a principal cycle and a principal cocycle associated with e, respectively, we have

$$e = (\gamma(e) + B) + (\omega(e) + B). \tag{4.4.2}$$

Let $O = \{e | \exists B \in \mathscr{B}, e \in B\}$. It can be verified that the decomposition form (4.4.2) of an edge is unique for $\gamma(e)\backslash O$ and $\omega(e)\backslash O$. Then, we have a tripartition of the edge set E of G as

$$E = M + N + O, \tag{4.4.3}$$

where $M = \{e | \exists \gamma \in \mathscr{C}, (e \in \gamma) \wedge (e+\gamma \in \mathscr{C}^{\perp})\}$ and $N = \{e | \exists \omega \in \mathscr{C}^{\perp}, (e \in \omega) \wedge (e+\omega \in \mathscr{C})\}$.

Theorem 4.4.1. $\forall e \in E, \forall B \in \mathscr{B}$,

(i) $e \in M \Rightarrow (|\gamma(e) + B| = 1 \pmod 2) \wedge (|\omega(e) + B| = 0 \pmod 2)$;
(ii) $e \in N \Rightarrow (|\gamma(e) + B| = 0 \pmod 2) \wedge (|\omega(e) + B| = 1 \pmod 2)$;
(iii) $|B| = 0 \pmod 2$.

Proof. First prove (iii). Because $\forall B \in \mathscr{B}$, $(B, B) = 0$, $|B| = 0 \pmod 2$. This is (iii). Then, prove (ii) and (i). From (iii) and eq. (4.4.2), $|\gamma(e)| + 1 = |\omega(e)| \pmod 2$. If $e \in M$, then $e \in \gamma(e)$. Since $\gamma(e) + e \in \mathscr{C}^{\perp}$, $(\gamma(e), \gamma(e) + e) = 0$, i.e. $|\gamma(e)| + 1 = 0 \pmod 2$. Therefore, $|\gamma(e)| = 1 \pmod 2$. From (iii), (i) is obtained. Similarly, (ii) holds. □

This theorem can be used to simplify the procedure of determining a tripartition of E. Generally, according to eq. (4.4.2), we may always find a tripartition of E by the following principles. For $e \in E$, let

$$\begin{cases} \mathscr{C}\langle e\rangle &= \{C | \forall C \in \mathscr{C}, (C, e) = 0\}; \\ \mathscr{C}[e] &= \{C | \forall C \in \mathscr{C}, (C, e) = 1\}. \end{cases} \tag{4.4.4}$$

Similarly, $\mathscr{C}^{\perp}\langle e\rangle$ and $\mathscr{C}^{\perp}[e]$ are known. Of course, $\mathscr{A}(\mathscr{C})$ and $\mathscr{A}(\mathscr{C}^{\perp})$ denote the alternating subspaces of $\mathscr{C}$ and $\mathscr{C}^{\perp}$, respectively.

Prin.1 $\forall e \in E$,

$$e \in M \Leftarrow \exists A \in \mathscr{C}\backslash\mathscr{A}(\mathscr{C}), (e \in A) \wedge (A \perp \mathscr{C}\langle e\rangle) \\ \wedge (\forall B \in \mathscr{C}[e], (A, B) = 1).$$

Prin.2 $\forall e \in E$,

$$e \in N \Leftarrow \exists A \in \mathscr{C}^{\perp}\backslash\mathscr{A}(\mathscr{C}^{\perp}), (e \in A) \wedge (A \perp \mathscr{C}^{\perp}\langle e\rangle) \\ \wedge (\forall B \in \mathscr{C}^{\perp}[e], (A, B) = 1).$$

Prin.3 $\forall e \in E$,

$$e \in O \Leftarrow \exists A \in \mathscr{A}(\mathscr{C}), (e \in A) \wedge (A \perp \mathscr{C}).$$

Lemma 4.4.1. *Let $a, b \in M + N$. Then,*

$$a \in \gamma(b) \Leftrightarrow b \in \gamma(a); \; a \in \omega(b) \Leftrightarrow b \in \omega(a).$$

Proof. Since $(\gamma(a), \gamma(b)) = (a + \omega(a), \gamma(b))$ and $(\gamma(a), \gamma(b)) = (\gamma(a), b + \omega(b))$ as well, by the orthogonality, we have $(a, \gamma(b)) = (\gamma(a), b)$, hence the first statement. Similarly, for the last statement. The lemma is done. □

From Theorem 4.1.4, we may also have the decomposition of $A \in \mathscr{B}^{\perp}$ as

$$A = \gamma(A) + \omega(A), \tag{4.4.5}$$

where $\gamma(A) \in \mathscr{C}$ and $\omega(A) \in \mathscr{C}^{\perp}$ are, respectively, called *principal cycle* and *principal cocycle* for A as well. Of course,

$$\gamma(A) = \sum_{e \in A} \gamma(e) \text{ and } \omega(A) = \sum_{e \in A} \omega(e)$$

for $A \cap O = \emptyset$ from eq. (4.4.1).

Theorem 4.4.2. $\forall A, B \in \mathscr{B}^{\perp}$,

$$(\gamma(A), B) = (A, \gamma(B)); \; (\omega(A), B) = (A, \omega(B)).$$

Proof. Similarly to the proof of Lemma 4.4.1, by the linearity of the spaces involved, the theorem is obtained. □

Further, we can find the criteria of determining principal cycles and principal cocycles for any $A \subset E$. For $A \subset E$, let

$$\begin{cases} \mathscr{C}\langle A\rangle = \{C \mid \forall C \in \mathscr{C}, A \perp C\}; \\ \mathscr{C}[A] = \{C \mid \forall C \in \mathscr{C}, (A, C) = 1\}. \end{cases} \tag{4.4.6}$$

Similarly, we get $\mathscr{C}^{\perp}\langle A\rangle$ and $\mathscr{C}^{\perp}[A]$ by using $\mathscr{C}^{\perp}$ instead of $\mathscr{C}$ in eq. (4.4.6).

Crit.1 $\forall A \subset E, C \in \mathscr{C}$,

$$C = \gamma(A) \Leftrightarrow (C \perp \mathscr{C}\langle A\rangle) \wedge (\forall D \in \mathscr{C}[A], (C, D) = 1).$$

Crit.2 $\forall A \subset E, C \in \mathscr{C}^\perp$,

$$C = \omega(A) \Leftrightarrow (C \perp \mathscr{C}^\perp\langle A\rangle) \wedge (\forall D \in \mathscr{C}^\perp[A], (C, D) = 1).$$

If for $a, b \in M + N, a \neq b$, we have $(a, \gamma(b)) = 1$ ($(a, \omega(b)) = (\gamma(a), b) = (\omega(a), b) = 1$ as well), then a and b are said to be *interlaced*, denoted by "a int b" or "b int a". Further, let

$$\text{Int } a = \{b | \ \forall b \in E, \ b \text{ int } a\}. \tag{4.4.7}$$

It is easily seen that for $a \in M + N$,

$$\text{Int } a = \gamma(a) \cap \omega(a) \cap (M + N). \tag{4.4.8}$$

Theorem 4.4.3. *If $\mathscr{B} = 0$, then for $e \in E$,*

$$|\text{Int } e| = 0 (\text{mod } 2). \tag{4.4.9}$$

Proof. Because $M + N = E$ in this case, eq. (4.4.8) becomes Int $e = \gamma(e) \cap \omega(e)$. By the orthogonality, $|\text{Int } e| = (\gamma(e), \omega(e)) = 0 \ (\text{mod } 2)$. The theorem is then obtained. □

Let $G \bullet e$ and $G : e$ be the resultant graphs of *contracting* e into a vertex and *bisectioning* e into two edges by introducing a new vertex for $e \in E$ from $G = (V, E)$.

Lemma 4.4.2. *Suppose G has the bicycle space of dimension r. If $e \in O$, then the bicycle spaces of $G - e$, $G \bullet e$ and $G : e$ are the same which is of dimension $r - 1$. If $e \in M$, then the dimensions of the bicycle spaces of $G - e$, $G \bullet e$ and $G : e$ are $r, r + 1$ (of course, e is not a loop) and $r + 1$, respectively. If $e \in N$, then the dimensions of the bicycle spaces of $G - e$, $G \bullet e$ and $G : e$ are $r + 1$ (when e is not a cut-edge), r and r, respectively.*

Proof. From inspection of that all the bicycles containing e in G do not produce bicycles in $G - e$, $G \bullet e$ and $G : e$ when $e \in O$, the first statement follows. The other two statements can be deduced by observing the decomposition by (4.4.2). □

Theorem 4.4.4. *For any graph G (connected of course), except only for the trivial graph, $\mathscr{B} = 0$ if, and only if, $\tau(G) = 1(\text{mod } 2)$, where $\tau(G)$ is the number of trees on G.*

Proof. By induction on the size of G, it is easily checked that for trees and graphs of small size, the theorem is true. Because contracting a cut-edge on a graph does not change both the number of trees on the graph and the number of bicycles, we may assume that G has no cut-edge. Since G has no loop as defined in Section 1.3, we always have that for $e \in E$, both $G - e$ and $G \bullet e$ are connected and $\tau(G) = \tau(G - e) + \tau(G \bullet e)$, where $\tau(G - e)$ and $\tau(G \bullet e)$ are the numbers of trees containing and not containing e

on G. Because for $e \in O$, $G-e$ and $G \bullet e$ have the same dimension of their bicycle spaces by Lemma 4.4.2, we have $\tau(G-e) = \tau(G \bullet e)$ by hypothesis and hence

$$\tau(G) = \tau(G-e) + \tau(G \bullet e) = 0 (\text{mod } 2).$$

It suffices to discuss two possible cases. (1) If $e \in M$, then $G \bullet e$ has a non-empty bicycle by Lemma 4.4.2 and hence $\tau(G \bullet e) = 0 (\text{mod } 2)$ by hypothesis. Moreover, $G-e$ has the same bicycle space as G by Lemma 4.4.2 and $\tau(G) = \tau(G-e) + \tau(G \bullet e) = \tau(G-e) (\text{mod } 2)$. (2) If $e \in N$, then $G \bullet e$ has the same bicycle space as G by Lemma 4.4.2 and $\tau(G) = \tau(G \bullet e) (\text{mod } 2)$ similarly. Therefore, by the hypothesis, the theorem is true. □

A *dendroid*, denoted by R^0, of $\mathscr{B}$ is a minimal joint set of $\mathscr{B}$ that is a subset of E whose any proper subset does never be a joint set of $\mathscr{B}$. Similarly to the cycle and cocycle spaces, a dendroid determines a basis of the bicycle space $\mathscr{B}$, which is called a *fundamental basis* of $\mathscr{B}$ as well.

Let $G-A$, $G \bullet A$ and $G : A$ be the respective graphs obtained by deleting, contracting and bisectioning all the edges in A from $G = (V, E)$.

Theorem 4.4.5. *If R^0 is a dendroid of G, then $G-R^0$, $G \bullet R^0$ and $G : R^0$ are graphs without bicycles.*

Proof. Because a dendroid determines a basis of the bicycle space of G, from Lemma 4.4.2, the theorem is obtained. □

It seems that no simpler graphic way has been found for finding a dendroid of a graph in general. However, for planar graphs, we have a much simpler procedure to do that. The crucial step is to find a set of travels which forms a polyhedron in a proper way.

T.T. Rule In a planar embedding of a planar graph, starting from a vertex travel along one edge on the boundary of a face f, whenever arriving at the midpoint of an edge, traverse the edge to the other side along the boundary of the face adjacent to f.

Suppose TTr be the set of travels obtained by the T.T. Rule such that any edge appears exactly twice in the members of TTr. For $t \in TTr$, let $q(t)$ be the set of edges which appear in t only once. Because TTr is a polyhedron, $\partial_2 t = q(t)$.

Lemma 4.4.3. $\forall t \in TTr$,

$$(q(t) \in \mathscr{B}) \wedge (q(t) = 0 \Leftrightarrow |TT_r| = 1.)$$

Proof. Because $q(t) = \partial_2 t \in \mathscr{C}$ and $(q(t), \partial_2 f) = 0$ for $f \in F$ from the planar duality, hence $q(t) \in \mathscr{C}^{\perp}$. Therefore, $q(t) \in \mathscr{B}$. The sufficiency of the last statement is obvious by the condition of TT_r. The necessity is obtained by the T.T. Rule itself. □

Lemma 4.4.4. *Let* $TTr = \{t_1, t_2, \ldots, t_r\}$. *Then the* r *vectors* $q(t_i), i = 1, \ldots, r$, *are linearly dependent but any* $r - 1$ *of them are linearly independent.*

Proof. The first statement is obvious because $\sum_{1<i\le r} q(t_i) = 0$. If otherwise on the last statement, then it is allowed to suppose $q(t_1) = \sum_{2\le i\le r-1} q(t_i)$ by probably reordering them. However, by the first statement, $\sum_{2\le i\le r} q(t_i) = 0$, which implies that $TTr = \{t_2, t_3, \ldots, t_r\}$. This is a contradiction to the condition of the lemma. □

Theorem 4.4.6. $\dim \mathscr{B} = |TTr| - 1$.

Proof. From Lemmas 4.4.3–4.4.4 and the discussions on the cycle and cocycle spaces before, it is allowed to determine a subset of E denoted by $\{e_1, \ldots, e_{r-1}\}$ with the property: $(q(t_i), e_i) = 1$, $(q(t_j), e_i) = 0$, $j \neq i$, $1 \le i, j \le r - 1$. That means $R^0 = \{e_1, \ldots, e_{r-1}\}$ is a minimal joint set of the space generated by $\{q(t_i) | \, i = 1, 2, \ldots, r\}$. Further, to prove R^0 is a dendroid. By induction on r, if $r = 1$, then $R^0 = \emptyset$. From Lemma 4.4.3, the conclusion is right. If $r > 1$, let e_1 be in $q(t_1)$ and $q(t_r)$ without loss of generality, then $TTr(G - e_1) = \{t_2, \ldots t_{r-1}, t_r'\}$, where t_r' is compounded from t_1 and t_r. By hypothesis, $R^0 \setminus \{e_1\}$ is a dendroid of $G - e_1$. By Lemma 4.4.2 for $e_1 \in O$, R^0 is a dendroid of G. Therefore, $\dim \mathscr{B} = |R^0| = r - 1 = |TTr| - 1$. □

4.5 Notes

4.5.1 The basic theory described in this chapter is from Liu [214, 216, 217] with some clarification.

4.5.2 Attention 4.1.1 and Example 4.1.1 are for clarifying the objects we are concerning with for avoiding some misunderstanding on surfaces and embeddings. As in Figure 4.5.1 shown in Abrams and Slilaty [1], the graph obtained by identifying x and y is not the skeleton of the polyhedron, determined by all the faces described there, because of Attention 4.1.1.

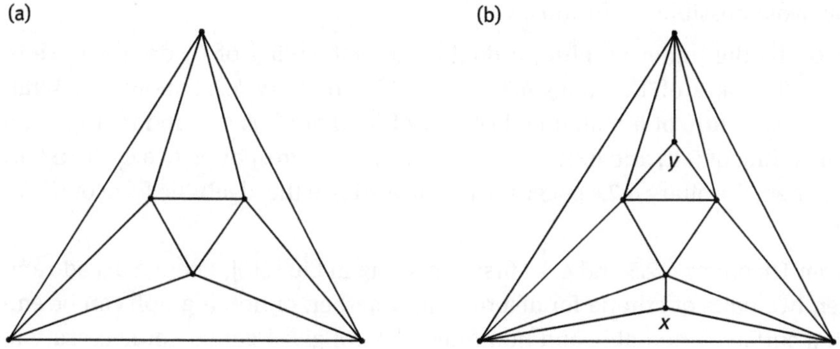

Figure 4.5.1: (a) The octahedron and (b) with new vertices x and y.

This is shown that what obtained by identifying x and y from Figure 4.5.1(b) is a type of surface not considered here, because the identified vertex is a boundary. Our surfaces are compact 2-manifolds without boundary. One might like to extend this theory to more general surfaces, e.g. those with boundaries.

4.5.3 Most papers on discussing spaces mainly cycle and cocycle spaces in a graph are over the real field for an orientation of edges, although the results obtained in this way are in fact independent of the choice of the orientation. The earliest paper should be due to Kirchhoff for the analysis of an electrical network. The main theory can be seen in Chen [43, 44]. Notably, Brooks, Smith, Stone and Tutte solved the problem of squaring the square by their method much related to the network theory shown in Brooks et al. [37]. On the other hand, the theory of chain groups on a graph is more or less influenced by the algebraication of topology. The first paper in this aspect seems to be the one by Tutte in Tutte [335].

4.5.4 Although the V-space and the E-space on a graph were introduced in Lefschetz [172], the F-space began to be used for investigating the structures of a number of types of quotient spaces including homological spaces since Liu [214, 216, 217].

4.5.5 Bicycle spaces were employed by investigating the spaces in a graph over GF(2) in Chen [44]. Then, several authors further studied, especially in Rosenstiehl [294], Rosenstiehl and Read [298]. In Berman [27], bicycles over the integer ring Z was discussed to find a kind of unique factorization of the tree number $\tau = t_1 \dots t_m = \prod_{1 \le i \le m} t_i$ on a graph, such that the group $\mathscr{B}$ of bicycles over Z is isomorphic to $Z(t_1) \times \dots \times Z(t_m)$, $Z(t_i) = \{a \in Z \mid t_i a = 0\}$, $i = 1, \dots, m$. More generally, in Breman and Liu [28, 29].

4.5.6 In Section 4.4, a much simpler method is provided for finding a basis of a bicycle space in a planar graph. However, it seems that this method cannot be developed directly for general graphs. The initial idea of this method appeared in Shank [310] and more early in Dehn [68] and then Mullin and Schellenberg [272].

4.5.7 Although several criteria of justifying if a graph is planar have been found in Section 4.3, they would not be directly helpful for designing efficient algorithms. In Chapter 8, we shall discuss further in order to make theorems efficient.

4.5.8 Although this chapter is from Liu [217] (Chapter 5), descriptions are much clarified to avoid possible confusions.

4.5.9 Let G_g^* be the under graph of a dual of an embedding of G, on a surface of genus g. On the basis of Theorems 4.2.5 and 4.3.2, Corollary 4.2.2 shows the Whitney algebraic structure of a planar embedding of G_0^* that G_0^* is the under graph of a planar embedding of G if, and only if, any cocycle in G_0^* corresponds to a cycle in G or vice versa. Then, Corollary 4.2.4 suggests us to investigate the algebraic form of G_g^* for any $g \neq 0$.

4.5.10 From Theorems 4.2.5 and 4.3.2 first appearing in Liu [216], about a decade and a half later, two pairs of criteria for determining whether, or not, a graph can be embedded in a surface, orientable and non-orientable, of given genus not necessary to be 0, such that the two theorems are the special case of genus 0 when orientable, shown in Liu [231]. The two pairs of general theorems are Theorems 13.2.1 and 13.2.2 in Liu [230] with their dual forms. Then, three pairs of criteria for recognizing the embeddability of graphs on a surface (orientable and non-orientable) of genus given not necessary 0, via three distinct ways, shown as in Liu [229, 232, 236].

4.5.11 On the structures of cycles, particularly circuits in a graph, another direction involves with cycle basis of a graph, on surfaces of genus not zero, shown in Ren et al. [283, 284]. Some others seen as in Cribb et al. [61], Deo et al. [72].

5 Polyhedra on the Sphere

5.1 Planar polyhedra

Suppose Σ to be a polyhedron whose underlying graph is $G = (V, E)$, with F being the set of faces. Let $\rho(v)$ and $\rho^*(f)$ be the valency of vertex v and the valency of face f respectively. Then, we have

$$\begin{cases} V = \sum_{i\geq 1} V_i, \ V_i = \{v | \forall v \in V, \ \rho(v) = i\}; \\ F = \sum_{i\geq 1} F_i, \ F_i = \{f | \forall f \in F, \ \rho^*(f) = i\}. \end{cases} \tag{5.1.1}$$

Further, let $\nu_i = |V_i|$, $1 \leq i \leq \nu$, $\varphi_i = |F_i|$, $1 \leq i \leq \varphi$, then

$$\begin{cases} \nu = |V| = \sum_{i\geq 1} \nu_i; \\ \varphi = |F| = \sum_{i\geq 1} \varphi_i; \\ 2\epsilon = \sum_{i\geq 1} i\nu_i = \sum_{i\geq 1} i\varphi_i. \end{cases} \tag{5.1.2}$$

In the dual case, the dual polyhedron Σ^* whose underlying graph $G^* = (V^*, E^*)$ has its face set F^* such that $V^* = F$ and $F^* = V$. For V^* and F^*, we also have the formulae similar to eqs. (5.1.1) and (5.1.2). Moreover, the following relations are satisfied:

$$\begin{cases} \nu^* = |V^*| = |F| = \varphi, \ \nu_i^* = \varphi_i, \ i \geq 1; \\ \varphi^* = |F^*| = |V| = \nu, \ \varphi_i^* = \nu_i, \ i \geq 1. \end{cases} \tag{5.1.3}$$

Lemma 5.1.1. *Let $\delta(G)$ and $\delta^*(G) = \delta(G^*)$ be, respectively, the minimum valencies of vertices in G and G^*. If $\delta(G), \delta^*(G) \geq 3$, then*

$$\left\lceil \frac{3\max(\nu, \varphi)}{2} \right\rceil \leq \epsilon \leq 3\left(\min(\nu, \epsilon) - Eul\,(\Sigma)\right). \tag{5.1.4}$$

where Eul(Σ) *is the Eulerian characteristic of Σ. And, the equality holds if, and only if, $V_i = \emptyset, i \geq 4$ or $F_i = \emptyset, i \geq 4$.*

Proof. From the third of eq. (5.1.2), we have

$$2\epsilon \geq \max\left(3\sum_{i\geq 1} \nu_i, 3\sum_{i\geq 1} \varphi_i\right) \tag{5.1.5}$$

and hence the left inequality of eq. (5.1.4). Then by the Eulerian characteristic mentioned in Section 2.5 and eq. (5.1.5),

DOI 10.1515/9783110479492-005

$$\begin{aligned} 3\text{Eul}(\Sigma) &\le \min(-\epsilon + 3v, -\epsilon + 3\varphi) \\ &= -\epsilon + 3\min(v, \varphi). \end{aligned} \tag{5.1.6}$$

Thus, $\epsilon \le 3\min(v, \varphi) - 3\text{Eul}(\Sigma)$ which is the right inequality of eq. (5.1.4).

Since the equality holds in eq. (5.1.5) or eq. (5.1.6) according to $|V_i| = 0, i \ge 4$, or $|F_i| = 0, i \ge 4$, the last statement of the lemma is soon obtained. □

From the lemma, a number of known results can be deduced directly in Liu [203, 204]. If $\Sigma \subset O_p$, then

$$\left\lceil \frac{3\max(v, \varphi)}{2} \right\rceil \le \epsilon \le 3(\min(v, \varphi) - 2 + 2p) \tag{5.1.7}$$

by eq. (3.6.2). Of course, for $p = 0$, i.e. the sphere or the plane, we have

$$\left\lceil \frac{3\max(v, \varphi)}{2} \right\rceil \le \epsilon \le 3\min(v, \varphi) - 6. \tag{5.1.8}$$

If a polyhedron Σ has all faces as triangles, then it is called a *triangulation*. A triangulation on the sphere is also called a *maximal planar graph*. For maximal planar graphs, the equality holds in eq. (5.1.8).

If $\Sigma \subset Q_q$, then

$$\left\lceil \frac{3\max(v, \varphi)}{2} \right\rceil \le \epsilon \le 3(\min(v, \epsilon) - 1 + q). \tag{5.1.9}$$

In a polyhedron Σ, if all the vertices are k-valent, then it is called *k-regular*. Similarly, if all the faces are k-valent, then it is called k^**-regular*. A 3-regular polyhedron is also said to be *cubic*. Of course, a triangulation is a 3^*-regular polyhedron. A polyhedron which is both s-valent and t^*-valent is called *totally regular*, or precisely *(s, t)-regular*, and written as *(s, t)-polyhedron*.

Let $\bar{\rho}$ and $\bar{\rho}^*$ be, respectively, the average valencies of vertices and faces. Then, we have

$$\bar{\rho} = \frac{1}{v}\sum_{i\ge1} i v_i;\ \bar{\rho}^* = \frac{1}{\varphi}\sum_{i\ge1} i\varphi_i. \tag{5.1.10}$$

Lemma 5.1.2. *For a polyhedron Σ whose underlying graph is $G = (V, E)$ with $\delta(G)$, $\delta^*(G) \ge 3$, we have*

$$\frac{2}{\bar{\rho}} - 1 + \frac{2}{\bar{\rho}^*} \begin{cases} > 0, & \textit{when } \Sigma \subset P_0 \textit{ or } Q_1\,; \\ = 0, & \textit{when } \Sigma \subset P_1 \textit{ or } Q_2\,; \\ < 0, & \textit{otherwise.} \end{cases} \tag{5.1.11}$$

Proof. From eq. (5.1.10), $v\bar{\rho} = 2\epsilon = \varphi\bar{\rho}^*$ and hence

$$\mathrm{Eul}(\Sigma) = \frac{2\epsilon}{\bar{\rho}} - \epsilon + \frac{2\epsilon}{\bar{\rho}^*} = \epsilon\left(\frac{2}{\bar{\rho}} - 1 + \frac{2}{\bar{\rho}^*}\right).$$

Further, by eq. (3.6.2), eqn. (5.1.11) is obtained. □

According to Lemma 5.1.2, the only possible (s, t)-polyhedra are as follows

$$(s, t) = (\bar{\rho}, \bar{\rho}^*) = (3, 3), (3, 4), (3, 5), (4, 3), (5, 3).$$

Theorem 5.1.1. *There are five* (s, t)*-polyhedra* $(s, t \geq 3)$ *on the sphere, and there are four* (s, t)*-polyhedra on the projective plane.*

Proof. In fact, they are the *Platonic solids*: the *tetrahedron* (the (3, 3)-polyhedron Π_4), the *cube* (the (3, 4)-polyhedron Π_6), the *octahedron* (the (4, 3)-polyhedron Π_8), the *dodecahedron* (the (3, 5)-polyhedron Π_{12}) and the *icosahedron* (the (5, 3)-polyhedron Π_{20}) on the sphere. From the discussion above, the first part of the theorem is obtained.

On the second part of the theorem, the four (s, t)-polyhedra are as shown in Figure 5.1.1. They are the (4, 3)-polyhedron (as in (a)) and its dual (3, 4)-polyhedron, (5, 3)-polyhedron (as in (b)) and its dual (3, 5)-polyhedron. However, (3, 3)-polyhedron does not exist on the projective plane (left to the reader). Therefore, the last part of the theorem is proved. □

Similarly, from eq. (5.1.11), there are only three possible kinds of (s, t)-polyhedra: $(s, t) = (\bar{\rho}, \bar{\rho}^*) = (3, 6), (4, 4), (6, 3)$ on the torus or the Klein bottle. In fact, Figure 5.1.2(a) and 5.1.2(b) show the (3, 6)- and the (4, 4)-polyhedra on the torus. The

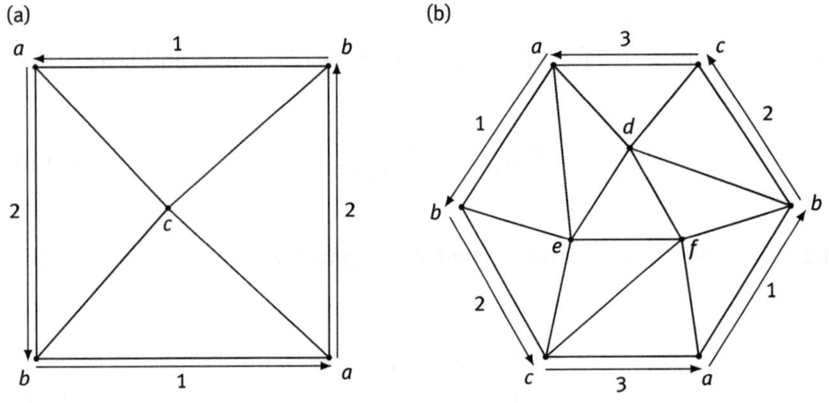

Figure 5.1.1: Four (s,t)-polyhedra.

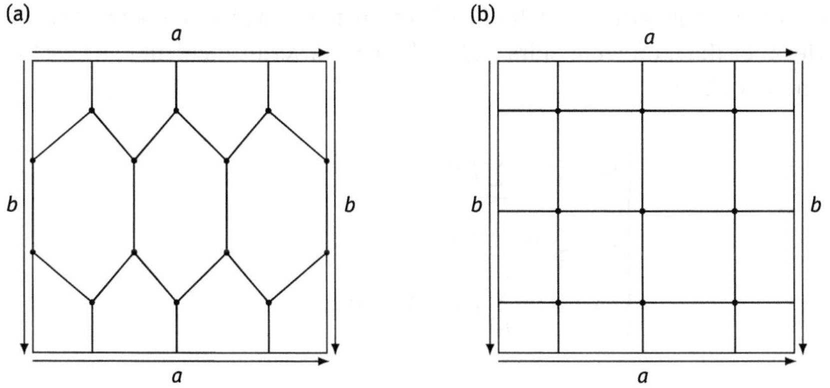

Figure 5.1.2: (3,6)- and (4,4)-polyhedra on torus.

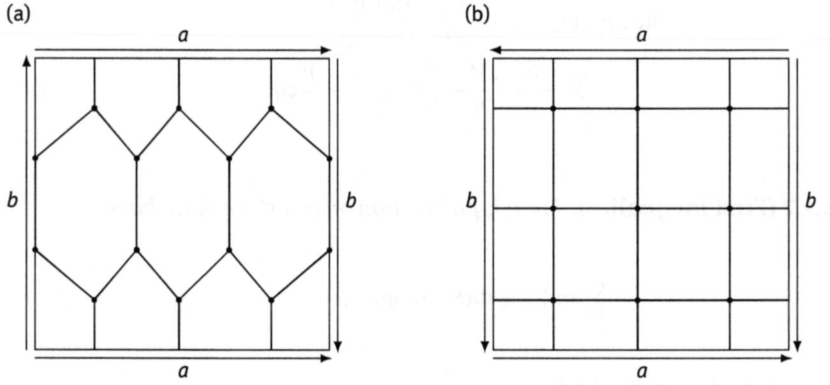

Figure 5.1.3: (3,6)- and (4,4)-polyhedra on Klein bottle.

(6, 3)-polyhedron is the dual of (3, 6)-polyhedron in Figure 5.1.2(a). In Figure 5.1.3, the corresponding kinds of polyhedra on the Klein bottle are also provided.

Lemma 5.1.3. *For a k-polyhedron* (i.e. *k-regular,* $k \geq 3$) *on a surface S, we have*

$$\sum_{j\geq 3}\Big(2k-(k-2)j\Big)\varphi_j = 2kEul(S). \tag{5.1.12}$$

Proof. From eq. (5.1.2),

$$\begin{aligned}\sum_{j\geq 3}(2k-(k-2)j)\varphi_j &= 2\sum_{j\geq 3} j\varphi_j - k\sum_{j\geq 3} j\varphi_j + 2k\sum_{j\geq 3}\varphi_j \\ &= 2k\sum_{j\geq 3} \nu_j - 2k\epsilon + 2k\varphi = 2k\mathrm{Eul}(S).\end{aligned}$$

Thus, the lemma is done. □

Let m_i^t be the number of vertices, each of which is incident to i faces of valency t, for $i, t \geq 3$ and let λ^t be the number of pairs $\{v, f\}$, where v is a vertex incident to a face f of valency t. Then, we have

$$\begin{cases} t\varphi_t = \lambda^t = \sum_{i=1}^{k} i m_i^t; \\ v = \sum_{i=0}^{k} m_i^t, \ t \geq 3; \\ t\varphi_t \leq lv + \sum_{j=1}^{k-l} j m_{l+j}^t, \ 3 \leq t \leq k. \end{cases} \tag{5.1.13}$$

In the inequality, the equality holds if, and only if, $m_0^t = m_1^t = \ldots = m_{l-1}^t = 0$. Further, let

$$\Phi(S; \varphi_3, \varphi_4, \ldots) = \frac{2(k-l)\mathrm{Eul}(S)}{2k-(k-2)t} + \sum_{\substack{j\geq 3 \\ j\neq t}} \frac{((k-2)t-2l)j - 2(k-l)t}{2k-(k-2)t}\varphi_j. \tag{5.1.14}$$

Theorem 5.1.2 (First inequality). *For a k-polyhedron on a surface S, we have*

$$\sum_{s=1}^{k-l} s m_{l+s}^t \geq \Phi(S; \varphi_3, \varphi_4, \ldots) \tag{5.1.15}$$

where $t \geq 3$, $t \neq 2k/(k-2)$, $3 \leq l < k$.

Proof. From the third of eq. (5.1.13), the third of eq. (5.1.2) for the k-regularity, and eq. (5.1.12), we have

$$\begin{aligned} \sum_{1\leq s\leq k-l} s m_{l+s}^t \geq t\varphi_t - lv &= \left(1-\frac{l}{k}\right)t\varphi_t - \frac{l}{k}\sum_{\substack{j\geq 3 \\ j\neq t}} j\varphi_j \\ &= \frac{k-l}{k}t\left(\frac{2k\mathrm{Eul}(S)}{2k-(k-2)t} - \sum_{\substack{j\geq 3 \\ j\neq t}} \frac{2k-(k-2)j}{2k-(k-2)t}\varphi_j\right) \\ &\quad - \frac{l}{k}\sum_{\substack{j\geq 3 \\ j\neq t}} j\varphi_j \\ &= \Phi(S; \varphi_3, \varphi_4, \ldots). \end{aligned}$$

The lemma is then obtained. □

From eq. (5.1.15) and the second of eq. (5.1.13), we have

$$m_k^t \geq -(k-l-1)v + \Phi(S; \varphi_3, \varphi_4, \ldots), \tag{5.1.16}$$

where $3 \leq l < k$. For the sake of brevity, let us write

$$\begin{aligned}\Psi(S; \varphi_3, \varphi_4, \ldots) &= \frac{2t\mathrm{Eul}(S)}{2k-(k-2)t} \\ &\quad + \sum_{\substack{j\geq 3 \\ j\neq t}} \frac{[(k-2)t-2(k-1)]j-2t}{2k-(k-2)t}\varphi_j.\end{aligned} \tag{5.1.17}$$

Theorem 5.1.3 (Second inequality). *For any k-polyhedron on a surface S, we have*

$$m_k^t \geq \Psi(S; \varphi_3, \varphi_4, \ldots), \tag{5.1.18}$$

where $t \geq 3,\ t \neq 2k/(k-2),\ k \geq 3.$

Proof. From eq. (5.1.12) and the third of eq. (5.1.2) for the k-regular case, we shall obtain

$$v = \frac{2t\mathrm{Eul}(S)}{2k-(k-2)t} + 2\sum_{\substack{j\geq 3 \\ j\neq t}} \frac{j-t}{2k-(k-2)t}\varphi_j. \tag{5.1.19}$$

Then, by substituting eq. (5.1.19) for v in eq. (5.1.16), after reduction, the inequality (5.1.18) is deduced. □

From the two inequalities described above, a number of results can be directly derived in Liu [203, 204]. Here, we only present a few commonly used results.

Corollary 5.1.1. *For k-polyhedron on a surface S, if Eul (S) > 0, i.e., S is the sphere or the projective plane, then it makes sense only when $k \leq 5$. Or, in other words, any polyhedron on a surface S, Eul (S) > 0 has a vertex and a face of valency at most 5. In particular, any planar graph has a vertex and a face of valency at most 5.*

Corollary 5.1.2. *For k-polyhedron on the torus and the Klein bottle, we have $k \leq 6$. If $k = 6$, then all the faces are triangles.*

Corollary 5.1.3. *On the sphere or on the projective plane, in any 4- or 5-polyhedron there exists a face which is a triangle. In particular, a planar graph with the minimum valency of vertices not less than 4 has a face which is a triangle. On the torus or the Klein bottle in any 4- or 5-polyhedron, there exists a face which is a triangle.*

Corollary 5.1.4. *On a surface with* Eul $(S) < 0$ *in any k-polyhedron there exists a face of valency not greater than*

$$\left\lfloor \frac{2k(1 - \text{Eul}\,(S))}{k-2} \right\rfloor.$$

5.2 Jordan closed-curve axiom

The classical version of Jordan curve theorem in topology states that a single closed curve C separates the sphere into two connected components, of which C is their common boundary. In this section, we investigate the polyhedral statements and proofs of the Jordan curve theorem.

Let $\Sigma = \Sigma\,(G; F)$ be a polyhedron whose underlying graph $G = (V, E)$ with F as the set of faces. If any circuit C of G not a face boundary of Σ has the property that there exist two proper subgraphs In and Ou of G such that

$$In \bigcup Ou = G;\ In \bigcap Ou = C, \tag{5.2.1}$$

then Σ is said to have the *first Jordan curve property*, or simply write as 1-JCP. For a graph G, if there is a polyhedron $\Sigma = \Sigma\,(G; F)$ which has the 1-JCP, then G is said to have the 1-JCP as well.

Of course, in order to make sense for the problems discussed in this section, we always suppose that all the members of F in the polyhedron $\Sigma = \Sigma\,(G; F)$ are circuits of G.

Theorem 5.2.1 (First Jordan curve theorem). *G has the* 1*-JCP if, and only if, G is planar.*

Proof. Because $\mathcal{H}_1(\Sigma) = 0, \Sigma = \Sigma\,(G; F)$, from Theorem 4.2.5, we know that Im $\partial_2 =$ Ker $\partial_1 = \mathcal{C}$, the cycle space of G and hence Im $\partial_2 \supseteq F$ which contains a basis of $\mathcal{C}$. Thus, for any circuit $C \notin F$, there exists a subset D of F such that

$$C = \sum_{f \in D} \partial_2 f;\ C = \sum_{f \in F \setminus D} \partial_2 f. \tag{5.2.2}$$

Moreover, if we write

$$Ou = G\left[\bigcup_{f \in D} f\right];\ In = G[\bigcup_{f \in F \setminus D} f],$$

then Ou and In satisfy the relations in eq. (5.2.1) since any edge of G appears exactly twice in the members of F. This is the sufficiency.

Conversely, if G is not planar, then G only has embedding on surfaces of genus not 0. Because of the existence of a non-contractible circuit, such a circuit does not satisfy the 1-JCP and hence G is without 1-JCP. This is the necessity. □

Let $\Sigma^* = \Sigma(G^*; F^*)$ be a dual polyhedron of $\Sigma = \Sigma(G; F)$. For a circuit C in G, let $C^* = \{e^* \mid \forall e \in C\}$, or say the corresponding vector in $\mathscr{G}_1^*$, of $C \in \mathscr{G}_1$.

Lemma 5.2.1. *Let C be a circuit in Σ. Then, $G^*\backslash C^*$ has at most two connected components.*

Proof. Suppose H^* be a connected component of $G^*\backslash C^*$ but not the only one. Let D be the subset of F corresponding to $V(H^*)$. Then,

$$C' = \sum_{f \in D} \partial_2 f \subseteq C.$$

However, if $\emptyset \neq C' \subset C$, then C itself is not a circuit. This is a contradiction to the condition of the lemma. From that any edge appears twice in the members of F, there is only one possibility that

$$C = \sum_{f \in F \backslash D} \partial_2 f.$$

Hence, $F\backslash D$ determines the other connected component of $G^*\backslash C^*$ when $C' = C$. □

Any circuit C in G which is the underlying graph of a polyhedron $\Sigma = \Sigma(G; F)$ is said to have the *second Jordan curve property*, or simply write 2-JCP for Σ with its dual $\Sigma^* = \Sigma(G^*; F^*)$ if $G^*\backslash C^*$ has exactly two connected components. A graph G is said to have the 2- JCP if all the circuits in G have the property.

Theorem 5.2.2 (Second Jordan curve theorem). *A graph G has the 2-JCP if, and only if, G is planar.*

Proof. To prove the necessity. Because for any circuit C in G, $G^*\backslash C^*$ has exactly two connected components, any C^* which corresponds to a circuit C in G is a cocircuit. Since any edge in G^* appears exactly twice in the elements of V^*, which are all cocircuits, from Lemma 5.2.1, V^* contains a basis of Ker δ_1^*. Moreover, V^* is a subset of Im δ_0^*. Hence, Ker $\delta_1 \subseteq$ Im δ_0. From Lemma 4.3.2, Im $\delta_0^* \subseteq$ Ker δ_1^*. Then, we have Ker $\delta_1^* =$ Im δ_0^*, i.e. $\widetilde{\mathscr{H}}_1(\Sigma^*) = 0$. From Theorem 4.3.2 in the dual case, G^* is planar and hence so is G. Conversely, to prove the sufficiency. From the planar duality, for any circuit C in G, C^* is a cocircuit in G^*. Then, $G^*\backslash C^*$ has two connected components and hence C has the 2-JCP. □

For a graph G, of course, connected without loop, associated with a polyhedron $\Sigma = \Sigma(G; F)$, let C be a circuit and E_C, the set of edges incident to, but not on C. We may define an equivalence on E_C, denoted by $\sim_C$ as the transitive closure of that $\forall a, b \in E_C$,

$$\begin{aligned} a \sim_C b \Leftrightarrow \exists f \in F, \ &(a^\alpha C(a,b) b^\beta \subset f) \\ &\vee (b^{-\beta} C(b,a) a^{-\alpha} \subset f), \end{aligned} \tag{5.2.3}$$

where $C(a, b)$ or $C(b, a)$ is the common path from a to b or from b to a in $C \cap f$, respectively. It can be seen that $|E_C/\sim_C| \le 2$ and the equality holds for any C not in F only if Σ is orientable.

In this case, the two equivalent classes are denoted by $E_{\mathscr{L}} = E_{\mathscr{L}}(C)$ and $E_{\mathscr{R}} = E_{\mathscr{R}}(C)$. Further, let $V_{\mathscr{L}}$ and $V_{\mathscr{R}}$ be the subsets of vertices by which a path between the two ends of two edges in $E_{\mathscr{L}}$ and $E_{\mathscr{R}}$ without a common vertex with C passes, respectively.

From the connectedness of G, it is clear that $V_{\mathscr{L}} \cup V_{\mathscr{R}} = V \backslash V(C)$. If $V_{\mathscr{L}} \cap V_{\mathscr{R}} = \emptyset$, then C is said to have the *third Jordan curve property*, or simply write 3-JCP. In particular, if C has the 3-JCP, then every path from $V_{\mathscr{L}}$ to $V_{\mathscr{R}}$ (or vice versa) crosses C and hence C has the 1-JCP. If every circuit which is not the boundary of a face f of $\Sigma(G)$, one of the underlain polyhedra of G has the 3-JCP, then G is said to have the 3-JCP as well.

Lemma 5.2.2. *Let C be a circuit of G which is associated with an orientable polyhedron $\Sigma = \Sigma(G; F)$. If C has the 2-JCP, then C has the 3-JCP. Conversely, if $V_{\mathscr{L}}(C) \neq \emptyset$, $V_{\mathscr{R}}(C) \neq \emptyset$ and C has the 3-JCP, then C has the 2-JCP.*

Proof. For a vertex $v^* \in V^* = V(G^*)$, let $f(v^*) \in F$ be the corresponding face of Σ. Suppose In^* and Ou^* are the two connected components of $G^* \backslash C^*$ by the 2-JCP of C. Then,

$$In = \bigcup_{v^* \in In^*} f(v^*) \text{ and } Ou = \bigcup_{v^* \in Ou^*} f(v^*)$$

are subgraphs of G such that $In \cup Ou = G$ and $In \cap Ou = C$ and also $E_{\mathscr{L}} \subset In$ and $E_{\mathscr{R}} \subset Ou$ (or vice versa). The only thing remaining is to show $V_{\mathscr{L}} \cap V_{\mathscr{R}} = \emptyset$. By contradiction, if $V_{\mathscr{L}} \cap V_{\mathscr{R}} \neq \emptyset$, then In and Ou have a vertex which is not on C in common and hence have an edge incident with the vertex, which is not on C, in common. This is a contradiction to $In \cap Ou = C$.

Conversely, from Lemma 5.2.1, we may assume that $G^* \backslash C^*$ is connected by contradiction. Then there exists a path P^* from v_1^* to v_2^* in $G^* \backslash C^*$ such that $V(f(v_1^*)) \cap V_{\mathscr{L}} \neq \emptyset$ and $V(f(v_2^*)) \cap V_{\mathscr{R}} \neq \emptyset$. Consider

$$H = \bigcup_{v^* \in P^*} f(v^*) \subseteq G.$$

Suppose $P = v_1 v_2 \ldots v_l$ is the shortest path in H from $V_{\mathscr{L}}$ to $V_{\mathscr{R}}$.

To show that P does not cross C. By contradiction, assume that v_{i+1} is the first vertex of P that crosses C. From the shortestness, v_i is not in $V_{\mathscr{R}}$. Suppose that subpath $v_{i+1} \ldots v_{j-1}$, $i + 2 \le j < l$, lies on C and that v_j does not lie on C. By the definition of $E_{\mathscr{L}}$, $(v_{j-1}, v_j) \in E_{\mathscr{L}}$ and hence $v_j \in V_{\mathscr{L}}$. This is a contradiction to the shortestness. However, from that P does not cross C, $V_{\mathscr{L}} \cap V_{\mathscr{R}} \neq \emptyset$. This is a contradiction to the 3-JCP. □

Theorem 5.2.3 (Third Jordan curve theorem). *Let $G = (V, E)$ be with an orientable polyhedron $\Sigma = \Sigma(G; F)$. Then, G has the 3-JCP if, and only if, G is planar.*

Proof. From Theorem 5.2.2 and Lemma 5.2.2, the sufficiency is obvious. Conversely, assume that G is not planar. By Lemma 4.2.6, $\mathrm{Im}\partial_2 \subseteq \mathrm{Ker}\partial_1 = \mathscr{C}$, the cycle space of G. By Theorem 4.2.5, $\mathrm{Im}\partial_2 \subset \mathrm{Ker}\partial_1$. Then, from Theorem 5.2.2, there exists a circuit $C \in \mathscr{C} \setminus \mathrm{Im}\partial_2$ without the 2-JCP. Moreover, we also have that $V_{\mathscr{L}} \neq \emptyset$ and $V_{\mathscr{R}} \neq \emptyset$. If otherwise $V_{\mathscr{L}} = \emptyset$, let

$$D = \{f | \exists e \in E_{\mathscr{L}}, e \in f\} \subset F.$$

Because $V_{\mathscr{L}} = \emptyset$, any $f \in D$ contains only edges and chords of C, we have

$$C = \sum_{f \in D} \partial_2 f$$

that contradicts to $C \notin \mathrm{Im}\,\partial_2$. Therefore, from Lemma 5.2.2, C does not have the 3-JCP. The necessity holds. □

5.3 Uniqueness

Suppose $\Sigma = \Sigma(G; F)$ to be a planar embedding of $G = (V, E)$. If there is a face f in Σ, which has the following form:

$$f = Aa^{\alpha} b^{\beta} Bc^{\delta} d^{\lambda} \tag{5.3.1}$$

such that $\Sigma \setminus \{f\}$ can be partitioned into two parts $\{f_1^{(1)}, \ldots, f_s^{(1)}\}$ and $\{f_1^{(2)}, \ldots, f_t^{(2)}\}$ with

$$\begin{cases} \Sigma_1 &= (b^{\beta} Bc^{\delta}, f_1^{(1)}, \ldots, f_s^{(1)}); \\ \Sigma_2 &= (d^{\lambda} Aa^{\alpha}, f_1^{(2)}, \ldots, f_t^{(2)}), \end{cases} \tag{5.3.2}$$

which are two polyhedra, of course, planar, then Σ is said to be *planarity 1-separable*. Apparently, if Σ is planarity 1-separable, then G is separable because $G = G_1 + G_2$, i.e. $G = G_1 \cup G_2$ provided $G_1 \cap G_2 = \{v\}, v \in V$, where G_1, G_2 are the underlying graphs of Σ_1, Σ_2, respectively.

Theorem 5.3.1. *Let G be a separable planar graph with $G = G_1 + G_2$, where both G_1, and G_2 are non-separable with at least two edges. Suppose Σ_1 and Σ_2 are a planar embedding of G_1 and G_2, respectively. Then, G has at least $2\rho_v(G_1)\rho_v(G_2)$ distinct planar embeddings of G within isomorphisms such that each of them is planarity 1-separated into Σ_1 and Σ_2, where $\rho_v(G_1)$ and $\rho_v(G_2)$ are the valencies of the common vertex v in G_1 and G_2, respectively.*

Proof. Suppose $E_v(G_1) = \{a_1, a_2, \ldots, a_{\rho_1}\}$ and

$$E_v(G_2) = \{b_1, b_2, \ldots, b_{\rho_2}\}, \rho_1 = \rho_v(G_1), \rho_2 = \rho_v(G_2).$$

And, suppose

$$\mathscr{A}_1 = \{A_i a_i^{-1} a_{i+1} | i = 1, 2, \ldots, \rho_1, a_{\rho_1+1} = a_1\}$$

and

$$\mathscr{A}_2 = \{B_j b_j^{-1} b_{j+1} | j = 1, 2, \ldots, \rho_2, b_{\rho_2+1} = b_1\}$$

are the sets of faces incident with v in Σ_1 and Σ_2, respectively. Then, we may find $\rho_1\rho_2$ planar embeddings of G from Σ_1 and Σ_2 given as follows: for $1 \le i \le \rho_1$, $1 \le j \le \rho_2$,

$$\begin{cases} \Sigma_{(i,j)} = (A_i a_i^{-1} b_j B_j^{-1} b_{j+1} a_{i+1}, \Sigma_1\langle i\rangle, \Sigma_2\langle j\rangle); \\ \Sigma_{(i,j)} = (A_i a_i^{-1} b_{j+1} B_j b_j^{-1} a_{i+1}, \Sigma_1\langle i\rangle, \Sigma_2\langle j\rangle), \end{cases} \tag{5.3.3}$$

where $\Sigma_1\langle i\rangle$ and $\Sigma_2\langle j\rangle$ stand for all the faces of Σ_1 without $A_i a_i^{-1} a_{i+1}$ and of Σ_2 without $B_j b_j^{-1} b_{j+1}$, respectively. Since there is a face in Σ, which is incident with v and is composed by a face in $\mathscr{A}_1$ or $\mathscr{A}_1^{-1} = \{a_{i-1}^{-1} a_i A_i^{-1} | i = 1, 2, \ldots, \rho_1, a_{\rho_1+1} = a_1\}$ of Σ_1 and the other in $\mathscr{A}_2$ of Σ_2, it is easily seen that the $2\rho_1\rho_2$ planar embeddings of Σ contain all the possible ones of G confirmed in this way by Σ_1 and Σ_2 from the definition of an isomorphism between two polyhedra. It suffices to show that it is allowed that for any $(i_1, j_1) \ne (i_2, j_2)$, $\Sigma_{(i_1,j_1)} \not\sim \Sigma_{(i_2,j_2)}$. However, it is true because all the faces in $\Sigma_{(i_1,j_1)}$ are the same as the corresponding ones in $\Sigma_{(i_2,j_2)}$ except for

$$\begin{aligned} f_{i_1,j_1} &= A_{i_1} a_{i_1}^{-1} b_{j_1+1} B_{j_1} b_{j_1}^{-1} a_{i_1+1} \\ \not\sim_{EL} f_{i_2,j_2} &= A_{i_2} a_{i_2}^{-1} b_{j_2+1} B_{j_2} b_{j_2}^{-1} a_{i_2+1}. \end{aligned}$$

□

If there are two faces:

$$f_1 = A_1 a_1^{\alpha_1} b_1^{\beta_1} B_1 c_1^{\delta_1} d_1^{\lambda_1}$$

and

$$f_2 = A a_2^{\alpha_2} b_2^{\beta_2} B_2 c_2^{\delta_2} d_2^{\lambda_2} \in F$$

in a planar polyhedron $\Sigma = \Sigma(G; F)$, $G = (V, E)$ such that $\Sigma\langle f_1, f_2\rangle = \Sigma \setminus \{f_1, f_2\}$ can be partitioned into $\Sigma_1\langle f_1, f_2\rangle$ and $\Sigma_2\langle f_1, f_2\rangle$ with the property:

$$\begin{cases} \Sigma_1 = (A_1 a_1^{\alpha_1} b_2^{\beta_2} B_2 c_2^{\delta_2} d_1^{\lambda_1}, \Sigma_1\langle f_1, f_2\rangle); \\ \Sigma_2 = (A_2 a_2^{\alpha_2} b_1^{\beta_1} B_1 c_1^{\delta_1} d_2^{\lambda_2}, \Sigma_2\langle f_1, f_2\rangle) \end{cases} \tag{5.3.4}$$

are polyhedra with at least two edges each, planar, of course, then Σ is said to be *2-separable*. The pair $\{f_1, f_2\}$ of faces is called a *separable pair* of Σ. Moreover, according to the definition of dual polyhedra in Section 1.5, there are two faces: $f_1^* = A_1^* a_1^{-\alpha_1} b_1^{\beta_1} B_1^* c_1^{-\delta_1} d_1^{\lambda_1}$ and $f_2^* = A_2^* a_2^{-\alpha_2} b_2^{\beta_2} B_2^* c_2^{-\delta_2} d_2^{\lambda_2} \in F^*$ in the dual Σ^* of Σ, with two vertices: v_1 and v_2 corresponding to f_1^* and f_2^* in Σ, respectively. Because the resultant graph of deleting v_1 and v_2 from the underlying graph G of Σ becomes disconnected, G is also called *2-separable*. The pair of vertices $\{v_1, v_2\}$ is said to be a *separating pair* of G. Naturally, a separable graph can also be called *1-separable* one. If a graph does not have a separating pair, then it is called *3-connected*. A planar polyhedron of a 3-connected graph (planar, of course) is called a *c-net*.

Theorem 5.3.2. *A non-separable planar graph G is 3-connected if, and only if, so is its planar dual G^*.*

Proof. From Corollary 4.2.2 and what discussed above we have known that a non-separable planar graph is 2-separable if, and only if, so is its planar dual. Therefore, from the definition of the 3-connectedness, the theorem follows. □

It is easily seen that if a planar graph is 2-separable, then it may have two non-isomorphic planar embeddings because, from eq. (5.3.4), we may find

$$\Sigma' = (f_1', f_2', \Sigma\langle f_1, f_2\rangle), \tag{5.3.5}$$

where $f_1' = B_1 c_1^{\delta_1} c_2^{-\delta_2} B_2^{-1} b_2^{-\beta_2} b_1^{\beta_1}$ and $f_2' = A_2 a_2^{\alpha_2} a_1^{-\alpha_1} A_1^{-1} d_1^{-\lambda_1} d_2^{\lambda_2}$. The two polyhedra Σ and Σ' are both planar with the underlying graph G but non-isomorphic in any case. Here, Σ' is said to be obtained by *the reflection* of Σ_1 with respect to the separating pair. If all the planar embeddings of a graph G are isomorphic, then G is said to have a unique planar embedding or simply *the uniqueness*.

Theorem 5.3.3. *A planar non-separable graph G with the valencies of vertices and the lengths of circuits not less than 3 has the uniqueness if, and only if, G is 3-connected.*

Proof. Since 2-separable planar graphs with the condition have at least two non-isomorphic planar embeddings each, the necessity is obvious.

Conversely, if G with the condition is 3-connected. By contradiction, suppose G has at least two non-isomorphic planar embeddings $\Sigma_1 = \Sigma_1(G; F_1)$ and $\Sigma_2 = \Sigma_2(G; F_2)$. Then, there is a face $f_1 \in F_1$ on Σ_1 such that $\partial_2 f_1$ is not the boundary of a face in F_2 on Σ_2. Because G is non-separable, $\partial_2 f_1$ is a circuit C in its own right. From the Jordan curve theorem in Section 5.2, the faces in F_2 are partitioned into two parts: $F_2^{(1)}$ and $F_2^{(2)}$ such that G_1 and G_2 are the respective underlying graphs of $\Sigma_2^{(1)}(G_1; F_2^{(1)} + \{C\})$ and $\Sigma_2^{(2)}(G_2; F_2^{(2)} + \{C\})$ with the property:

$$G_1 \bigcup G_2 = G; \; G_1 \bigcap G_2 = C. \tag{5.3.6}$$

However, $C = \partial_2 f_1$ on Σ_1 . Then, G_1 and G_2 are on the same side of f_1 and hence there is a face f on this side such that $\{f_1, f\}$ forms a separating pair on Σ_1. That implies G is 2-separable from the planarity. This is a contradiction to the 3-connectedness of G . □

For a planar graph $G = (V, E), \Sigma = \Sigma(G; F)$ is a planar embedding of G. If G has a separating pair $\{u, v\}, u, v \in V$, then Σ has a separable pair $\{f_1, f_2\}, f_1, f_2 \in F$ corresponding to $\{u, v\}$. The operation of transforming Σ into $\Sigma_1 = \Sigma(G_1; F_1)$ and $\Sigma_2 = \Sigma(G_2; F_2)$ as shown in eq. (5.3.4) is called *splitting* G into G_1 and G_2, which are said to be *splitting blocks* of G. The separating pair is also called a *splitting pair.* If G_1 is a splitting block of G and G_1 does not have a separating pair, then G_1 is said to be a *basic splitting block*. Notice that if $\{u, v\}$ is a separating pair of G and $e = (u, v) \in E$, then e is a splitting block in its own right, of course, a basic one. It is also easily seen that if by contracting all the edges each of which has its two ends, which do not form a splitting pair themselves, in distinct splitting pairs of G, then the resultant graph, probably with multi-edges, has the set of all its basic splitting blocks well defined. We call the resultant graph denoted by $\tilde{G}$ a *basic core* of G. It is also easily shown that any non-separable planar graph G has its basic core $\tilde{G}$ unique and that the cardinality of the set of all basic splitting blocks of G is the same as that of the set of all those of $\tilde{G}$. The number of basic splitting blocks, none of which is an edge of G, is denoted by $\alpha_{\mathrm{Bl}}(G)$. A splitting pair $\{u, v\}$ of G is said to be *simple* if $G = G_1 \cup G_2,\ G_1 \cap G_2 = \{u, v\}$ but $\{u, v\}$ is not a splitting pair of G_1, or G_2 anymore. If G has all the splitting pairs simple, then it is called *simply 2-separable.*

We may construct a graph which is called the *splitting block graph* denoted by $\mathrm{Bl}(\tilde{G})$ with vertices as basic splitting blocks of $\tilde{G}$. An edge of $\mathrm{Bl}(\tilde{G})$ represents that the two splitting blocks corresponding its two ends have a common splitting pair in $\tilde{G}$. Easy to show that $\mathrm{Bl}(\tilde{G})$ is a forest if G is simply 2-separable.

Theorem 5.3.4. *A planar non-separable graph G (simply 2-separable) with all the valencies of vertices and the lengths of circuits not less than three has* $2^{\alpha} Bl^{(G)-1}$ *topologically non-equivalent planar embeddings.*

Proof. By induction on $\alpha_{\mathrm{Bl}}(G)$. If G has no separating pair, that is to say, G is 3-connected. Then, $\alpha_{\mathrm{Bl}}(G) = 1$. From Theorem 5.3.3, the theorem is true. In general, because $\mathrm{Bl}(\tilde{G})$ is a forest, G has a basic splitting block G_1 with only one separating pair, which corresponds to an articulate or isolated vertex. Suppose $G = G_1 \cup G_2$ with $G_1 \cap G_2 = \{u, v\}$. By the hypothesis, G_2 has $2^{\alpha}\mathrm{Bl}^{(G)-2}$ non-isomorphic planar embeddings since $\alpha_{\mathrm{Bl}}(G_2) = \alpha_{\mathrm{Bl}}(G) - 1$. Then, by reflection with respect to the separating pair, we can find two topologically non-equivalent planar embeddings; each of which produces $2^{\alpha}\mathrm{Bl}^{(G)-2}$ distinct embeddings. Therefore, G has $2 \cdot 2^{\alpha}\mathrm{Bl}^{(G)-2} = 2^{\alpha}\mathrm{Bl}^{(G)-1}$ topologically non-equivalent planar embeddings. The theorem is soon obtained. □

Of course, if G is not simply 2-separable, then because a splitting pair $\{u_i, v_i\}$ with s_i basic splitting blocks produces $(s_i - 1)!$ distinct planar embeddings, from the theorem, the total number of distinct(topologically nonequivalent) planar embeddings of G is

$$2^{\alpha}\mathrm{Bl}^{(G)-1}\prod_{i\geq 1}((s_i-1)!)^{l_i},$$

where l_i is the number of splitting pairs; each of which is incident with s_i basic splitting blocks.

5.4 Straight-line representations

Let $\Sigma = \Sigma(G; F)$ be a planar polyhedron whose underlying graph is $G = (V, E)$ with F as the set of faces. The problem is how to represent Σ on the plane. That is, how to draw Σ on the plane such that a vertex of Σ is represented by a point and an edge, by a segment of a curve, or a straight line with the property that no two segments cross at an interior point of a segment. We are particularly interested in the representation with straight-line segments, which is called a *straight-line embedding* of G. If a straight-line embedding has all its vertices on the grid lattice which consists of all the integer points if the coordinates are introduced on the plane, then it is called a *grid point embedding* of G. Because graphs here have neither loop nor multi-edge, the planar polyhedra of them, if any, have neither 1-valent nor 2-valent face. Actually, 1-valent and 2-valent vertices are not essential for our purpose. We are allowed to discuss planar polyhedra with all the valencies of vertices and faces not less than 3 only.

Lemma 5.4.1. *If a graph has a grid point embedding, then it always has a grid point embedding such that for any face and any edge, there is an inner point which is on the grid lattice.*

Proof. Suppose G has a grid point embedding $G(\Sigma)$ in which there is a face f (or an edge e) without a grid point as an inner point. Let d_f (or d_e) be the shortest distance (or the length) between two points corresponding to vertices on the boundary of f (or e). Then on the subdivision of the grid lattice in the Euclidean plane such that two adjacent grid points have their distance less than $d_f/2$ (or $d_e/2$), the embedding $G(\Sigma)$ has the property that the lemma requires. □

Lemma 5.4.2. *Any planar graph G has a grid point embedding.*

Proof. Because the underlying graph of any polyhedron is a subgraph of that of a triangulation with the same order, we are allowed to observe triangulations only. By induction on the order of G, when the order is small, it is easy to check with whatever is chosen to be the infinite face on the plane. Suppose v to be a vertex whose valency is less than or equal to 5 according to Corollary 5.1.1. Then, it suffices to discuss the three possible cases: $\rho(v) = 3$, 4 or 5. Because any face of a planar embedding of G can be chosen as the infinite face, it is allowed to assume that vertex v is on the outer face boundary.

When $\rho(v) = 3$, let $G' = G - v$. Then by the hypothesis, G' has a grid point embedding $\Sigma(G', F\langle v\rangle)$ with all the vertices adjacent to v on the boundary of the infinite face of $\Sigma(G-v, F\langle v\rangle)$, where $F\langle v\rangle$ consists of all the faces in F not incident with v and a new face which is the infinite face. Suppose that v_1, v_2 and v_3 are the vertices adjacent to v in G, and that the longest segment among v_1v_2, v_2v_3 and v_3v_4 is v_2v_3 on a horizontal line in the embedding. Because the length of v_2v_3 is greater than 1, there is a grid point p between v_2 and v_3. Let v be a grid point which is not v_1, it does exist, on or near enough to the ray from v_1 along the line v_1p without passing through p. Then, $\Sigma(G, F)$ is obtained by connecting the segment vv_1, vv_2 and vv_3 in the infinite face of $\Sigma(G', F\langle v\rangle)$.

When $\rho(v) = 4$, let v_i, $i = 1, 2, 3, 4$, be the vertices adjacent to v in G with the circuit $v_1v_2v_3v_4v_1$ and G' be the resultant graph of connecting a new edge (v_1, v_3) on $G - v$. By hypothesis, G' has a grid point embedding $\Sigma(G', F\langle v\rangle)$ with the infinite face boundary $v_1v_4v_3v_1$. Suppose the line v_1v_2 meets the segment v_4v_3, which is assumed to be on a horizontal line, at p. By Lemma 5.4.1, let q be a grid point between p and v_4. Thus, v can be chosen to be a grid point, it does exists, on or near enough to the ray from v_1 along the line qv_1 without passing through q to obtain a grid point embedding of G by joining the segments: vv_i, $i = 1, 2, 3, 4$ in the infinite face of the embedding obtained by deleting the edge v_1v_3 from $\Sigma(G', F\langle v\rangle)$.

When $\rho(v) = 5$, similar to the cases of $\rho(v) = 4$ above, a grid point embedding of G can also be found. □

For a planar embedding $G(\Sigma)$ of G, if there is a grid point embedding of G, which is isomorphic to $G(\Sigma)$ with their infinite faces fixed to be in correspondence, then $G(\Sigma)$ is said to be *grid point compatible*.

Theorem 5.4.1. *Any planar embedding of a graph is grid point compatible.*

Proof. We are allowed only to discuss that the embedding is a triangulation as well here. By induction on the order of G, for the order is small, it is easy to check. In general, by Corollary 5.1.1, only the three cases: $\exists v, \rho(v) = 3, 4$ and 5 have to be investigated. If v is on the boundary of the infinite face of the embedding $G(\Sigma)$, it is just the case as shown in Lemma 5.4.2. Only that v is in the inner domain(contractible) of the infinite face has to be considered.

Case 1. $\rho(v) = 3$. Because of v in the inner domain of the infinite face, the triangle $v_1v_2v_3$, $V_v = \{v_1, v_2, v_3\}$, in G is a inner face in the grid point embedding of $G' = G - v$ by the hypothesis. By Lemma 5.4.1, let v be a grid point in the inner domain of the face $v_1v_2v_3$. Thus, a grid point embedding of G is obtained by jointing the straight segments vv_1, vv_2 and vv_3 in the inner domain of $v_1v_2v_3$.

Case 2. $\rho(v) = 4$. In a similar way to case 1 for G' whose embedding is the resultant one of joining an edge v_1v_3 on $G(\Sigma - v)$, where $\Sigma - v = \Sigma(G - v)$, $V_v = \{v_1, v_2, v_3, v_4\}$ in G.

By Lemma 5.4.1, let the inner grid point of the segment v_1v_3 be v, then a grid point embedding of G is obtained by joining straight segments vv_2 and vv_4 on $\Sigma - v$.

Case 3. $\rho(v) = 5$. Let $V_v = \{v_i | i = 1, 2, \ldots, 5\}$ in G. Because of v in the inner domain of the infinite face, let G' be the resultant embedding of jointing edges: v_1v_3 and v_1v_4 in the embedding $G(\Sigma - v)$. By the hypothesis, G' has a grid point embedding. From Lemma 5.4.1, there is a grid point p close enough to v_1 in the inner domain of the triangle $v_1v_3v_4$ such that all the straight segments pv_i, $i = 1, 2, \ldots, 5$ are in the inner domain of the closed Jordan curve $v_1v_2v_3v_4v_5v_1$. Let $v = p$. Then, a grid point embedding of G is obtained from that of G' in a obvious way. □

5.5 Convex representation

A polygon is said to be *convex* in the plane if the set of common points of the polygon with any straight line never has two or more connected components, at least one of which is a finite segment. By definition, a polygon is convex if, and only if, its complement in the plane is convex. A planar embedding of a graph, if any, is said to be *convex* if all its faces are convex polygons.

If a convex embedding has all its vertices on the grid points of the grid lattice which consists of all the points with integer coordinates in the Euclidean plane, then it is called a *grid point convex embedding.*

A face of a planar embedding of a graph G is said to be *grid point convex extensible* if it can be chosen as the outer (infinite) face of a grid point convex embedding isomorphic to the planar one of G.

An inner face f of a grid point convex embedding of G is said to have the *strongly convex property* if there is an interior grid point p of f (or the infinite face f_0 if f adj f_0) such that any ray from p to a vertex on the boundary of f never meets an inner point of any inner face adjacent to f.

By geometric observation, it enables to make a grid point convex embedding have a face with the strongly convex property.

Lemma 5.5.1. *A face f of a planar embedding of G is grid point convex extensible if, and only if, any vertex v which is not on the boundary of f has three vertex-disjoint paths from v to three distinct vertices on the boundary of f.*

Proof. To prove the sufficiency by induction on the order of G. Because G has all the valencies of vertices and faces not less than 3, the graph of least order is K_4 whose planar embedding is tetrahedron. It is easily seen that the lemma holds for the tetrahedron. In general, from Corollary 5.1.1, only three possible cases have to be considered. Moreover, we may without loss of generality assume f is the infinite face.

Case 1. There exists a vertex v, $\rho(v) = 3$ in G. Suppose $V_v = \{v_1, v_2, v_3\}$. And, let

$$\begin{aligned} G' &= (V - v, E(G - v) \bigcup \{(v_1, v_2), (v_2, v_3), (v_1, v_3)\}) \\ &= (G - v) \bigcup \{(v_1, v_2), (v_2, v_3), (v_3, v_1)\}. \end{aligned}$$

If $v \in \partial_2 f$, then let $v_2 \notin \partial_2 f$. It is easily shown that G' satisfies the required condition for f (if $v \notin \partial_2 f$), or f' (if $v \in \partial_2 f$) such that

$$\partial_2 f' = \partial_2 f - (v, v_1) - (v, v_3) + (v_1, v_2) + (v_2, v_3).$$

By the hypothesis, G' has a grid point convex embedding $G'(\Sigma)$. When $v \in \partial_2 f$, suppose v is a grid point, by Lemma 5.4.1, in the triangle which is formed by the three straight lines: the segment v_1v_3 and its successor and predecessor edges in which only v_1v_3 is on the infinite face boundary of the embedding of G'. Otherwise, that is, $v \notin \partial_2 f$, let v be a grid point in the inner domain of the triangle $v_1v_2v_3$ by Lemma 5.4.1 as well. Then, joining the straight segments: vv_1, vv_2 and vv_3 on the resultant one of deleting the edges not in G from $G'(\Sigma)$ to find a grid point convex embedding of G.

Case 2. There is a vertex $v, \rho(v) = 4$. Suppose $V_v = \{v_1, v_2, v_3, v_4\}$. Consider $G' = (G - v) \cup \{(v_1, v_2), (v_2, v_3), (v_3, v_4), (v_1, v_4)\}$. Because G' is also with the condition for the face f when $v \notin \partial_2 f$ (at most by suitably adding an edge (v_1, v_3) or (v_2, v_4)) or for the face f' otherwise, $\partial_2 f' = \partial f + e_1 + e_2 + e_3$ in vector operation (or symmetric difference on sets), where $e_1 = (v, v_1), e_2 = (v, v_4) \in E$, and $e_3 = (v_1, v_4) \notin \partial_2 f$. By hypothesis, G' has a grid point convex embedding. Then, by the similar treatment, on account of the strongly convex property a grid point one of G is obtained from the one of G'.

Case 3. There is a vertex $v, \rho(v) = 5$. Similarly to the two cases above by considering $G' = (G - v) \cup \{(v_i, v_{i+1}) | \, i = 1, 2, \ldots, 5, \ v_6 = v_1\}$ for $V_v = \{v_i | \, i = 1, \ldots, 5\}$ with the natural cyclic order along a face boundary in G' (at most by suitably adding edges among the vertices in V_v).

The necessity is obvious by contradiction for the typical configuration as shown in Figure 5.5.1(a), where f_1 and f_2 cannot be represented by convex polygons as the outer faces of convex embeddings of G. □

A planar embedding of G which has all faces grid point convex extensible is said to be *totally grid point convex extensible.* Figure 5.5.1(b) shows a planar embedding of G is convex extensible for the faces f_1 and f_2 but it is not totally convex extensible. If a graph G has a planar embedding which is totally grid point convex extensible, then G is said to be *totally grid point convex.*

Theorem 5.5.1. *A planar graph G is totally grid point convex if, and only if, G has no separating pair of vertices.*

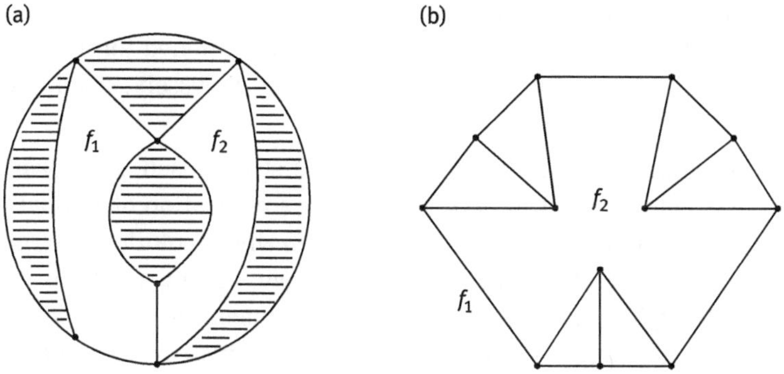

Figure 5.5.1: Two configurations.

Proof. Because G has all the valencies of vertices and faces not less than 3, the necessity is obvious by Lemma 5.5.1. For the sufficiency, we are allowed to observe only one planar embedding of G by Theorem 5.3.3. Because G has no separating pair of vertices, for any face f and a vertex v not on the boundary of f, there are at least three vertex-disjoint paths from v to three distinct vertices on the boundary of f (otherwise, there has to be a configuration as shown in Figure 5.5.1(a)). Then, by Lemma 5.5.1, f is grid point convex extensible. From the arbitrariness of f chosen, G is totally grid point convex. □

All descriptions of the proofs of theorems here are only existential type without considering the design of algorithms. Some algorithmic accounts of them can be discussed based on the OD-tree technique.

5.6 Notes

5.6.1 The Euler formula for planar polyhedra is a very important tool in the construction of a variety of polyhedra, with some kinds of symmetry in Coxeter [58], and in tiling of the plane, seen in Grunbaum and Shephard [111]. However, it seems that problems discussed in Section 5.1 for general surfaces had not been observed until the beginning of the last decade in Liu [204].

5.6.2 Several combinatorial versions of the Jordan curve theorem has been discussed since 1979 in Tutte [349] and Vince and Little [354]. In Tutte [349], a theory of general combinatorial maps are also provided.

5.6.3 One thing should be mentioned in Section 5.2 is that Theorem 5.2.1 provides a necessary and sufficient condition of the 1-JCP instead of only a necessity before. Further results for general surfaces of genus not 0 can be seen in Liu and Liu [248]. However, because the order relation on boundary should be considered, the revised form appears in Liu [229].

5.6.4 The straight-line representation of a planar graph was first proposed and solved by Fary [84]. Of course, Theorem 5.4.1 here is a somewhat strengthened form of the result of Fary. The algorithmic account was first discussed by Fraysseix, Pach and Pollack [91].

5.6.5 The convex representations of a planar graph was first proposed and solved by Tutte [344]. Then, it is also due to Tutte [339] to present a realization of a convex embedding of a graph. He presented a linear equation system whose solution determines the location of the vertices in the plane for the convex embedding at a time. He called this kind of embedding *barycentric*. Although he did not mention of grid points as shown by Theorem 5.5.1, the vertices of a barycentric embedding can be made on grid points by Lemma 4.4.1 here. More optimization problems will be investigated in Chapter 13.

5.6.6 In fact, we are allowed to introduce positive weights on vertices of a graph for finding a convex embedding of the graph, such that the total length of all inner edges is minimum, by considering the weights under the boundary condition, i.e. the outer-face boundary is a given convex polygon on the plane. The detail will be discussed in Chapter 13 as well.

5.6.7 This chapter is based on Liu [217] (Chapter 6) with Theorem 5.2.1 completed for the new sufficient condition, in spite of the section added on straight-line embedding, and a number of minor revisions.

5.6.8 Since the surface closed curve axiom occurs in Liu [218] (page 8), or Liu [219] (page 7), we were awoken to investigate its polyhedral form, for establishing a criterion of recognizing whether, or not, a graph can be embedded onto a surface of a given genus not 0.

6 Automorphisms of a Polyhedron

6.1 Automorphisms of polyhedra

Let δ be an operation on an edge from one end to the other of the edge, and σ, another from one side to the other of the edge, called the *end operation* and *side operation*, respectively.

Let $\Sigma = \Sigma(V, E, F)$ be a polyhedron with $G = (V, E)$ as its underlying graph and F as the set of faces. For $e = (u, v) \in E$, let

$$X(e; \Sigma) = \{e_{u,r}, e_{u,l}, e_{v,r}, e_{v,l}\} \tag{6.1.1}$$

be called the *quadricell* of e where $e_{u,r}$, $e_{u,l}$ and $e_{v,r}$, $e_{v,l}$ stand for the four directed *quart-edges* associated with e on the right-, left-hand sides of going away from u and v, respectively, and let

$$\mathscr{X}(\Sigma) = \bigcup_{e \in E} X(e; \Sigma) \tag{6.1.2}$$

be called the *universal set* of Σ. Further, for $v \in V$, suppose $\varrho(v) = (a, b, c, \ldots)$ be the rotation at v on Σ. Let

$$\begin{cases} \varrho_r(v) = (a_{v,r}, b_{v,r}, c_{v,r}, \ldots); \\ \varrho_l(v) = (\ldots, c_{v,l}, b_{v,l}, a_{v,l}) \end{cases} \tag{6.1.3}$$

be, respectively, called the *right*; *left* rotation at v in Σ. Then, we may find a permutation $\pi(\Sigma)$ on $\mathscr{X}(\Sigma)$ associated with Σ as

$$\pi(\Sigma) = \prod_{v \in V} \varrho_r(v)\varrho_l(v) \tag{6.1.4}$$

which is called a Σ-*permutation*. What has to be especially noticed here is that the direction of $\varrho_r(v)$ is the inverse of that of $\varrho_l(v)$ and vice versa. We occasionally write $e_{0,r}$, $e_{0,l}$ and $e_{1,r}$, $e_{1,l}$ instead of $e_{u,r}$, $e_{u,l}$ and $e_{v,r}$, $e_{v,l}$ for $e \in E$. Hence, 0 represents the tail and 1, the head of e when the orientation of e on Σ has to be considered.

Furthermore, for convenience we introduce two operations σ and δ, which are in fact two special kinds of permutations on $\mathscr{X}(\Sigma)$ as for $x \in \mathscr{X}$, σx (or x^σ in some case) represents the element which is incident with the same vertex as x but on the other side of x and δx (or x^δ in some case), the element incident with the different vertex from that with x but on the same side as x in the same quadricell. Because for any $e \in E$, we have

DOI 10.1515/9783110479492-006

$$\begin{cases} \sigma(e_{0,r}) = e_{0,l}, & \sigma(e_{0,l}) = e_{0,r}; \\ \delta(e_{0,r}) = e_{1,r}, & \delta(e_{0,l}) = e_{1,l} \end{cases} \tag{6.1.5}$$

and

$$\begin{aligned} \sigma\delta(e_{0,r}) &= \delta\sigma(e_{0,r}), \\ \sigma\delta(e_{0,l}) &= \delta\sigma(e_{0,l}). \end{aligned} \tag{6.1.6}$$

And, in a similar way, for $e_{1,r}$ and $e_{1,l}$. In other words, for any $x \in \mathscr{X}(\Sigma)$, σ and δ satisfy the following relations:

$$\begin{aligned} \sigma^2(x) = \sigma(\sigma(x)) = \delta^2(x) &= x, \\ \sigma\delta(x) &= \delta\sigma(x), \end{aligned} \tag{6.1.7}$$

where σ and δ are, respectively, called the *side switch* and the *end switch* as shown σ and δ in Chapter 2.

Lemma 6.1.1. *For two polyhedra $\Sigma_1 = \Sigma(V_1, E_1, F_1)$ and $\Sigma_2 = \Sigma(V_2, E_2, F_2)$, a bijection $\tau : E_1 \longrightarrow E_2$ is an isomorphism between Σ_1 and Σ_2 if, and only if, τ induces a bijection, denoted by τ as well : $\mathscr{X}(\Sigma_1) \longrightarrow \mathscr{X}(\Sigma_2)$ such that the following diagrams :*

$$\begin{array}{ccc} \mathscr{X}(\Sigma_1) & \xrightarrow{\quad\tau\quad} & \mathscr{X}(\Sigma_2) \\ {\scriptstyle\gamma_1}\downarrow & & \downarrow{\scriptstyle\gamma_2} \\ \mathscr{X}(\Sigma_1) & \xrightarrow{\quad\tau\quad} & \mathscr{X}(\Sigma_2) \end{array} \tag{6.1.8}$$

are all commutative for $(\gamma_1, \gamma_2) = (\sigma, \sigma)$, (δ, δ) and (π_1, π_2) where $\pi_i = \pi(\Sigma_i)$, $i = 1, 2$.

Proof. From τ being an isomorphism between Σ_1 and Σ_2, for $f = (a_f, b_f, c_f, \ldots) \in F_1$, we have

$$\tau(f) = (\tau(a_f), \tau(b_f), \tau(c_f), \ldots) \tag{6.1.9}$$

and hence from the discussion in Chapter 2 for $v \in V$, $\varrho(v) = (a_v, b_v, c_v, \ldots)$, we have

$$\tau(\varrho(v)) = (\tau(a_v), \tau(b_v), \tau(c_v), \ldots). \tag{6.1.10}$$

We may find the induced bijection $\tau : \mathscr{X}(\Sigma_1) \longrightarrow \mathscr{X}(\Sigma_2)$ from $\tau : E_1 \longrightarrow E_2$ by the following procedure.

First, choose a vertex $o \in V_1$ as the starting point. Suppose $\varrho(o) = (a_o, b_o, c_o, \ldots)$ and hence

$$\tau(\varrho(o)) = (\tau(a_o), \tau(b_o), \tau(c_o), \ldots).$$

Let $x(a_o) \in \mathscr{X}(\Sigma_1)$ be one of $a_{o,r}$ and $a_{o,l}$. Define

$$\tau(x(a_o)) = \begin{cases} x(\tau(a_o)), & \text{if } \varrho(o) \text{ and } \tau(\varrho(o)) \text{ have} \\ & \quad \text{the same direction;} \\ \sigma x(\tau(a_o)), & \text{otherwise.} \end{cases} \tag{6.1.11}$$

Then, extend it into the whole $\mathscr{X}(\Sigma_1)$ by the rule: if for $x \in \mathscr{X}(\Sigma_1)$, $\tau(x)$ has been defined, then define

$$\begin{cases} \tau(\sigma x) = \sigma\tau(x); \\ \tau(\delta x) = \delta\tau(x); \\ \tau(\pi_1 x) = \pi_1\tau(x). \end{cases} \tag{6.1.12}$$

Because of the connectedness of $G_1 = (V_1, E_1)$, the induced bijection τ which is known from the properties of σ, δ and π_1 is well defined on $\mathscr{X}(\Sigma_1)$. Moreover, from eq. (6.1.12), the diagrams as shown in eq. (6.1.8) are commutative. This is the necessity.

Conversely, from $\tau : \mathscr{X}(\Sigma_1) \longrightarrow \mathscr{X}(\Sigma_2)$, we can find the induced one $\tau : E_1 \longrightarrow E_2$ by the rule: for $e \in E_1$, $\tau(e) \in E_2$ corresponds to the quadricell

$$\{x(\tau e), \sigma x(\tau e), \delta x(\tau e), \sigma\delta x(\tau e)\},$$

which is well defined by eq. (6.1.8) for $(\gamma_1, \gamma_2) = (\sigma, \sigma)$ and (δ, δ). Of course, τ is a bijection between E_1 and E as τ does between $\mathscr{X}(\Sigma_1)$ and $\mathscr{X}(\Sigma)$. Further, the rotation at a vertex v on Σ_1 is determined to be in agreement with the right one on $\pi(\Sigma_1)$. From diagram (6.1.8) for $(\gamma_1, \gamma_2) = (\pi_1, \pi_2)$, we have eq. (6.1.10) for $v \in V_1$, $\varrho(v) = (a_v, b_v, c_v, \ldots)$. By the same reason discussed Chapter 2, we have eq. (6.1.9) for $f = (a_f, b_f, c_f, \ldots) \in F_1$. Therefore, τ is an isomorphism between Σ_1 and Σ_2. This is the sufficiency. □

Lemma 6.1.2. *Let $x, y \in \mathscr{X}(\Sigma)$ given for a polyhedron $\Sigma = \Sigma(V, E, F)$. Suppose τ_1 and τ_2 are two bijections on $\mathscr{X}(\Sigma)$ which determine two automorphisms of Σ. If $\tau_1(x) = \tau_2(x) = y$, then $\tau_1 = \tau_2$.*

Proof. Because $\tau_1(x) = \tau_2(x) = y$, from Lemma 6.1.1, we know that

$$\begin{cases} \tau_1(\sigma x) = \sigma y = \tau_2(\sigma x); \\ \tau_1(\delta x) = \delta y = \tau_2(\delta x) \end{cases} \tag{6.1.13}$$

and by eqn. (6.1.5),

$$\tau_1(\pi x) = \pi(y) = \tau_2(\pi x). \tag{6.1.14}$$

Further, because the underlying graph $G = (V, E)$ of Σ is connected, for any $z \in \mathscr{X}(\Sigma)$ $(z \neq y)$ there is a path from the edge associated with y to that with z and hence there are integers: $k > 0$, $r_i, s_i, t_i \geq 0$, $i = 1, 2, \ldots, k$, such that

$$z = \prod_{1 \leq i \leq k} \sigma^{r_i} \delta^{s_i} \pi^{t_i}(y). \tag{6.1.15}$$

From eqs (6.1.13) and (6.1.14), we find that for any $z \in \mathscr{X}(\Sigma)$,

$$\tau_1(\tilde{x}) = z = \tau_2(\tilde{x}),$$

where

$$\tilde{x} = \prod_{1 \leq i \leq k} \sigma^{r_i} \delta^{s_i} \pi^{t_i}(x).$$

Therefore, $\tau_1 = \tau_2$. The lemma is obtained. □

Corollary 6.1.1. *If τ is a bijection on $\mathscr{X}(\Sigma)$, which induces an automorphism of Σ such that there is an $x \in \mathscr{X}(\Sigma)$, $\tau(x) = x$, then $\tau = 1$, the identity.*

Theorem 6.1.1. *Let $Aut(\Sigma)$ be the automorphism group of a polyhedron $\Sigma = \Sigma(V, E, F)$, $aut(\Sigma) = |Aut(\Sigma)|$, the order of the group, and $\epsilon(\Sigma) = |E|$. Then,*

$$aut(\Sigma) \mid 4\epsilon(\Sigma). \tag{6.1.16}$$

Proof. First, we classify the elements on $\mathscr{X}(\Sigma)$ by the equivalence denoted by $\sim_{\text{Aut}}$: $\forall x, y \in \mathscr{X}(\Sigma)$,

$$x \sim_{\text{Aut}} y \Longleftrightarrow \exists \tau \in \text{Aut}(\Sigma),\ \ x = \tau y. \tag{6.1.17}$$

From the equivalence, the classification is found as

$$\mathscr{X}(\Sigma) = \sum_{i=1}^{s} X(x_i) \tag{6.1.18}$$

such that $x_i \in X(x_j)$ if, and only if, $i = j$. Then, we show that

$$|X(x_i)| = \text{aut}(\Sigma), \qquad 1 \leq i \leq s. \tag{6.1.19}$$

In fact, for any τ_1, $\tau_2 \in \text{Aut}(\Sigma)$, if $\tau_1 x_i = \tau_2 x_i$, then $\tau_2^{-1}\tau_1 x_i = x_i$. Because $\tau_2^{-1}\tau_1 \in \text{Aut}(\Sigma)$, from Corollary 6.1.1, we have $\tau_1 = \tau_2$. This implies eq. (6.1.19). Moreover,

$$|\mathscr{X}(\Sigma)| = 4\epsilon(\Sigma)$$

from eqs. (6.1.1) and (6.1.2). The theorem follows from eq. (6.1.19). □

Corollary 6.1.2. *For any polyhedron* Σ*, we have*

$$aut(\Sigma) \leq 4\epsilon(\Sigma).$$

For a planar polyhedron Σ*, the equality holds if, and only if,* Σ *is one of the Platonic polyhedra,* C_n *and* C_n^**. Here,* C_n *itself is a circuit of length* n*,* C_n^**, a cocircuit of* n *muli-edges* $n \geq 1$.

Proof. The first statement is obvious from Theorem 6.1.1. From eqs (6.1.9) and (6.1.10), the equality holds if, and only if, Σ is total regular and hence the corollary by a similar discussion of Theorem 5.1.1. □

In the same way as the discussions in section 5.1, we can easily find all the polyhedra Σ whose automorphism groups are of order $4\epsilon(\Sigma)$ on the projective plane, the torus or the Klein bottle.

Corollary 6.1.3. *For a 3-connected planar graph* $G = (V, E)$*, we have*

$$aut(G) \leq 4\epsilon(G), \tag{6.1.20}$$

where $aut(G)$ *is the order of the automorphism group* $Aut(G)$ *and* $\epsilon(G)$ *is the size of* G.

Proof. From the uniqueness of planar embedding of a 3-connected graph, we know that $\mathrm{Aut}(G) = \mathrm{Aut}(\mu G)$, where $\mu(G)$ is a planar embedding of G. The corollary follows from Corollary 6.1.2. □

However, it is allowed to have aut $(G) > 4\epsilon(G)$ for a graph $G = (V, E)$ even a planar (of course, not 3-connected) one.

Theorem 6.1.2. *Let* v_i*,* φ_i *be the respective number of vertices, faces of valency* i*,* $i \geq 1$*, on a polyhedron* Σ*. Then*

$$aut(\Sigma) \,\big|\, \big(2iv_i, 2j\varphi_j \,|\, \forall i,\ i \geq 1,\ \forall j,\ j \geq 1\big), \tag{6.1.21}$$

where $(2iv_i, 2j\varphi_j \,|\, \forall i,\ i \geq 1,\ \forall j,\ j \geq 1)$ *is the greatest common divisor of all the numbers in the parenthesis.*

Proof. From eq. (6.1.8), an automorphism τ on Σ has to have the property that for $x \in \mathscr{X}(\Sigma)$, which is incident to a vertex of valency i, $i \geq 1$, or a face of valency j, $j \geq 1$, $\tau(x)$ has to be incident to a vertex of valency i or a face of valency j as well. We are also allowed to classify the elements in $\mathscr{X}(\Sigma)$, which are incident to a vertex of valency i by the rule (6.1.17). And then, it is seen that all the classes obtained in this way have the same cardinality, which is the order of the automorphism group Aut (Σ). Because the number of the elements incident to a vertex of valency i is $2iv_i$, we have $\text{aut}(\Sigma) \mid 2iv_i$. Similarly, we may also find $\text{aut}(\Sigma) \mid 2j\varphi_j$. From the arbitrariness of the choice of i, $i \geq 1$, and j, $j \geq 1$, eq. (6.1.21) is found. □

Following the idea as shown in the proof of the theorem, we are further allowed to consider the elements in $\mathscr{X}(\Sigma)$, which are incident to an i-valent vertex and a j-valent face and so forth to improve eq. (6.1.21). We would rather like to leave it for the reader.

Corollary 6.1.4. *For any polyhedron* Σ,

$$aut(\Sigma) \leq (2iv_i,\ 2j\varphi_j \mid \forall i,\ i \geq 1,\ \forall j,\ j \geq 1). \tag{6.1.22}$$

Proof. A direct result of Theorem 6.1.2. □

Corollary 6.1.5. *For a 3-connected planar graph* $G = (V, E)$,

$$aut(G) \leq (2iv_i,\ 2j\varphi_j \mid \forall i,\ i \geq 1,\ \forall j,\ j \geq 1). \tag{6.1.23}$$

Proof. Similar to the proof of Corollary 6.1.3. □

6.2 Eulerian and non-Eulerian codes

In this section, we discuss the representations of a polyhedron $\Sigma = \Sigma(V, E, F)$ by a number of $2\epsilon + 1$ digits to the base v or simply called a v-nary number of $2\epsilon + 1$ digits, where $v = v(\Sigma) = |V|$, $\epsilon = \epsilon(\Sigma) = |E|$.

For a polyhedron Σ given, we may find a number of ways to produce a v-nary number of $2\epsilon + 1$ digits, which represents the polyhedron by a variety of orientations for an OD-tree on the underlying graph $G = (V, G)$ of Σ. Here, only two of them are investigated.

The first kind of ways for orientation of an OD-tree is according to the cyclic order of the face boundary as far as possible.

Let us call a vertex being on the state: *new* or *old*, denoted by N or $\overline{N}$, respectively, accordingly as it has not yet been or has been visited. An edge is arranged to have three states: O, I and II representing that the edge has been visited zeroth, first and second time (or say *used*), respectively. Of course, we always assume the underlying graph $G = (V, E)$ of Σ is simple for the brevity. However, all the statements in what follows can easily be extended to more general cases.

FOD-procedure: Let $\Sigma = \Sigma(V, E, F)$ be represented by a permutation as shown in eq. (6.1.4) on $\mathscr{X}(\Sigma)$. For $v \in V$, $e \in E$, let

$$S(v, e) = (S(v), S(e))$$

be the states v and e are on. Write $\Lambda = \{0, 1, 2, \ldots, v - 1\}$.

Starting rule. Choose an initial vertex v on an element $x(e)$ of $\mathscr{X}(\Sigma)$, which is incident with v. Set $S(v, e) = (N, O)$.

F-rule 1. When a vertex v is reached on an element $x(e) \in \mathscr{X}(\Sigma)$ associated with $S(v, e) = (N, O)$, record the minimum number at v among those that have not been recorded in Λ, set $S(v, e) = (\overline{N}, I)$. Travel the edge $e = (v, u)$ from v along $x(e)$ to u on $x(e_1) = \pi\sigma\delta x(e)$ and set $S(u) = N$, $S(e) = I$, $S(e_1) = O$, if $S(u) \neq \overline{N}$; on $\sigma\delta x(e)$ and set $S(e) = I$, otherwise.

F-rule 2. When a vertex v is reached on an element $x(e) \in \mathscr{X}(\Sigma)$ associated with

$$S(v, e) = (\overline{N}, I),$$

record the number that has already been at v and set $S(e) = II$. Travel the edge $e = (v, u)$ from v along $x(e)$ to u on $x(e_1) = \pi^i\sigma\delta x(e)$, where i is the minimum number of j such that the edge incident with $\pi^j\sigma\delta x(e)$ has not been used, and set $S(e_1) = O$ if e_1 is without a state; II, otherwise.

Finishing rule. Finish doing the procedure by recording the number 0 at v as soon as all edges are on the state II and the starting vertex v is reached.

Theorem 6.2.1. *For a vertex v and $x(e) \in \mathscr{X}(\Sigma)$ incident with v given, the FOD-procedure is well defined and what is recorded on vertices is a ν-nary number of $2\epsilon + 1$ digits with both of the first and the last digits being* 0.

Proof. Because the FOD-procedure executed is just following the OD-tree of the orientation defined by the cyclic orders of faces on Σ, from the discussion in Section 2.2 in Liu [217], the first statement is obtained. Furthermore, what is obtained by the FOD-procedure is a Eulerian tour or a Eulerian travel on the underlying graph $G = (V, E)$ of Σ with the consideration of each edge doubled. Because each step of moving from a vertex to another along an edge records the number at the vertex once and the last time of the recording has to be the initial vertex whose number is 0. All have been recorded in agreement with the order of the Eulerian tour is in fact a number of $2\epsilon + 1$ digits. Since the starting vertex is numbered by 0, both of the first and last digits have to be 0. Because each digit recorded is a number in Λ, it is ν-nary of $2\epsilon + 1$ digits. □

The ν-nary number of $2\epsilon + 1$ digits, denoted by

$$X = X(\Sigma) = x_1x_2\ldots x_{2\epsilon+1},$$

obtained by the FOD-procedure is called an *F-code* of Σ.

First, from the FOD-procedure we may see that the sequence which consists of the numbers occurring for the first time in the order of their position in X from left to right is just the natural one: $0, 1, 2, \ldots, \nu - 1$. A ν-nary number as a sequence of its digit with this property is called *well ordered.*

Second, because each edge is travelled exactly twice in the FOD-procedure, the set of all the 2ϵ successive pairs of digits in X can be partitioned into two parts: X_1 and X_2, $|X_1| = |X_2| = \epsilon$, with a bijection between X_1 and X_2 such that the corresponding pairs only have different orders if they are not the same. A ν-nary number with this property is said to be *well polyhedral*.

Third, from the F-rule 1 in FOD-procedure, each successive pair $x_i x_{i+1}$ in X with x_{i+1} occurring not for the first time and its corresponding successive pair never occurred (or occurred, otherwise) before having its successor $x_{i+2} = x_i$ $(x_{i+2} \neq x_i)$, $2 \leq i \leq 2\epsilon$. That is a reflection at x_i. Of course, $x_{i+1} < x_i$. Further, let $f_1 = x_1 x_2 \ldots x_{i_1}$ such that $x_{i_1} = x_1$ and all x_i, $1 \leq i \leq i_1 - 1$ are not reflective. From F-rule 1, $x_i = i - 1$, $1 \leq i \leq i_1 - 1$. If $f_1, f_2, \ldots, f_k$ are found, then let

$$X\langle f_1, f_2, \ldots, f_k \rangle = x_1^{(k)} x_2^{(k)} \ldots \tag{6.2.1}$$

consists of all the successive pairs that have not been used in $f_1, f_2, \ldots, f_k$ with the order determined by X and let

$$f_{k+1} = \prod_{i=0}^{l} X^{(k)}(t_i, s_{i+1}), \tag{6.2.2}$$

where

$$X^{(k)}(t_i, s_{i+1}) = x_{t_i}^{(k)} x_{t_i+1}^{(k)} \ldots x_{s_{i+1}}^{(k)}, \quad t_0 = 1, \tag{6.2.3}$$

all $x_{t_i+1}^{(k)}, x_{t_i+2}^{(k)}, \ldots, x_{s_{i+1}-1}^{(k)}$ are not reflective, for $0 \leq i \leq l$, $x_{s_i}^{(k)} = x_{t_i}^{(k)}$, $1 \leq i \leq l$, $x_{s_{l+1}}^{(k)} = x_{t_0}^{(k)}$ and the operations in eq. (6.2.2) are done in agreement with the order in $X\langle f_1, \ldots, f_k \rangle$. It is easy to check that an F-code of a polyhedron has the property that there exists an integer φ such that $f_1, f_2, \ldots, f_\varphi$ satisfy

$$X\langle f_1, f_2, \ldots, f_\varphi \rangle = \emptyset.$$

A ν-nary number of $2\epsilon + 1$ digits which has this property is said to be *well reflective*.

A ν-nary number of $2\epsilon + 1$ digit which has each of $0, 1, 2, \ldots, \nu - 1$ occurring at least once and is well ordered, well polyhedral and well reflective is said to be *well satisfied*. Of course, any F-code of a polyhedron is well satisfied from the discussion above.

Fourth, if a segment $x_i x_{i+1} \ldots x_j$, $1 \leq i < j \leq 2\epsilon$, in X with all x_l, $l = i, i+1, \ldots, j$, occurring for the first time, then it is a segment of the natural sequence $0, 1, 2, \ldots, \nu - 1$ and if $x_l x_{l+1}$ has $x_l = x_j$ and $x_{l+1} = x_j - 1$ for $1 \leq x_l \leq \nu - 1$, then x_j never occurs again in X after the occurrence of $x_l x_{l+1}$, i.e., $\forall i\ (i > l)$, $x_i \neq x_j$. A ν-nary number of $2\epsilon + 1$ digits, $\epsilon \geq \nu - 1$, of course, having this property is called *well assigned*.

Lemma 6.2.1. *Any ν-nary number of $2\epsilon + 1$ digits*

$$Z = z_1 z_2 \ldots z_{2\epsilon+1}$$

which is well satisfied is well assigned.

Proof. Because each of $0, 1, 2, \ldots, \nu - 1$ occurs at least once in Z, from the well orderedness, any segment $z_i z_{i+1} \ldots z_j$ with all z_l, $i < l < j$, occurring for the first time has to be a segment of the natural sequence. Moreover, from the well reflectiveness and the well polyhedralness, the well assignedness of Z is obtained. □

A ν-nary number

$$X = x_1 x_2 \ldots x_{2\epsilon+1}$$

of $2\epsilon + 1$ digits is said to be *completely contractible* if it is well reflective and can be transformed into a ν-nary number of $2\nu - 1$ digits, which forms a planar embedding of an OD-tree by contracting all the possible segments:

$$x_{i-1} x_i x_{i+1},\ x_i < x_{i-1} = x_{i+1},\ 2 \le i \le 2\epsilon,$$

into a digit $x (= x_{i-1} = x_{i+1})$ that corresponds to deleting all the reflective edges in X, which is considered as a single-face polyhedron.

Lemma 6.2.2. *A ν-nary number of $2\epsilon + 1$ digits is well satisfied if, and only if, it is completely contractible.*

Proof. By the well reflectiveness, what is obtained by contracting all $x_{i-1} x_i x_{i+1}$, $x_i < x_{i-1} = x_{i+1}$, $2 \le i \le 2\epsilon$, into a digit with the value $x = x_{i-1}$ ($= x_{i+1}$ as well) is a ν-nary number of $2\nu - 1$ digits, which is well ordered and well polyhedral. In fact, it is a planar embedding of a tree from the well polyhedralness and hence one of an OD-tree from the well orderedness. The necessity is true. Conversely, from the completely contractibility, it is easy to check that the sufficiency is true as well. □

Theorem 6.2.2. *A ν-nary number of $2\epsilon+1$ digits X is an F-code of a polyhedron with each edge incident to exactly two faces if, and only if, it is well satisfied.*

Proof. The necessity is obvious from the discussion above. Conversely, according to the well satisfiedness, we may find that

$$\Sigma = (f_1, f_2, \ldots, f_\varphi) \tag{6.2.4}$$

determined by eqs (6.2.1)–(6.2.3) is the polyhedron whose F-code is X obtained by following the FOD-procedure step by step on Σ. □

Further, from Lemma 6.2.2, we can soon find the following characterization of a ν-nary number of $2\epsilon + 1$ digits to be an F-code of a polyhedron, which has the advantage of designing a linear time algorithm for the recognition on computers.

Theorem 6.2.3. *A ν-nary number of $2\epsilon+1$ digits X is an F-code of a polyhedron with each edge incident to exactly two faces if, and only if, it is completely contractible.*

The 8-nary number of 25 digits

$$X_1 = \underbrace{0\,1\,2\,3\,4\,5\,(0}_{f_1}\overbrace{0)\,5\,6\,7\,(4}^{f_2}\underbrace{4)\,7\,(2}_{f_3}\overbrace{2)\,7\,6\,(1}^{f_4}$$

$$\underbrace{1)\,6\,5\,(4}_{f_3}\overbrace{4)\,3\,(0}^{f_2}\overbrace{0)\,3\,(2}^{f_4}\underbrace{2)\,(1}_{f_3}\overbrace{1)\,0}^{f_4}$$

in which (xx) for $x = 0, 1, 2, \ldots$ represents a single digit x itself from now and then is seen to be well satisfied with (f_1, f_2, f_3, f_4) as

$$f_1 = (0, 1, 2, 3, 4, 5); \qquad f_2 = (0, 5, 6, 7, 4, 3);$$
$$f_3 = (4, 7, 2, 1, 6, 5); \qquad f_4 = (2, 7, 6, 1, 0, 3).$$

From Theorem 6.2.2, X_1 is an F-code of a polyhedron. In fact, it is an F-code of the (3,6)-polyhedron on the torus as shown in Figure 6.2.1(a). And, the 8-nary number of 25 digits

$$X_2 = \underbrace{0\,1\,2\,3\,4\,5\,(0}_{f_1\to}\overbrace{0)\,5\,6\,(2}^{f_2\to}\underbrace{2)\,6\,7\,(3}_{f_4\to}\overbrace{3)\,7\,(0}^{f_2\to}$$

$$\underbrace{0)\,7\,6\,5\,4\,(1}_{f_3\to}\underbrace{1)\,4\,(3}_{f_4\leftarrow}\overbrace{3)\,(2}^{f_2\leftarrow}\underbrace{2)\,(1}_{f_4\leftarrow}\underbrace{1)\,0}_{f_3\to}$$

is also seen to be well satisfied with (f_1, f_2, f_3, f_4) as

$$f_1 = (0, 1, 2, 3, 4, 5); \qquad f_2 = (0, 5, 6, 2, 3, 7);$$
$$f_3 = (0, 7, 6, 5, 4, 1); \qquad f_4 = (2, 6, 7, 3, 4, 1).$$

From Theorem 6.2.2, X_2 is an F-code of a polyhedron as well. In fact, it is also an F-code of the (3,6)-polyhedron but on the Klein bottle as shown in Figure 6.2.1(b). Here, the arrows on X_2 as shown above have to be noticed for the non-orientable case.

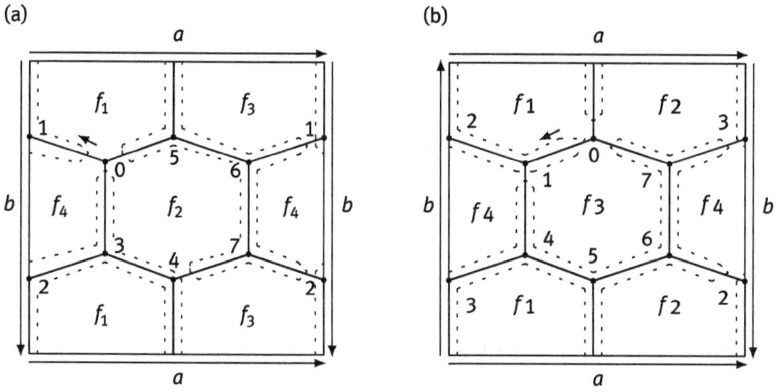

Figure 6.2.1: (a) X_1 and (b) X_2.

On the other hand, if Theorem 6.2.3 is applied to recognize whether or not X_1 or X_2 is an F-code of a polyhedron, it is soon found that both X_1 and X_2 are completely contractible and hence are F-codes of polyhedra.

A code of a polyhedron $\Sigma = \Sigma(V, E, F)$, which is determined by a Eulerian tour of $G = (V, E)$ with $e \in E$ doubled, is said to be *Eulerian*; otherwise, *non-Eulerian*. Of course, an F-code of a polyhedron is a Eulerian one. We may also be allowed to introduce a non-Eulerian code of a polyhedron.

One might think of the dual form of the FOD-procedure, which would better be called *VOD-procedure* by considering the duality of polyhedra to find another code, which would better be called a *V-code*. However, we construct another kind of procedure which is called *St-procedure* to find a so-called *St-code* of a polyhedron.

Let $\Sigma = \Sigma(V, E, F)$ be a polyhedron, with $G = (V, E)$ being its underlying graph and F, its face set. For $v \in V$, suppose

$$\sigma_v = ((v, u_1), (v, u_2), \ldots, (v, u_h)) \tag{6.2.5}$$

is the rotation of the edges at v. Then, we write

$$\pi(v) = (u_1, u_2, \ldots, u_h) \tag{6.2.6}$$

with the cyclic order, or

$$u_i \pi(v) = u_{i+1}$$

for $i = 1, 2, \ldots, h$ and

$$u_{h+1} = u_1.$$

Of course,

$$N(v) = \{u_1, u_2, \ldots, u_h\},$$

the neighbourhood of v. If v is represented by a number i, then we may always write

$$\pi(v) = \pi_i \tag{6.2.7}$$

for $i \in \Lambda = \{0, 1, 2, \ldots, \nu - 1\}$.

St-procedure By defining an order in which the vertices are numbered from Λ step by step, on the process, an edge e which has been visited twice is denoted by $S(e) = \overline{N}$ and a vertex v is with $S(v) = \overline{N}$ whenever all the edges incident to v are visited at least once.

Starting rule Choose a vertex v,

$$\pi(v) = (u_1, u_2, \ldots, u_k),$$

and let v be the vertex which is numbered by 0, i.e. $n(v) = 0$. Set

$$\begin{cases} \pi_0 = (1, 2, \ldots, k); \\ f(i) = 0, \qquad i = 1, 2, \ldots, k, \quad k = \rho(v); \\ n(u_i) = i, \qquad i = 1, 2, \ldots, k; \\ S(0) = \overline{N}; \qquad \Lambda \Leftarrow \Lambda - \{0, 1, 2, \ldots, k\}. \end{cases}$$

Record $0\pi_0 = 0\,(1\,2\,3\,4 \ldots k)$.

Proceeding rule Suppose $S(i) = \overline{N},\ i = 0, 1, 2, \ldots, l - 1$ and

$$\pi(l) = (u_1, u_2, \ldots, u_s).$$

Then, set

$$\pi_l = (f(l), i_1, i_2, \ldots, i_{s-1}), \qquad S(l) = \overline{N}$$

where for $j = 1, 2, \ldots, s - 1$,

$$i_j = \begin{cases} n(u_{j-1}),\ f(i_j) = l \text{ and } S((l, i_j)) = \overline{N}, \\ \qquad \text{if } u_{j-1} \text{ has been numbered;} \\ m = \min\{i \mid \forall i \in \Lambda\},\ n(i_j) = m \\ \qquad \text{and} \Lambda \Leftarrow \Lambda - \{m\}, \quad \text{otherwise.} \end{cases}$$

Record π_l to obtain $0\,\pi_0\,\pi_1\,\ldots\,\pi_{l-1}\,\pi_l$.

Finishing rule When $v-1$ has been used to number a vertex and all the edges and all the vertices are associated with $\overline{N}$, the procedure stops.

Moreover, it can be shown that the resultant sequence of numbers recorded or say the St-code obtained by the St-procedure is a v-nary number of $2\epsilon+1$ digits as well. In fact, it is

$$0\pi_0\pi_1\pi_2\ldots\pi_{v-1}. \tag{6.2.8}$$

By a similar clue as shown for F-code, we may also find the characterization of a v-nary number of $2\epsilon+1$ digits to be an St-code of a polyhedron. Because of the limitation of space, the detail should be left to the reader. One might like to see what happens to the St-code of the polyhedron in Figure 6.2.1(a) or 6.2.1 (b). The St-code of the former is as follows:

$$0(123)(045)(064)(065)(172)(173)(237)(456)$$

from which we may find the faces of the polyhedron as

$$f_1 = (1, 0, 2, 6, 3, 5); \quad f_2 = (2, 0, 3, 6, 7, 4);$$
$$f_3 = (3, 0, 1, 4, 7, 5); \quad f_4 = (4, 1, 5, 7, 6, 2)$$

as shown in Figure 6.2.2(a).

The St-code of the latter is as follows:

$$0(123)(045)(046)(076)(127)(167)(2\overbrace{35}^{\leftarrow})(\underbrace{3\overbrace{54}^{\leftarrow}}_{\rightarrow})$$

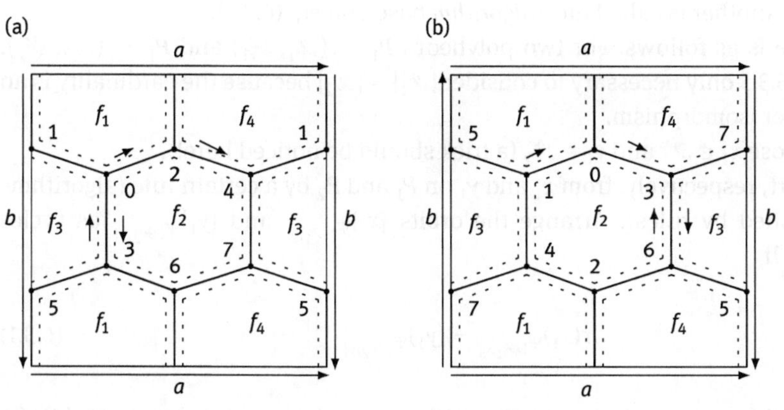

Figure 6.2.2: Faces of polyhedron.

from which the following faces should be derived:

$$f_1 = (1, 0, 2, 4, 7, 5); \qquad f_2 = (3, 0, 1, 4, 2, 6);$$
$$f_3 = (3, 7, 4, 1, 5, 6); \qquad f_4 = (6, 2, 0, 3, 7, 5)$$

as shown in Figure 6.2.2(b). However, in the non-orientable case indicated as above, anti-orders have to appear in the rotations at some vertices along the boundaries of faces as shown by the arrows on the St-code above.

6.3 Determination of automorphisms

Although some invariants are provided, they are still far from determining an isomorphism between two polyhedra in section 6.1.

In fact, it is seen in section 6.2 that an automorphism of a polyhedron can be determined by codes, i.e. a sequence of invariants in a number as a function of their size. This section concerns with clarification and simplification.

In order to do this, algorithms are established for justifying and recognizing if two polyhedra are isomorphic. In other words, an isomorphism can be found between two polyhedra if any; or no isomorphism exists at all otherwise.

Generally speaking, since the ground set of a polyhedron is finite, i.e. its cardinality is 4ϵ, ϵ is the size of polyhedron, by a theoretical point of view, there exists a permutation which corresponds to an isomorphism among all the $(4\epsilon)!$ permutations if any, or no isomorphism at all between two polyhedra otherwise. However, this is an impractical way even on a modern computer.

Our purpose is to establish an algorithm directly with the amount of computation as small as possible without counting all the permutations.

Here, two types of algorithms are presented. One is called *vertex-algorithm* based on eq. (6.3.2). Another is called *face-algorithm* based on eq. (6.3.5).

Their clue is as follows. For two polyhedra $P_1 = (\mathcal{X}_1, \mathcal{P}_1)$ and $P_1 = (\mathcal{X}_2, \mathcal{P}_2)$, from Lemma 6.3.3 only necessary to consider $|\mathcal{X}_1| = |\mathcal{X}_2|$ because the cardinality is an invariant under isomorphism.

First, choose $x_1 \in \mathcal{X}_1$ and $y_1 \in \mathcal{X}_2$ (a trick should be noticed here!).

Then, start, respectively, from x_1 and y_1 on P_1 and P_2 by a certain rule (algorithms are distinguished by rules). Arrange the orbits $\{x_1\}_{\Psi_{\{\mathcal{P}_1,\gamma\}}}$ and $\{y_1\}_{\Psi_{\{\mathcal{P}_2,\gamma\}}}$ as cycles where $\gamma = \sigma\delta$. If

$$\tau(x_1)_{\Psi_{\{\mathcal{P}_1,\gamma\}}} = (y_1)_{\Psi_{\{\mathcal{P}_2,\gamma\}}} \tag{6.3.1}$$

can be induced from $y_1 = \tau(x_1)$, then stop. Otherwise, choose another y_1(a trick!). Go to the procedure on P_2 until every possible y_1 has be chosen.

Finally, if it stops at the latter, then it is shown that P_1 and P_2 are not isomorphic, and denoted by $P_1 \neq P_2$; otherwise, an isomorphism between P_1 and P_2 is done from eq. (6.3.1), denoted by $P_1 = P_2$.

Algorithm 6.3.1. Based on vertices, determine if two maps are isomorphic.

Given two polyhedra $P = (\mathscr{X}, \mathscr{P})$ and $Q = (\mathscr{Y}, \mathscr{Q})$, and their order, size and coorder are all equal (otherwise, not isomorphic!). For convenience, for any $x \in \mathscr{X}$, let $|x| = |\{x\}_{\mathscr{P}}|$, i.e. the valency of vertex $(x)_{\mathscr{P}}$.

Initiation Given $x \in \mathscr{X}$, choose $y \in \mathscr{Y}$. Let $\tau(x) = y$ and $\tau Kx = Ky$. Label both x and y by 1. Naturally, $Kx = Ky = K1 = \{1, \sigma 1, \delta 1, \gamma 1\}$ (Here, the number 1 deals with a symbol!). Label $(x)_{\mathscr{P}}$ by 0, then $x = 1$ is the first element coming to vertex 0. By (v, t_v) denote that t_v is the first element coming to vertex v.

Let S be a sequence of symbols storing numbers and symbols and l, the maximum of labels on all the edges with a label. Here, $S = \emptyset$, $l = 1$ and the minimum of labels among all labelled but not passed vertices $n = 0$. If vertex $(\gamma 1)_{\mathscr{P}} = (1)_{\mathscr{P}}$, the maximum vertex label $m = 0$; otherwise, label vertex $(\gamma 1)_{\mathscr{P}}$ by $1m = 1$.

Proceeding When all vertices are labelled as used, then go to Halt (1).

For n, let s_P and s_Q be, respectively, the number of edges without label on $(\gamma t_n)_{\mathscr{P}}$ and $(\gamma t_n)_{\mathscr{Q}}$.

If $s_P \neq s_Q$, when no y can be chosen, then go to Halt (2); otherwise, choose another y and then go to Initiation.

In the direction starting from γt_n, label those edges by $l+1, \ldots, l+s, s = s_P = s_Q \geq 0$ in order. Thus, two linear orders of elements with numbers labelled

$$\langle \gamma t_n, \mathscr{P}\gamma t_n, \ldots, \mathscr{P}^{-1}\gamma t_n \rangle$$

and

$$\langle \gamma t_n, \mathscr{Q}\gamma t_n, \ldots, \mathscr{Q}^{-1}\gamma t_n \rangle$$

are obtained.

If the two are not equal, when no y is available to choose, then go to Halt (2); otherwise, choose another y and then go to Initiation.

Put this linear order into S as a last part and then substitute the extended sequence for S. In the meantime, label $K(l+1), K(l+2), \ldots, K(l+s)$ on P and Q. Substitute $l+s$ for l. Mark vertex n as used. Substitute $n+1$ for n. Let r be the number of vertices without label in

$$(\gamma(l+1))_{\mathscr{P}}, \ldots, (\gamma(l+s))_{\mathscr{P}},$$

and label them as $m+1, \ldots, m+r$ in order. Substitute $m+r$ for m. Go on the Proceeding.

Halt (1) Output S. (2) P and Q are not isomorphic.

With regard to Algorithm 6.3.1, from the method of choosing y, each element in the ground set is passed through at most once. So there exists a constant c such that the amount of computation is at most $c|\mathscr{X}|$. Since the worst case is for y chosen all over the ground set $\mathscr{Y}$, the total amount of computation is at most $c|\mathscr{X}|^2$. Because of $|\mathscr{X}| = 4\epsilon$, where ϵ is the size of the map, this amount is with its order as the size squared, i.e. $O(\epsilon^2)$.

As described above, if on checking all possibilities of $|\mathscr{Y}|!$, by Stirling formula,

$$\begin{aligned}|\mathscr{Y}|! &\sim \sqrt{2\pi}e^{-|\mathscr{Y}|}|\mathscr{Y}|^{|\mathscr{Y}|-\frac{1}{2}} \\ &>> O(e^{|\mathscr{Y}|}) >> O(\epsilon^{\epsilon}) \\ &>> O(\epsilon^2)\end{aligned}$$

when $|\mathscr{Y}| = |\mathscr{X}| = 4\epsilon$ is large enough. Thus, this algorithm is much efficient.

Algorithm 6.3.2. Based on faces, determine if two polyhedra are isomorphic.

Given two polyhedra $P = (\mathscr{X}_{\sigma,\delta}, \mathscr{P})$ and $Q = (\mathscr{Y}_{\sigma,\delta}, \mathscr{Q})$, and their order, size and co-order are all equal (otherwise, not isomorphic!). For convenience, let $\mathscr{X} = \mathscr{X}_{\sigma,\delta}$, $\mathscr{Y} = \mathscr{Y}_{\sigma,\delta}$ and for any $x \in \mathscr{X}$, let $|x| = |\{x\}_{\mathscr{P}\gamma}|$, i.e. the valency of face $(x)_{\mathscr{P}\gamma}$ where $\gamma = \sigma\delta$.

Initiation Given $x \in \mathscr{X}$, choose $y \in \mathscr{Y}$. Let $\tau(x) = y$ and $\tau Kx = Ky$. Label both x and y by 1. Naturally, $Kx = Ky = K1 = \{1, \sigma 1, \delta 1, \gamma 1\}$ (Here, the number 1 deals with a symbol!). Label $(x)_{\mathscr{P}|ga}$ by 0, then $x = 1$ is the first element coming to face 0. By (f, t_f) denote that t_f is the first element coming to face f.

Let T be a sequence of symbols storing numbers and symbols and l, the maximum of labels over all the edges with a label. Here, $T = \emptyset$, $l = 1$ and the minimum of labels among all labelled but not passed faces $n = 0$. If face $(\gamma 1)_{\mathscr{P}\gamma} = (1)_{\mathscr{P}\gamma}$, the maximum face label $m = 0$; otherwise, label face $(\gamma 1)_{\mathscr{P}\gamma}$ by $1m = 1$.

Proceeding When all faces are labelled as used, then go to Halt (1).

For n, let s_P and s_Q be, respectively, the number of edges without label on $(\gamma t_n)_{\mathscr{P}\gamma}$ and $(\gamma t_n)_{\mathscr{Q}\gamma}$.

If $s_P \neq s_Q$, when no y can be chosen, then go to Halt (2); otherwise, choose another y and then go to Initiation.

In the direction starting from γt_n, label those edges by $l+1, \ldots, l+s, s = s_P = s_Q \geq 0$ in order. Thus, two linear orders of elements with numbers labelled

$$\langle \gamma t_n, \mathscr{P}\gamma\gamma t_n, \ldots, \mathscr{P}\gamma^{-1}\gamma t_n \rangle$$

and

$$\langle \gamma t_n, \mathscr{Q}\gamma\gamma t_n, \ldots, \mathscr{Q}\gamma^{-1}\gamma t_n \rangle$$

are obtained.

If the two are not equal, when no y is available to choose, then go to Halt (2); otherwise, choose another y and then go to Initiation.

Put this linear order into S as last part and then substitute the extended sequence for S. In the meantime, label $K(l+1), K(l+1), \ldots, K(l+s)$ on P and Q. Substitute $l+s$ for l. Mark n as used. Substitute $n+1$ for n. Let r be the number of vertices without label in

$$(\gamma(l+1))_{\mathscr{P}}, \ldots, (\gamma(l+s))_{\mathscr{P}},$$

and label them as $m+1, \ldots, m+r$ in order. Substitute $m+r$ for m. Go on the Proceeding.

Put this linear order into T as last part and then substitute the extended sequence for T. In the meantime, label $K(l+1), K(l+2), \ldots, K(l+s)$ on P and Q. Substitute $l+s$ for l. Mark face n as used. Substitute $n+1$ for n. Let r be the number of faces without label in

$$(\gamma(l+1))_{\mathscr{P}}, \ldots, (\gamma(l+s))_{\mathscr{P}},$$

and label them as $m+1, \ldots, m+r$ in order. Substitute $m+r$ for m. Go on the Proceeding.

Halt (1) Output T. (2) P and Q are not isomorphic.

With regard to Algorithm 6.3.2, it can be seen as the dual of Algorithm 6.3.1. The amount of its computation is also estimated as $O(\epsilon^2)$.

Note 6.3.1. These two algorithms suggest that whenever a cyclic order of edges at each vertex is given, an efficient algorithm for justifying and recognizing if two graphs are isomorphic within the cyclic order at each vertex can be established. By calling an algorithm *efficient*, it is meant that there exists an constant c such that the amount of its computation is about $O(\epsilon^c)$, ϵ is the size of the graphs.

If without considering the limitation of a cyclic order at each vertex, no efficient algorithm for an isomorphism of two graphs has been found yet up to now. However, a new approach is, from what has been discussed here, provided for further investigation of an isomorphism between two graphs.

In this section, it is shown that the two algorithms described in the last section can be used for justifying and recognizing whether, or not, two polyhedra are isomorphic.

Lemma 6.3.1. *Let S and T are, respectively, the outputs of Algorithms 6.3.1 and 6.3.2 at Halt (1), then*

(i) *Elements in S and T are all in the same orbit of group $\Psi_{\{\mathscr{P},\gamma\}}$ on $\mathscr{X}$.*

(ii) *S forms an orbit of group $\Psi_{\{\mathscr{P},\gamma\}}$ on $\mathscr{X}$ if, and only if, T forms an orbit of group $\Psi_{\{\mathscr{P},\gamma\}}$ on $\mathscr{X}$;*

(iii) *S forms an orbit of group $\Psi_{\{\mathscr{P},\gamma\}}$ on $\mathscr{X}$ if, and only if, for any $x \in S$, $\gamma x \in S$.*

Proof.

(i) From the proceedings of the two algorithms, it is seen that from an element only passes through γ and $\mathscr{P}$ (Algorithm 6.3.1), or γ and $\mathscr{P}\gamma$ (Algorithm 6.3.2) for getting an element in S, or T. Because $\gamma, \mathscr{P}, \mathscr{P}\gamma \in \Psi_{\{\mathscr{P},\gamma\}}$ and $\gamma^2 = 1$, elements in S and T are all in the same orbit of group $\Psi_{\{\mathscr{P},\gamma\}}$ on $\mathscr{X}$.

(ii) Necessity. Because S forms an orbit of group $\Psi_{\{\mathscr{P},\gamma\}}$ on $\mathscr{X}$, and from Algorithm 6.3.1, S contains half the elements of $\mathscr{X}$, by Lemma 2.3.4, group $\Psi_{\{\mathscr{P},\gamma\}}$ has two orbits on $\mathscr{X}$. This implies in the orientable case. Thus, from (i), T forms an orbit of group $\Psi_{\{\mathscr{P},\gamma\}}$ on $\mathscr{X}$ as well.

Sufficiency. On the basis of duality, it is deduced from the necessity.

(iii) Necessity. Since S forms an orbit of group $\Psi_{\{\mathscr{P},\gamma\}}$ on $\mathscr{X}$ and S contains only half the elements of $\mathscr{X}$, by Lemma 2.3.4, group $\Psi_{\{\mathscr{P},\gamma\}}$ has two orbits on $\mathscr{X}$. From the orientability, for any $x \in S$, $\gamma x \in S$.

Sufficiency. Since for any $x \in S$, $\gamma x \in S$, and S only contains half the elements of $\mathscr{X}$, by Theorem 2.3.2, it is only possible that S itself forms an orbit of group $\Psi_{\{\mathscr{P},\gamma\}}$ on $\mathscr{X}$. □

For non-orientable maps, such two algorithms have their outputs S and T also containing half the elements of $\mathscr{X}$ but not forming an orbit of group $\Psi_{\{\mathscr{P},\gamma\}}$.

Lemma 6.3.2. *Let S and T are, respectively, the outputs of Algorithms* 6.3.1 *and* 6.3.2 *at Halt* (1). And, let G_S and G_T be, respectively, the graphs induced by elements in S and T, then $G_S = G_T = G(P)$.

Proof. From Lemma 6.3.1(i), by the procedures of the two algorithms, because the intersection of each of S and T with any quadricell consists of two elements incident to the two ends of the edge, S, T as well, is incident to all edges with two ends of each edge in polyhedron P.

Therefore, $G_S = G_T = G(P)$. □

Theorem 6.3.1. *The output S of Algorithm* 6.3.1 *at Halt (1) induces an isomorphism between polyhedra P and Q. Halt (2) shows that polyhedra P and Q are not isomorphic.*

Proof. Let τ be a mapping from $\mathscr{X}$ to $\mathscr{Y}$ such that the image and the co-image are with the same label. From the transitivity of a map, τ is a bijection. Because $\tau K x = K \tau x$, $x \in \mathscr{X}$, then $\tau\sigma\tau^{-1} = \sigma$ and $\tau\delta\tau^{-1} = \delta$. And in the Proceeding, for labelling a vertex $(x)_{\mathscr{P}}$, $\tau(x)_{\mathscr{P}} = (\tau x)_{\mathscr{Q}}$. From Lemma 6.3.2, this implies that $\tau\mathscr{P}\tau^{-1} = \mathscr{Q}$. Based on Lemma 6.1.1, τ is an isomorphism between P and Q. This is the first statement.

By contradiction to prove the second statement. Assume that there is an isomorphism τ between P and Q. If $\tau(x) = y$, then by Algorithm 6.3.1 the procedure should terminate at Halt (1). However, a termination at Halt (2) shows that for any $x \in \mathscr{X}$, there is no elements in $\mathscr{Y}$ corresponding to x in an isomorphism between polyhedra P and Q, and hence it is impossible to terminate at Halt (1). This is a contradiction.

Therefore, the theorem is true. □

Although the following theorem has its proof with a similar reasoning, in order to understand the precise differences the proof is still in a detailed explanation.

Theorem 6.3.2. *The output T of Algorithm 6.3.2 at Halt (1) induces an isomorphism between polyhedra P and Q. Halt (2) shows that polyhedra P and Q are not isomorphic.*

Proof. Let τ be a mapping from $\mathscr{X}$ to $\mathscr{Y}$ such that the image and the co-image are with the same label. From the transitivity of a map, τ is a bijection. Because $\tau Kx = K\tau x$, $x \in \mathscr{X}$, then $\tau\sigma\tau^{-1} = \sigma$ and $\tau\delta\tau^{-1} = \delta$. And in the Proceeding, for labelling a face $(x)_{\mathscr{P}\gamma}$, $\tau(x)_{\mathscr{P}\gamma} = (\tau x)_{\mathscr{Q}\gamma}$. From Lemma 6.3.2, this implies that $\tau\mathscr{P}\gamma\tau^{-1} = \mathscr{Q}\gamma$. Based on Lemma 6.1.1, τ is an isomorphism between P and Q. This is the first statement.

By contradiction to prove the second statement. Assume that there is an isomorphism τ between P and Q. If $\tau(x) = y$, then by Algorithm 6.3.2 the procedure should terminate at Halt (1). However, a termination at Halt (2) shows that for any $x \in \mathscr{X}$, there are no elements in $\mathscr{Y}$ corresponding to x in an isomorphism between polyhedra P and Q, and hence it is impossible to terminate at Halt (1). This is a contradiction.

Therefore, the theorem is true. □

If missing all those related to y in Algorithm 6.3.1 and 6.3.2 (i.e., P=Q=M), then for any polyhedron $M = (\mathscr{X}, \mathscr{P})$, the procedures will always terminate at Halt (1). Thus, their outputs S and T are, respectively, called a *primal trail code* and a *dual trail code* of M. When an element x and a polyhedron P should be indicated, they are denoted by, respectively, $S_x(P)$ and $T_x(P)$.

Theorem 6.3.3. *Let $P = (\mathscr{X}, \mathscr{P})$ and $Q = (\mathscr{Y}, \mathscr{Q})$ be two given maps. Then, they are isomorphic if, and only if, for any $x \in \mathscr{X}$ chosen, there exists an element $y \in \mathscr{Y}$ such that $S_x(P) = S_y(Q)$, or $T_x(P) = T_y(Q)$.*

Proof. Necessity. Suppose τ is an isomorphism between polyhedra $P = (\mathscr{X}, \mathscr{P})$ and $Q = (\mathscr{Y}, \mathscr{Q})$. For the given element $x \in \mathscr{X}$, let $y = \tau(x)$. From Theorem 6.3.1 or 6.3.2, $S_x(P) = S_y(Q)$, or $T_x(P) = T_y(Q)$.

Sufficiency. From Theorem 6.3.1 or 6.3.2, it is known that by Algorithm 6.3.1, or by Algorithm 6.3.2, their outputs induce an isomorphism between $P = (\mathscr{X}, \mathscr{P})$ and $Q = (\mathscr{Y}, \mathscr{Q})$. □

Note 6.3.2. In justifying whether, or not, two polyhedra are isomorphic, the initial element x can be chosen arbitrarily in one of the two polyhedra to see if there is an element y in the other such that $S_x(P) = S_y(Q)$, or $T_x(P) = T_y(Q)$. This enables us to do for some convenience.

In addition, based on Theorem 6.3.3, all isomorphisms between two polyhedra can be found if any.

Here, two examples are provided for further understanding the procedures of the two algorithms described in the last section.

Example 6.3.1. Justify whether, or not, two polyhedra $P_1 = (\mathcal{X}_1, \mathcal{P}_1)$ and $P_2 = (\mathcal{X}_2, \mathcal{P}_2)$ are isomorphic where

$$\mathcal{X}_1 = Kx_1 + Ky_1, \ \mathcal{P}_1 = (x_1, y_1, \delta y_1)(\gamma x_1)$$

and

$$\mathcal{X}_2 = Kx_2 + Ky_2, \ \mathcal{P}_2 = (y_2, x_2, \delta y_2)(\gamma x_2).$$

First, for P_1, choose $x = x_1$. By Algorithm 6.3.1, find $S_x(P_1)$. Let

$$\mathcal{P}_1 = (x_1, y_1, \delta y_1)(\gamma x_1) = uv.$$

Initiation

$$x_1 = 1, \ Kx_1 = \{1, \sigma 1, \delta 1, \gamma 1\}, \ u = 0, v = 1,$$
$$S = \emptyset, \ l = 0, \ m = 1.$$

Proceeding

Step 1 $\mathcal{P}_1 = (1, y_1, \delta y_1)(\gamma 1)$.

$$y_1 = 2, \ Ky_1 = \{2, \sigma 2, \delta 2, \gamma 2\}, \ u = 0, \ v = 1,$$
$$S = \langle 1, 2, \delta 2 \rangle, \ l = 2, \ n = 1, \ m = 1.$$

Step 2 $\mathcal{P}_1 = (1, 2, \delta 2)(\gamma 1)$.

$$u = 0, \ v = 1,$$
$$S = \langle 1, 2, \delta 2, \gamma 1 \rangle, \ l = 2, \ n = 1, \ m = 1.$$

Halt (1) Output: $S_x(P_1) = S = \langle 1, 2, \delta 2, \gamma 1 \rangle$.

Then, for P_2, because a link should correspond to a link and a vertex should correspond to a vertex with the same valency, y has only two possibilities for choice, i.e. x_2 and σx_2. Choose $y = x_2$. By Algorithm 6.3.1, find $S_y(P_2)$. Let

$$\mathcal{P}_2 = (y_2, x_2, \delta y_2)(\gamma x_2) = uv.$$

Initiation

$$x_2 = 1, \ Kx_2 = \{1, \sigma 1, \delta 1, \gamma 1\}, \ u = 0, v = 1,$$
$$S = \emptyset, \ l = 0, \ m = 1.$$

Proceeding

Step 1 $\mathscr{P}_2 = (y_1, 1, \delta y_2)(\gamma 1)$.

$$\delta y_2 = 2,\ K\delta y_2 = \{2, \sigma 2, \delta 2, \gamma 2\},\ u = 0,\ v = 1,$$
$$S = \langle 1, 2, \delta 2\rangle,\ l = 2,\ n = 1,\ m = 1.$$

Step 2 $\mathscr{P}_2 = (2, 1, \delta 2)(\gamma 1)$.

$$u = 0,\ v = 1,$$
$$S = \langle 1, 2, \delta 2, \gamma 1\rangle,\ l = 2,\ n = 1,\ m = 1.$$

Halt (1) Output: $S_y(P_2) = S = \langle 1, 2, \delta 2, \gamma 1\rangle$.
Since $S_x(P_1) = S_y(P_2)$ and $y = x_2$, an isomorphism from P_1 to P_2 is found as τ_1:

$$\tau_1 K x_1 = K x_2,\ \tau_1 K y_1 = K y_2.$$

Then, choose $y = \sigma x_2$. By Algorithm 6.3.1, find $S_y(P_2)$. Let

$$\mathscr{P}_2 = (\sigma x_2, \sigma y_2, \gamma y_2)(\delta x_2) = uv.$$

Initiation

$$\sigma x_2 = 1,\ K\sigma x_2 = \{1, \sigma 1, \delta 1, \gamma 1\},\ u = 0, v = 1,$$
$$S = \emptyset,\ l = 0,\ m = 1.$$

Proceeding

Step 1 $\mathscr{P}_2 = (1, \sigma y_2, \gamma y_2)(\gamma 1)$.

$$\sigma y_2 = 2,\ K\sigma y_2 = \{2, \sigma 2, \delta 2, \gamma 2\},\ u = 0,\ v = 1,$$
$$S = \langle 1, 2, \delta 2\rangle,\ l = 2,\ n = 1,\ m = 1.$$

Step 2 $\mathscr{P}_2 = (1, 2, \delta 2)(\gamma 1)$.

$$u = 0,\ v = 1,$$
$$S = \langle 1, 2, \delta 2, \gamma 1\rangle,\ l = 2,\ n = 1,\ m = 1.$$

Halt (1) Output: $S_y(P_2) = S = \langle 1, 2, \delta 2, \gamma 1\rangle$.
Since $S_x(P_1) = S_y(P_2)$ and $y = \sigma x_2$, an isomorphism from P_1 to P_2 is found as τ_2:

$$\tau_2 K x_1 = K\sigma x_2,\ \tau_2 K y_1 = K\sigma y_2.$$

In consequence, there are two isomorphisms between P_1 and P_2 above in all. Since $2 \in S_x(P_1)$ but $\gamma 2 \notin S_x(P_1)$, by Lemma 6.3.1(iii), P_1, P_2 as well, is non-orientable.

Example 6.3.2. Justify whether, or not, $P_1 = (\mathscr{X}_1, \mathscr{P}_1)$ and $P_2 = (\mathscr{X}_2, \mathscr{P}_2)$ are isomorphic where

$$\mathscr{X}_1 = Kx_1 + Ky_1, \ \mathscr{P}_1 = (x_1, y_1, \gamma y_1)(\gamma x_1)$$

and

$$\mathscr{X}_2 = Kx_2 + Ky_2, \ \mathscr{P}_2 = (y_2, x_2, \gamma y_2)(\gamma x_2).$$

First, for P_1, choose $x = x_1$. By Algorithm 6.3.2, find $T_x(P_1)$. Let

$$\mathscr{P}_1\gamma = (x_1, \gamma x_1, y_1)(\gamma y_1) = fg.$$

Initiation

$$x_1 = 1, \ Kx_1 = \{1, \sigma 1, \delta 1, \gamma 1\}, \ f = 0, g,$$
$$T = \emptyset, \ l = 0, \ m = 0.$$

Proceeding

Step 1 $\mathscr{P}_1\gamma = (1, \gamma 1, y_1)(\gamma y_1)$.

$$y_1 = 2, \ Ky_1 = \{2, \sigma 2, \delta 2, \gamma 2\}, \ f = 0, \ g = 1,$$
$$T = \langle 1, \gamma 1, 2\rangle, \ l = 2, \ n = 1, \ m = 1.$$

Step 2 $\mathscr{P}_1\gamma = (1, \gamma 1, 2)(\gamma 2)$.

$$f = 0, \ g = 1,$$
$$T = \langle 1, \gamma 1, 2, \gamma 2\rangle, \ l = 2, \ n = 1, \ m = 1.$$

Halt (1) Output: $T_x(P_1) = T = \langle 1, \gamma 1, 2, \gamma 2\rangle$.
Then, for P_2, because a link should be corresponding to a link and a vertex should be corresponding to a vertex with the same valency, y only has two possibilities for choosing, i.e. x_2 and σx_2. Choose $y = x_2$. By Algorithm 6.3.2, find $T_y(P_2)$. Let

$$\mathscr{P}_2\gamma = (x_2, \gamma x_2, \gamma y_2)(y_2) = fg.$$

Initiation

$$x_2 = 1, \ Kx_2 = \{1, \sigma 1, \delta 1, \gamma 1\}, \ f = 0, g = 0,$$
$$T = \emptyset, \ l = 0, \ m = 0.$$

Proceeding

Step 1 $\mathscr{P}_2\gamma = (1, \gamma 1, \gamma y_2)(y_2)$.

$$\gamma y_2 = 2,\ K\gamma y_2 = \{2, \sigma 2, \delta 2, \gamma 2\},\ f = 0,\ g = 1,$$
$$T = \langle 1, \gamma 1, 2\rangle,\ l = 2,\ n = 1,\ m = 1.$$

Step 2 $\mathscr{P}_2\gamma = (1, \gamma 1, 2)(\gamma 2)$.

$$f = 0,\ g = 1,$$
$$T = \langle 1, \gamma 1, 2, \gamma 2\rangle,\ l = 2,\ n = 1,\ m = 1.$$

Halt (1) Output: $T_y(P_2) = T = \langle 1, \gamma 1, 2, \gamma 2\rangle$.
Since $T_x(P_1) = T_y(P_2)$ and $y = x_2$, an isomorphism from P_1 to P_2 is found as τ_1:

$$\tau_1 K x_1 = K x_2,\ \tau_1 K y_1 = K\gamma y_2.$$

Then, choose $y = \sigma x_2$. By Algorithm 7.2, find $T_y(P_2)$. Let

$$\mathscr{P}_2\gamma = (\sigma x_2, \delta x_2, \sigma y_2)(\delta y_2) = fg.$$

Initiation

$$\sigma x_2 = 1,\ K\sigma x_2 = \{1, \sigma 1, \delta 1, \gamma 1\},\ f = 0, g = 0,$$
$$T = \emptyset,\ l = 0,\ m = 0.$$

Proceeding

Step 1 $\mathscr{P}_2\gamma = (1, \gamma 1, \sigma y_2)(\delta y_2)$.

$$\sigma y_2 = 2,\ K\gamma y_2 = \{2, \sigma 2, \delta 2, \gamma 2\},\ f = 0,\ g = 1,$$
$$T = \langle 1, \gamma 1, 2\rangle,\ l = 2,\ n = 1,\ m = 1.$$

Step 2 $\mathscr{P}_2\gamma = (1, \gamma 1, 2)(\gamma 2)$.

$$f = 0,\ g = 1,$$
$$T = \langle 1, \gamma 1, 2, \gamma 2\rangle,\ l = 2,\ n = 1,\ m = 1.$$

Halt (1) Output: $T_y(P_2) = T = \langle 1, \gamma 1, 2, \gamma 2\rangle$.
Since $T_x(P_1) = T_y(P_2)$ and $y = \sigma x_2$, an isomorphism from P_1 to P_2 is found as τ_2:

$$\tau_2 K x_1 = K\sigma x_2,\ \tau_2 K y_1 = K\sigma y_2.$$

In consequence, there are two isomorphisms between P_1 and P_2 in all. By Lemma 6.3.1(iii), P_1, P_2, as well, is orientable.

Now, the automorphism groups of standard polyhedra, i.e. simplified butterflies and simplified barflies, are discussed.

First, observe the orientable case. For

$$O_1 = (\mathscr{X}_1, \mathscr{J}_1) = (Kx_1 + Ky_1, (x_1, y_1, \gamma x_1, \gamma y_1)),$$

by Algorithm 6.3.1,

$$\begin{aligned}
S_{x_1}(O_1) &= 1, y_1, \gamma 1, \gamma y_1 = 1, 2, \gamma 1, \gamma 2;\\
S_{\sigma x_1}(O_1) &= 1, \delta y_1, \gamma 1, \sigma y_1 = 1, 2, \gamma 1, \gamma 2;\\
S_{\delta x_1}(O_1) &= 1, \sigma y_1, \gamma 1, \delta y_1 = 1, 2, \gamma 1, \gamma 2;\\
S_{\gamma x_1}(O_1) &= 1, \gamma y_1, \gamma 1, y_1 = 1, 2, \gamma 1, \gamma 2;\\
S_{y_1}(O_1) &= 1, \gamma x_1, \gamma 1, x_1 = 1, 2, \gamma 1, \gamma 2;\\
S_{\sigma y_1}(O_1) &= 1, \sigma x_1, \gamma 1, \delta x_1 = 1, 2, \gamma 1, \gamma 2;\\
S_{\delta y_1}(O_1) &= 1, \delta x_1, \gamma 1, \sigma x_1 = 1, 2, \gamma 1, \gamma 2;\\
S_{\gamma y_1}(O_1) &= 1, x_1, \gamma 1, \gamma x_1 = 1, 2, \gamma 1, \gamma 2,
\end{aligned}$$

i.e.

$$\begin{aligned}
S_{x_1}(O_1) &= S_{\sigma x_1}(O_1) = S_{\delta x_1}(O_1) = S_{\gamma x_1}(O_1) = S_{y_1}(O_1)\\
&= S_{\sigma y_1}(O_1) = S_{\delta y_1}(O_1) = S_{\gamma y_1}(O_1)\\
&= 1, 2, \gamma 1, \gamma 2.
\end{aligned}$$

Thus, O_1 has its automorphism group of order 8, i.e.

$$\operatorname{aut}(O_1) = 4 \times (2 \times 1) = 8.$$

A polyhedron with a non-trivial automorphism group is said to be *symmetrical*. If a polyhedron with its automorphism group of order 4 times its size, then it is said to be *completely symmetrical*. It can be seen that O_1 is completely symmetrical. However, none of O_k, $k \geq 2$, is completely symmetrical although they are all symmetrical.

Theorem 6.3.4. *For simplified butterflies (orientable standard polyhedra)* $O_k = (\mathscr{X}_k, \mathscr{J}_k)$, $k \geq 1$, *where*

$$\mathscr{X}_k = \sum_{i=1}^{k} (Kx_i + Ky_i)$$

and

$$\mathscr{J}_k = \left(\prod_{i=1}^{k} \langle x_i, y_i, \gamma x_i, \gamma y_i \rangle \right),$$

we have

$$aut(O_k) = \begin{cases} 2k, & if\, k \geq 2; \\ 8, & if\, k = 1. \end{cases} \qquad (6.3.2)$$

Proof. From the symmetry between $\langle x_i, y_i, \gamma x_i, \gamma y_i\rangle, i \geq 2$, and $\langle x_1, y_1, \gamma x_1, \gamma y_1\rangle$ in $\mathcal{J}_k$, $k \geq 2$, only necessary to calculate $S_{x_1}(O_k)$, $S_{\gamma x_1}(O_k)$, $S_{y_1}(O_k)$, $S_{\gamma y_1}(O_k)$, $S_{\sigma x_1}(O_k)$, $S_{\delta x_1}(O_k)$, $S_{\sigma y_1}(O_k)$, and $S_{\delta y_1}(O_k)$ by Algorithm 6.3.1.

From Algorithm 6.3.1,

$$\begin{aligned} S_{x_1}(O_k) &= 1, y_1, \gamma 1, \gamma y_1, \prod_{i=1}^{k} \langle x_i, y_i, \gamma x_i, \gamma y_i\rangle \\ &= 1, 2, \gamma 1, \gamma 2, \prod_{i=2}^{k} \langle (2i-1), 2i, \gamma(2i-1), \gamma 2i\rangle \\ &= \prod_{i=1}^{k} \langle (2i-1), 2i, \gamma(2i-1), \gamma 2i\rangle, \end{aligned}$$

$$\begin{aligned} S_{\sigma x_1}(O_k) &= 1, \sigma\left(\prod_{i=1}^{k} \langle x_i, y_i, \gamma x_i, \gamma y_i\rangle\right)^{-1}, \delta y_1, \gamma 1, \sigma y_1 \\ &\neq S_{x_1}(O_k), \end{aligned}$$

$$\begin{aligned} S_{\gamma x_1}(O_k) &= 1, \gamma y_1, \prod_{i=1}^{k} \langle x_i, y_i, \gamma x_i, \gamma y_i\rangle, \gamma 1, y_1 \\ &\neq S_{x_1}(O_k), \end{aligned}$$

$$\begin{aligned} S_{\delta x_1}(O_k) &= 1, \sigma y_1, \gamma 1, \sigma\left(\prod_{i=1}^{k} \langle x_i, y_i, \gamma x_i, \gamma y_i\rangle\right)^{-1}, \delta y_1 \\ &\neq S_{x_1}(O_k), \end{aligned}$$

$$\begin{aligned} S_{y_1}(O_k) &= 1, \gamma x_1, \gamma 1, \prod_{i=1}^{k} \langle x_i, y_i, \gamma x_i, \gamma y_i\rangle, x_1 \\ &\neq S_{x_1}(O_k), \end{aligned}$$

$$\begin{aligned} S_{\sigma y_1}(O_k) &= 1, \sigma x_1, \sigma\left(\prod_{i=1}^{k} \langle x_i, y_i, \gamma x_i, \gamma y_i\rangle\right)^{-1}, \gamma 1, \delta x_1 \\ &\neq S_{x_1}(O_k), \end{aligned}$$

$$\begin{aligned}S_{\gamma y_1}(O_k) &= 1, \prod_{i=1}^{k}\langle x_i, y_i, \gamma x_i, \gamma y_i\rangle, x_1, \gamma 1, \gamma x_1\\ &\neq S_{x_1}(O_k),\end{aligned}$$

$$\begin{aligned}S_{\delta y_1}(O_k) &= 1, \delta x_1, \sigma 1, \sigma x_1, \sigma\left(\prod_{i=1}^{k}\langle x_i, y_i, \gamma x_i, \gamma y_i\rangle\right)^{-1}\\ &= 1, 2.\gamma 1, \gamma 2, \prod_{i=2}^{k}\langle(2i-1), 2i, \gamma(2i-1), \gamma 2i\rangle\\ &= S_{x_1}(O_k).\end{aligned}$$

Because of two automorphisms from $S_{\delta y_1}(O_k) = S_{x_1}(O_k)$, O_k have $2 \times k = 2k$ automorphisms altogether. Hence, when $k \geq 2$,

$$\text{aut}(O_k) = 2k.$$

When $k = 1$, $\text{aut}(O_1) = 8$ is known. □

Then, observe the non-orientable case. for

$$Q_1 = (\mathscr{X}_1, \mathscr{I}_1) = (Kx_1, (x_1, \delta x_1)),$$

by Algorithm 6.3.1,

$$\begin{aligned}S_{x_1}(Q_1) &= 1, \delta_1;\ S_{\sigma x_1}(Q_1) = 1, \delta 1;\\ S_{\delta x_1}(Q_1) &= 1, \delta 1;\ S_{\gamma x_1}(Q_1) = 1, \delta 1,\end{aligned}$$

i.e. $\text{aut}(Q_1) = 4$.

Theorem 6.3.5. *For simplified barflies $Q_l = (\mathscr{X}_l, \mathscr{I}_l)$, $l \geq 1$, where*

$$\mathscr{X}_l = \sum_{i=1}^{l} Kx_i$$

and

$$\mathscr{I}_l = \prod_{i=1}^{l}\langle x_i, \delta x_i\rangle,$$

we have

$$aut(Q_l) = \begin{cases} 2l, & l \geq 2; \\ 4, & l = 1. \end{cases} \tag{6.3.3}$$

Proof. From the symmetry of $\langle x_i, \delta x_i \rangle$, $i \geq 2$, and $\langle x_1, \delta x_1 \rangle$ in $\mathscr{I}_l$, $l \geq 2$, only necessary to calculate

$$S_{x_1}(Q_l), S_{\sigma x_1}(Q_l), S_{\delta x_1}(Q_l) \text{ and } S_{\gamma x_1}(Q_l)$$

by employing Algorithm 6.3.1. □

From Algorithm 6.3.1,

$$\begin{aligned} S_{x_1}(Q_l) &= 1, \delta 1, \prod_{i=2}^{l} \langle x_i, \delta x_i \rangle \\ &= 1, \delta 1, 2, \delta 2, \prod_{i=2}^{l} \langle x_i, \delta x_i \rangle \\ &= \prod_{i=1}^{l} \langle i, \delta i \rangle, \end{aligned}$$

$$\begin{aligned} S_{\sigma x_1}(Q_l) &= 1, \sigma \left(\prod_{i=2}^{l} \langle x_i, \delta x_i \rangle \right)^{-1}, \delta 1 \\ &\neq S_{x_1}(Q_l), \end{aligned}$$

$$\begin{aligned} S_{\delta x_1}(Q_l) &= 1, \prod_{i=2}^{l} \langle x_i, \delta x_i \rangle, \delta 1 \\ &\neq S_{x_1}(Q_l), \end{aligned}$$

$$\begin{aligned} S_{\gamma x_1}(Q_l) &= 1, \delta 1, \sigma \left(\prod_{i=2}^{l} \langle x_i, \delta x_i \rangle \right)^{-1} \\ &= 1, \delta 1, l, \delta x_l, \prod_{i=l-1}^{2} \langle \gamma x_i, \sigma x_i \rangle \\ &= \prod_{i=1}^{l} \langle i, \delta i \rangle \\ &= S_{x_1}(Q_l). \end{aligned}$$

Because of two automorphisms from $S_{\gamma x_1}(Q_l) = S_{x_1}(Q_l)$, Q_l has $2 \times l = 2l$ automorphisms altogether. Hence, when $l \geq 2$,

$$\text{aut}(Q_l) = 2l.$$

When $l = 1$, $\text{aut}(Q_1) = 4$ is known. ■

Similarly, the two theorems can also be proved by employing Algorithm 6.3.2.

6.4 Asymmetrization

For a given polyhedron $P = (\mathscr{X}, \mathscr{P})$, if a subset $R \subseteq \mathscr{X}$ is chosen such that an automorphism of P with R fixed, i.e. an element of R only corresponds to an element of R, then P is called a *set rooted* map. The subset R is called the *rooted set* of P, and an element of R is called a *rooted element*.

Theorem 6.4.1. *For a set rooted polyhedron $P^R = (\mathscr{X}, \mathscr{P})$, R is the rooted set,*

$$aut(P^R) \mid |R|. \tag{6.4.1}$$

Proof. Assume that all elements in R are partitioned into equivalent classes under the group $\text{Aut}(P^R)$. From Theorem 6.4.1, each class has $\text{aut}(P^R)$ elements. Therefore, eq. (6.4.1) is satisfied. □

Corollary 6.4.1. *For a set rooted polyhedron $P^R = (\mathscr{X}, \mathscr{P})$, R is the rooted set,*

$$aut(P^R) \leq |R|. \tag{6.4.2}$$

Proof. A direct result of eq. (6.4.1). □

For a given polyhedron $P = (\mathscr{X}, \mathscr{P})$, if a vertex v_x, $x \in \mathscr{X}$, is chosen such that an automorphism of P has to be with v_x fixed, i.e. an element incident with v_x has to correspond to an element incident with v_x, then P is called a *vertex rooted* map. The vertex v_x is called the *rooted vertex* of P, and an element incident with v_x, the *rooted element*.

Corollary 6.4.2. *For a vertex rooted polyhedron $P^{vr} = (\mathscr{X}, \mathscr{P})$, v_x is the rooted vertex,*

$$aut(P^{vr}) \mid 2|\{x\}_{\mathscr{P}}|. \tag{6.4.3}$$

Proof. This is eq. (6.4.1) when $R = \{x\}_{\mathscr{P}} \cup \{\sigma x\}_{\mathscr{P}}$. □

For a given polyhedron $P = (\mathscr{X}, \mathscr{P})$, if face f_x, $x \in \mathscr{X}$, is chosen such that an automorphism of P has f_x fixed, i.e. an element incident with f_x should be corresponding to an element incident with f_x, then P is said to be a *face-rooted* map. The face f_x is called the *rooted face* of P. An element in rooted face is called a *rooted element*.

Corollary 6.4.3. *For a face-rooted polyhedron $P^{fr} = (\mathcal{X}, \mathcal{P})$ with rooted face f_x,*

$$aut(P^{fr}) \mid 2|\{x\}_{\mathcal{P}\gamma}|. \tag{6.4.4}$$

Proof. This is eq. (6.4.1) when $R = \{x\}_{\mathcal{P}\gamma} \cup \{\sigma x\}_{\mathcal{P}\gamma}$. □

For given polyhedron $P = (\mathcal{X}, \mathcal{P})$, if edge e_x, $x \in \mathcal{X}$, is chosen such that an automorphism of P is with e_x fixed, i.e. an element in e_x is always corresponding to an element in e_x, then P is called an *edge-rooted* map. Edge e_x is the *rooted edge* of P. An element in the rooted edge is also called a *rooted element*.

Corollary 6.4.4. *For an edge-rooted polyhedron $P^{er} = (\mathcal{X}, \mathcal{P})$ with the rooted edge e_x,*

$$aut(P^{er}) \mid |Kx|. \tag{6.4.5}$$

Proof. The case of eq. (6.4.1) when $R = Kx$. □

For a given polyhedron $P = (\mathcal{X}, \mathcal{P})$, an element $x \in \mathcal{X}$ is chosen such that an automorphism of P is with x as a fixed point, then P is called a *rooted polyhedron*. The element x is the *root* of P. The vertex, the edge and the face incident to the root are, respectively, called the *root vertex*, the *root edge* and the *root face*.

Corollary 6.4.5. *For a rooted polyhedron $P^r = (\mathcal{X}, \mathcal{P})$ with its root x,*

$$aut(P^r) = 1. \tag{6.4.6}$$

Proof. The case of eq. (6.4.1) when $R = \{x\}$. □

This tells us that a rooted polyhedron does not have the symmetry at all. The method mentioned above shows such a general clue for transforming a problem with symmetry to a problem without symmetry and then doing the reverse.

Example 6.4.1. Polyhedron

$$P_1 = (Kx + Ky, (x)(\gamma x, y, \gamma y))$$

has four distinct ways for choosing the root. Because P_1 has the following four primal trail codes

$$S_x = \underline{1}_0, \underline{\gamma 1, 2, \gamma 2}_1 = S_{\sigma x}, S_{\gamma x} = \underline{1, 2, \gamma 2}_0, \underline{\gamma 1}_1 = S_{\delta x},$$
$$S_y = \underline{1, \gamma 1, 2}_0, \underline{\gamma 2}_1 = S_{\delta y}, S_{\gamma y} = \underline{1, 2, \gamma 1}_0 \underline{\gamma 2}_1 = S_{\sigma y},$$

the four ways of rooting are shown in Figure 6.4.1(a–d), where the root is marked at its tail.

Example 6.4.2. Polyhedron $P_2 = (Kx + Ky, (x)(\gamma x, y, \delta y))$ has four distinct ways for choosing the root. Because P_2 has the following four primal trail codes:

$$S_x = \underline{1}_0, \underline{\gamma 1, 2, \delta 2}_1 = S_{\sigma x}, S_{\gamma x} = \underline{1, 2, \delta 2}_0, \underline{\gamma 1}_1 = S_{\delta x},$$
$$S_y = \underline{1, \delta 1, 2}_0, \underline{\gamma 2}_1 = S_{\gamma y}, S_{\delta y} = \underline{1, 2, \delta 1}_0 \underline{\gamma 2}_1 = S_{\sigma y},$$

the four ways of rooting are shown in Figure 6.4.2(a–d), where the root is marked at its tail.

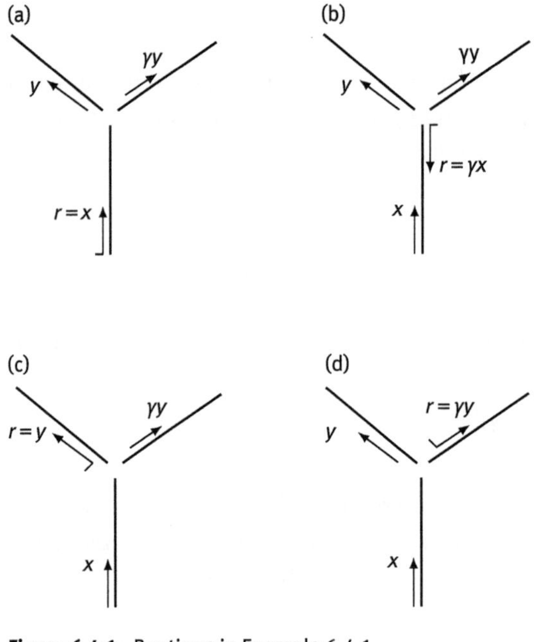

Figure 6.4.1: Rootings in Example 6.4.1.

6.5 Notes

6.5.1 Codes of planar graphs were introduced by Weinberg in the 1960s in Weinberg [374, 376]. According to those, he found the maximum order of the automorphism group of a 3-connected planar graphs in Weinberg [372], and an efficient algorithm for justifying isomorphisms of 3-connected planar graphs in Weinberg [375, 377] (1965–1966). However, the discussions in this chapter enable us to find the corresponding results for general planar graphs instead of 3-connected ones.

6.5.2 We are also allowed to justify isomorphisms of general graphs, instead of planar ones, by considering the embeddings on a certain kind of surfaces. However, the efficiency looks much related to the number of combinatorially distinct embeddings on a surface. FOD-procedure (VOD-procedure as well) and St-procedure enables us to find an isomorphism of a polyhedron not only planar, if any.

6.5.3 Hopcroft and Tarjan provided algorithms for the isomorphism of 3-connected planar graphs, which were mainly related to data structure in the 1970s in Hopcroft and Tarjan [137, 138] (1071–1972).

6.5.4 Hopcroft and Wong suggested to introduce a series of operations for simplifying the procedure of testing isomorphisms of 3-connected planar graphs, in order to find more efficient algorithm in Hopcroft and Wong [141].

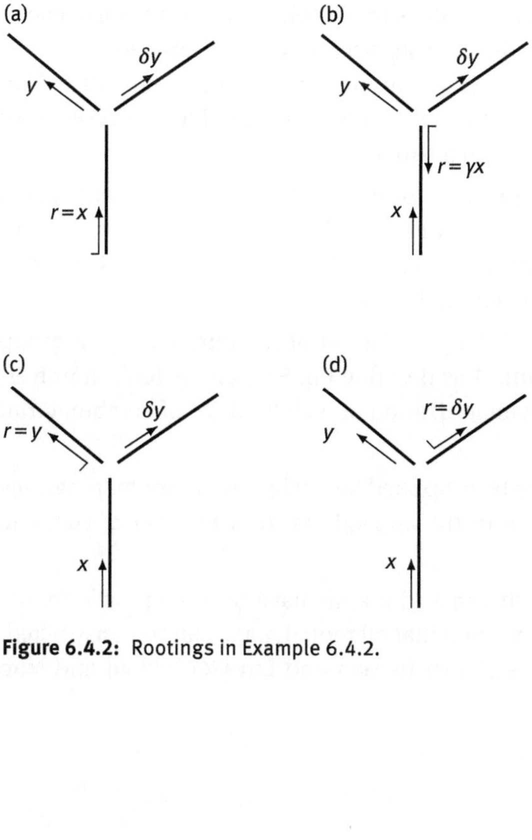

Figure 6.4.2: Rootings in Example 6.4.2.

6.5.5 For planar case, we may also be allowed to test the isomorphism, by using the planarity auxiliary graphs and the solutions of modulo 2 equations, or Boolean equations in Liu [192, 206, 207] and Wu [403].

6.5.6 Codes have been used to enumerate planar maps in Cori [55], Cori and Machi [56].

6.5.7 How to determine an isomorphism between two graphs is a very difficult problem.

Conjecture 6.5.1. *(Ulam) Two graphs $G_1 = (V_1, E_1)$ and $G_2 = (V_2, E_2)$ are isomorphic if, and only if, there is a bijection $\tau\colon V_1 \to V_2$ such that $G_1 - v$ and $G_2 - \tau(v)$ are isomorphic for $v \in V_1$.*

Such a conjecture has not yet been resolved. Then, a weak conjecture can be posed for maps.

Conjecture 6.5.2. *Two polyhedra $P_1 = (\mathscr{X}_1, \mathscr{J}_1)$ and $P_2 = (\mathscr{X}_2, \mathscr{J}_2)$ are isomorphic if, and only if, there is a bijection τ from a vertex v in P_1 to a vertex u in P_2 such that $P_1 - v$ and $P_2 - u$ are all isomorphic.*

The latter conjecture looks hopeful to be proved, much earlier than the former.

6.5.8 Because of the efficiency for determining an isomorphism between two polyhedra, if any, as shown in this chapter, it enables us to seek an isomorphism between two graphs via their underlain polyhedra on a surface, or types of surfaces.

Because of the known result that any graph but tree has an underlain polyhedron on a non-orientable surface of genus as its Betti number shown in Liu [195], it enables us to observe underlain polyhedra on such a surface.

6.5.9 The upper bound of the automorphism group order of any map is generalized from the result for planar map in Harary and Tutte [125]. More about the bound, one might read the asymmetrization in Liu [218]. This is the theoretical basis of recognizing an isomorphism of two maps efficiently.

6.5.10 The procedure for decreasing the upper bound of the automorphism group order of a polyhedron step by step until 1 as described in Section 6.4 looks a universal way which can be applied for asymmetrization of a wide range of combinatorial objects.

6.5.11 Algorithms in this chapter have been applied for designing computer programs to get results. Calculations are shown in the appendices for a number of concrete cases in Liu [219].

6.5.12 The symmetries of maps, polyhedra and graphs have been employed for exploiting the relationships among them, such that all rooted maps can be constructed. The reader is referred to read articles shown in Mao and Liu [264]–[266] and Mao et al. [267].

7 Gauss Crossing Sequences

7.1 Crossing polyhegons

For a Jordan closed curve or a circle C, let f be a continuous function from C into the plane such that $f(C)$ has finite self-intersection points, more precisely simply crossing points at each of which $r \in f(C)$ only simply crossing, or say without tangent points is allowed and there are exactly two points p and $q \in C$ satisfying $f(p) = f(q) = r$ or say *doubly crossing*. If all the self-intersection points of $f(C)$ are denoted by distinct letters: $a, b, c, \ldots$ then by travelling from a point p on $Z = f(C)$ and recording the letters which denote the intersection points passed by along the corresponding direction determined on the circle C to p, we can obtain a sequence of letters, which is called *the crossing sequence* of Z in the cyclic order.

It is easily seen that each letter appears in the sequence exactly twice because each intersection point is an image of exactly two points on C. A sequence of letters with the property as just mentioned is said to be *polyhedral*. In Figure 7.1.1(a), the crossing sequence is *abcabc*, which has three crossing points. And, in Figure 7.1.1(b), the crossing sequence is *abcdbadc* with four crossing points.

Both of them have even *length* which is the number of occurrences of letters recorded in the sequence. A sequence of letters is said to be a *crossing sequence* if there is a closed curve Z in the plane such that it is the crossing sequence of Z.

Of course, any crossing sequence is polyhedral. The *Gauss crossing problem* is that of determining if a given sequence of letters is a crossing sequence, or in other words, of characterizing the crossing sequences. Let Seq be a sequence of letters and $\Xi = \Xi(\text{Seq})$ be the set of all the letters in Seq. If the sequence Seq has the form

$$\text{Seq} = AaBbCaDb, \tag{7.1.1}$$

then a and b are said to be *interlaced*, or write a int b.

By an *interlaced graph* of a sequence Seq, we shall mean such a graph denoted by $G(I) = G(\text{Int}) = (V(\text{Int}), E(\text{Int}))$ that

$$V(\text{Int}) = \Xi;\ E(\text{Int}) = \text{Int}, \tag{7.1.2}$$

where $\text{Int} = \{(a, b) | \forall a, b \in \Xi,\ a \text{ int } b\}$.

One might see that the interlaced graph of the crossing sequence of the curve shown in Figure 7.1.1(a) is a triangle and that in Figure 7.1.1(b) a quadrangle. Both of them are Eulerian. It is easily seen that $G(\text{Int})$ has neither a loop nor a multi-edge. However, isolated vertices are allowed. Let

$$\text{Int}\, a = \{b | \forall b \in V(\text{Int}),\ (a, b) \in E(\text{Int})\}$$

for $a \in V(\text{Int})$ as in Section 3.4.

DOI 10.1515/9783110479492-007

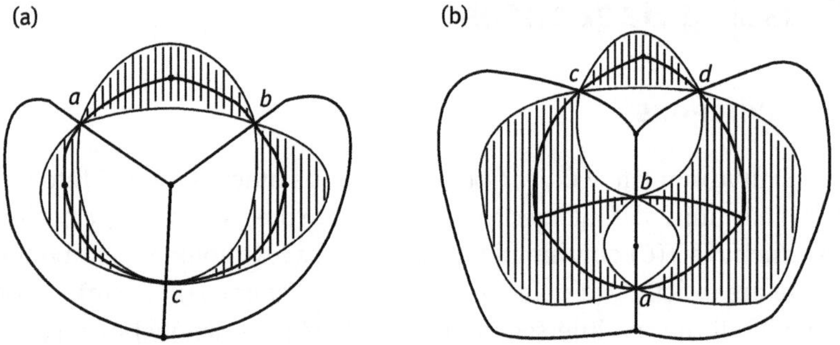

Figure 7.1.1: Two interlaced graphs.

Lemma 7.1.1. *For any crossing sequence* Seq, *each component of the interlaced graph* $G(\mathrm{Int})$ *is Eulerian.*

Proof. For $a \in \Xi$, assume that the sequence Seq has the form Seq $= AaBa$. Then, for $b \in \Xi$ $(b \neq a)$, $(a, b) \in E$ (Int) if, only if, b appears in both A and B. Because of simply and doubly crossing in Seq, both the point going way from and the point coming to a on aAa are in the inner or outer domain of aBa. From the Jordan curve theorem, the number of crossing points is even. Therefore,

$$|\mathrm{Int}\ a| = 0(\mathrm{mod}\ \ 2).$$

That implies the lemma from Theorem 1.3.6. □

Lemma 7.1.2. *For a crossing sequence Seq,* $\forall a, b \in \Xi \left((a, b) \notin E(Int)\right)$,

$$|Int\ a \bigcap Int\ b| = 0(mod\, 2).$$

Proof. Because $(a, b) \notin E(\mathrm{Int})$, Seq has the form

$$\mathrm{Seq} = AaBbCbDa.$$

Thus,

$$c \in \mathrm{Int}\ a \bigcap \mathrm{Int}\ b$$

if, and only if, c is a crossing point of aAa with bCb, by the same reason used in the proof of Lemma 7.1.1, from the Jordan curve theorem, we have the statement of the lemma. □

If Seq is the crossing sequence of a curve Z. Then, we may find that the graph whose vertices are crossing points and whose edges are the segments between two vertices without passing through other vertex of Z is a 4-valent planar graph denoted by $G(Z)$. From Theorem 1.3.7, the dual $G^*(Z)$ of $G(Z)$ is bipartite. This implies that the set of faces of $G(Z)$ can be partitioned into two parts in each of which, there are no two faces adjacent. Of course, at each vertex, its four incident faces are partitioned into two parts as well; each of which consists of two faces opposite one to another.

A pair of faces opposite one to another at a vertex is said to be *opposite*. For the two parts of faces of $G(Z)$, we may define a graph each by treating faces in the same part as vertices with that two vertices are adjacent if, only if, their corresponding faces form an opposite pair. It is easily seen that the two graphs produced by the two parts of faces are planar dual one of another.

Let $G_S(Z)$ be the one produced by the part of faces without the infinite face. Of course, the other is denoted by $G_S^*(Z)$, the planar dual of $G_S(Z)$. They are called the *associated graphs* of the crossing sequence. In Figure 7.1.1, $G_S(Z)$ and $G_S^*(Z)$ are represented by thin lines and heavy lines, respectively.

Lemma 7.1.3. *For a crossing sequence* Seq, *both* $G_S(Z)$ *and* $G_S^*(Z)$ *are without bicycle.*

Proof. Because both $G_S(Z)$ and $G_S^*(Z)$ are planar and the sequence Seq is the unique travel obtained by the T.T. rule in Section 3.4 on both of them, from Theorem 4.4.6 the lemma follows. □

Lemma 7.1.4. *For a crossing sequence* Seq, *let*

$$\begin{cases} U(\mathrm{Int}) = \{(a,b) | \forall (a,b) \in E(\mathrm{Int}); \\ |Int\ a \bigcap Int\ b| = 0 (mod\ 2)\}. \end{cases}$$

Then, $U(\mathrm{Int}) \in \mathscr{C}^{\perp}(\mathrm{Int})$, *i.e. a 1-coboundary or cocycle of* $G(I)$.

Proof. From Lemma 7.1.3, $G_S(Z)$ is without bicycle, i.e. $O = \emptyset$ in eq. (4.4.3). Thus, we have

$$E(G_S(Z)) = \Xi = M + N, \tag{7.1.3}$$

where $e \in M$ or N is determined by $e \in \gamma(e)$ or $\omega(e)$ in eq. (4.3.11), respectively.

From Theorem 4.3.6, we know that $\{\gamma(e) | \forall e \in \Xi\}$ and $\{\omega(e) | \forall e \in \Xi\}$ generate the cycle space $\mathscr{C}(G_S(Z))$, and the cocycle space $\mathscr{C}^{\perp}(G_S(Z))$, respectively. Because of eq. (4.4.8), Int $a = \gamma(a) \cap \omega(a)$. Then, we have

$$\gamma(a) = \begin{cases} \mathrm{Int}\ a \bigcup \{a\}, & \text{if } a \in M; \\ \mathrm{Int}\ a, & \text{if } a \in N, \end{cases} \tag{7.1.4}$$

and

$$\omega(a) = \begin{cases} \text{Int } a \bigcup \{a\}, & \text{if } a \in N; \\ \text{Int } a, & \text{if } a \in M. \end{cases} \tag{7.1.5}$$

Because of $(a, b) \in E(\text{Int})$, i.e. $b \in \text{Int } a$, we have

$$|\text{Int } a \bigcap (\text{Int } b \bigcup \{b\})| = |\text{Int } a \bigcap \text{Int } b| + 1 (\text{mod } 2).$$

From eqs. (7.1.4) and (7.1.5),

$$|\text{Int } a \bigcap \text{Int } b| = \begin{cases} (\gamma(a), \omega(b)) (\text{mod } 2) \\ \qquad \Leftrightarrow a \in N, b \in M; \\ (\omega(a), \gamma(b)) (\text{mod } 2) \\ \qquad \Leftrightarrow a \in M, b \in N. \end{cases}$$

By the orthogonality of the spaces $\mathscr{C}(G_S(Z))$ and $\mathscr{C}^{\perp}(G_S(Z))$,

$$|\text{Int } a \bigcap \text{Int } b| = 0 (\text{mod } 2) \Leftrightarrow a \in M, \\ b \notin M \text{ or } a \in N, b \notin N.$$

This implies that $U(\text{Int})$ is a 1-coboundary of $G(I)$. □

In fact, Lemmas 7.1.1 and 7.1.2 can also be shown directly from $G_S(Z)$ because $G_S(Z)$ has no bicycle. From eq. (4.4.9), Lemma 7.1.1 is obtained. For $(a, b) \notin E(\text{Int})$, i.e. $b \notin \text{Int } a$ and $a \notin \text{Int } b$ as well, we have

$$\begin{aligned} \text{Int } b \bigcap (\text{Int } a \bigcup \{a\}) &= \text{Int } a \bigcap (\text{Int } b \bigcup \{b\}) \\ &= \text{Int } a \bigcap \text{Int } b. \end{aligned}$$

However, according to eqs. (7.1.4) and (7.1.5), we have

$$(\gamma(a), \omega(b)) = \begin{cases} |\text{Int } a \bigcap (\text{Int } b \bigcup \{b\})| (\text{mod } 2), \\ \qquad \text{if } a \in N, \ b \in N; \\ |(\text{Int } a \bigcup \{a\}) \bigcap \text{Int } b| (\text{mod } 2), \\ \qquad \text{if } a \in M, \ b \in M \end{cases}$$

and hence

$$|\text{Int }a \bigcap \text{Int }b| = \begin{cases} (\gamma(a), \omega(b))(\text{mod } 2), \\ \qquad \text{if } a \in N,\ b \in M; \\ (\gamma(b), \omega(a))(\text{mod } 2), \\ \qquad \text{if } a \in M,\ b \in N. \end{cases}$$

By the orthogonality between $\mathscr{C}(G_S(Z))$ and $\mathscr{C}^{\perp}(G_S(Z))$, Lemma 7.1.2 is obtained.

Further, we shall verify that the conditions described in Lemmas 7.1.1, 7.1.2 and 7.1.4 are sufficient for a polyhedral sequence being a crossing sequence. This is an answer of what is called the *Gauss conjecture*.

7.2 Dehn's transformation

Let Seq be a cyclic sequence of letters. For a crossing point $a \in \Xi$, the set of letters which appear in Seq, the transformation denoted by λ_a from Seq = $aAaB$ to

$$\text{Seq}' = a'A^{-1}a'B$$

is said to be *splitting at the letter a*. Of course, if Seq is the crossing sequence of a curve Z, then splitting at a letter is splitting a crossing point a such that the resultant curve Z' has the sequence Seq$'$ as the crossing sequence whenever without considering the letter with a prime. Because a' is not a crossing point of the closed curve Z' anymore, we may imagine that a' is a label of two points on Z' as shown in Figure 7.2.1. Thus,

$$\text{Seq}' = \lambda_a(\text{Seq}')$$

is called the *labelled crossing sequence* of the curve Z' obtained by splitting the crossing point a on Z if Seq is the crossing sequence of Z; or a *labelled sequence*, otherwise. Because,

$$\lambda_a\lambda_b \neq \lambda_b\lambda_a$$

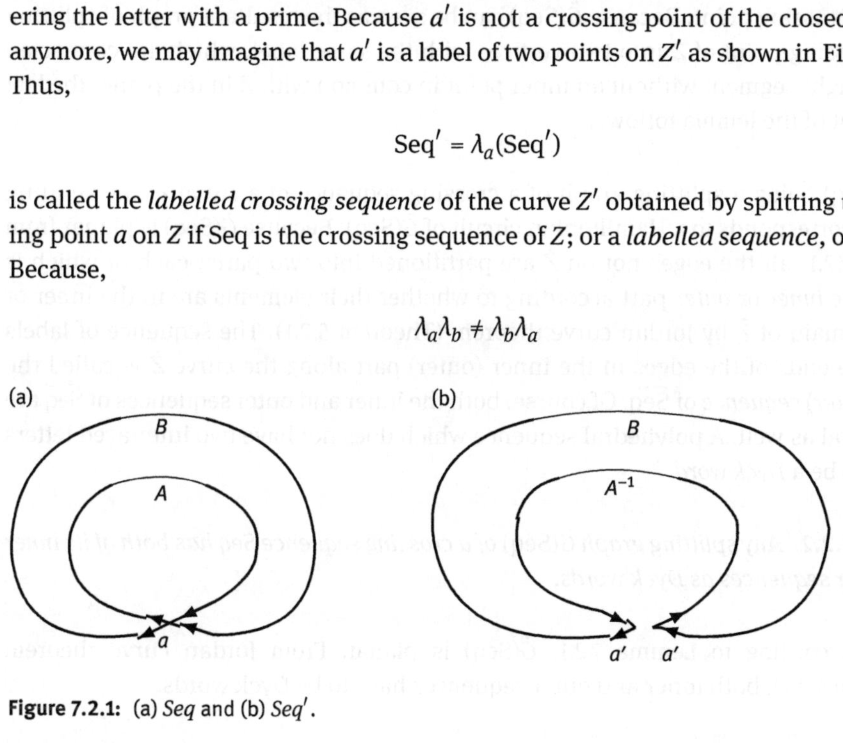

Figure 7.2.1: (a) *Seq* and (b) *Seq*$'$.

in general, for a linear order of all the letters on Ξ given, let

$$\widetilde{\text{Seq}} = \prod_{a \in \Xi} \lambda_a(\text{Seq}) \tag{7.2.1}$$

which is called a *splitting sequence* of Seq. Naturally, $\widetilde{\text{Seq}}$ is a labelled sequence with all the letters labelled.

A graph is said to be a *splitting graph* denoted by $\tilde{G}(\text{Seq})$ of a polyhedral sequence Seq if its vertices are all the labelled letters, each of which appears exactly twice, as well, of $\widetilde{\text{Seq}}$ and two vertices are adjacent if, and only if, they have the same label or are successive to another on $\widetilde{\text{Seq}}$. The closed curve $\tilde{Z}$ which corresponds to $\widetilde{\text{Seq}}$ is called the *splitting curve* of Z; of which, Seq is the crossing sequence.

Lemma 7.2.1. *For any polyhedral sequence* Seq*, all the splitting graphs* $\tilde{G}(\text{Seq})$ *are cubic. If* Seq *is a crossing sequence, then for any linear order on* Ξ *given, the splitting graph* $\tilde{G}(\text{Seq})$ *is planar.*

Proof. The first statement is obvious because each vertex in $\tilde{G}(\text{Seq})$ has exactly three incident edges: one is with the two ends of the same label and each of the other two is with the two ends successive on $\widetilde{\text{Seq}}$.

Suppose Seq is the crossing sequence of a closed curve Z. Let $\widetilde{\text{Seq}}$ be the splitting sequence of Seq. In fact, $\widetilde{\text{Seq}}$ is the splitting sequence of a closed Jordan curve $\tilde{Z}$ without crossing point. Because $\tilde{G}(\text{Seq})$ can be seen as the resultant graph of splitting each crossing point of Z into two vertices with the same label, which are connected by a straight segment without an inner point in common with Z in the plane, the last statement of the lemma follows. □

For $\tilde{G}(\text{Seq})$ being a splitting graph of a crossing sequence of a curve Z, the splitting curve $\tilde{Z}$ corresponds to a Hamiltonian circuit of $\tilde{G}(\text{Seq})$. Because $\tilde{G}(\text{Seq})$ is planar from Lemma 7.2.1, all the edges not on $\tilde{Z}$ are partitioned into two parts; each of which is called the *inner* or *outer* part according to whether their elements are in the inner or outer domain of $\tilde{Z}$ by Jordan curve theorem (Theorem 5.2.1). The sequence of labels of all the ends of the edges in the inner (outer) part along the curve $\tilde{Z}$ is called the *inner* (*outer*) *sequence* of $\widetilde{\text{Seq}}$. Of course, both the inner and outer sequences of $\widetilde{\text{Seq}}$ are polyhedral as well. A polyhedral sequence which does not have two interlaced letters is said to be a *Dyck word.*

Lemma 7.2.2. *Any splitting graph* $\tilde{G}(\text{Seq})$ *of a crossing sequence Seq has both of its inner and outer sequences as Dyck words.*

Proof. According to Lemma 7.2.1, $\tilde{G}(\text{Seq})$ is planar. From Jordan curve theorem (Theorem 5.2.1), both inner and outer sequences have to be Dyck words. □

Now, we are allowed to reverse the operation of splitting at a letter by *producing* a crossing point denoted by the letter without the prime which labels the two ends of an edge not on $\tilde{Z}$ in $\tilde{G}(\text{Seq})$ in the following way: first replace each edge not on $\tilde{Z}$ by a pair of parallel segments, sufficiently close together, then at each end of the edge delete the portion of $\tilde{Z}$ between the two ends of the new parallel segments, and finally introduce a crossing point, labelled by the same letter without the prime as that the two ends of the edge have in the middle of each pair of parallel segments as shown in Figure 7.2.2.

If this is done for every edge not on $\tilde{Z}$ and if the result is a single closed curve in agreement with the order of the sequence Seq, then $\tilde{G}(\text{Seq})$ is said to be *traceable.*

Lemma 7.2.3. *If* $\tilde{G}(\text{Seq})$ *is a splitting graph of a crossing sequence* Seq, *then* $\tilde{G}(\text{Seq})$ *is traceable.*

Proof. By the procedure of producing crossing points from $\tilde{G}(\text{Seq})$ as mentioned above, the result of it being done for every edge not on $\tilde{Z}$ is just a closed curve Z which realizes the sequence Seq and hence $\tilde{G}(\text{Seq})$ is traceable. □

Notice that splitting sequences and splitting graphs make sense for any polyhedral sequence not necessary to be a crossing sequence and that they are not unique for a sequence given as dependent on the choice of the linear order on Ξ. However, no matter what is chosen for the linear order on Ξ, as soon as a splitting sequence can be partitioned into two Dyck words and further the corresponding splitting graph is traceable, so can any splitting sequence and further so is any splitting graph.

Theorem 7.2.1. *A polyhedral sequence* Seq *is a crossing sequence if, and only if, its splitting sequence* $\widetilde{\text{Seq}}$ *can be partitioned into two Dyck words and its splitting graph is traceable.*

Proof. The necessity is derived from Lemmas 7.2.1 and 7.2.3. Because $\widetilde{\text{Seq}}$ is partitioned into two Dyck words, $\tilde{G}(\text{Seq})$ can be embedded into the plane such that one Dyck word

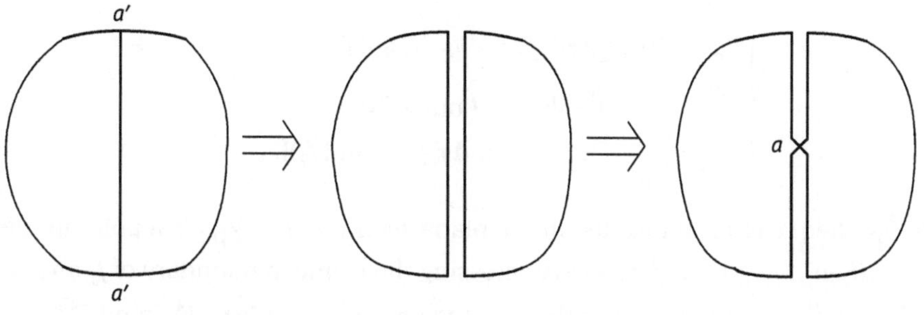

Figure 7.2.2: Splitting and crossing.

(the other) determines all the edges whose two ends are not successive on $\widetilde{\mathrm{Seq}}$ in the inner (outer) domain of the circle C which corresponds to $\widetilde{\mathrm{Seq}}$ in the plane.

Moreover, since $\tilde{G}(\mathrm{Seq})$ is traceable, the single closed curve Z obtained by the procedure of producing all the crossing points from the planar embedding of $\tilde{G}(\mathrm{Seq})$ realizes the sequence as the crossing sequence. □

Lemma 7.2.4. *For a polyhedral sequence* Seq, *the interlaced graph* $\tilde{G}(I)$ *of its splitting sequence* $\widetilde{\mathrm{Seq}}$ *is bipartite if, and only if,* $\widetilde{\mathrm{Seq}}$ *can be partitioned into two Dyck words.*

Proof. If $\tilde{G}(I)$ is bipartite, then its vertex set $\tilde{V}(\mathrm{Int}) = \tilde{V}_1 + \tilde{V}_2$. That implies any pair of vertices in $\tilde{V}_1$ (or $\tilde{V}_2$) is not interlaced. $\widetilde{\mathrm{Seq}}$ is partitioned into two subsequences; each of which consists of all the letters in $\widetilde{V}_1$ (or $\widetilde{V}_2$). In consequence, they determine two Dyck words into which the splitting sequence $\widetilde{\mathrm{Seq}}$ is partitioned. The necessary is obtained.

Conversely, the two Dyck words of $\widetilde{\mathrm{Seq}}$ determine a bipartition of $\widetilde{V}(\mathrm{Int})$ by their sets of letters. □

Theorem 7.2.2. *A polyhedral sequence* Seq *is a crossing sequence if, and only if, the interlaced graph* $\tilde{G}(I)$ *is bipartite and the splitting graph* $\tilde{G}(\mathrm{Seq})$ *is traceable.*

Proof. From Lemma 7.2.4 and Theorem 7.2.1, the theorem is obtained. □

For a polyhedral sequence Seq, if its splitting graph $\tilde{G}(\mathrm{Seq})$ is planar, let $\tilde{Z}$ be the Hamiltonian circuit corresponding to the splitting sequence $\widetilde{\mathrm{Seq}}$, $F_{\mathrm{in}}(\mathrm{Seq})$ and $F_{\mathrm{out}}(\mathrm{Seq})$ be the sets of all the faces in the inner and outer domains of $\tilde{Z}$, respectively, and $E_{\mathrm{in}}(\mathrm{Seq})$ (or $E_{\mathrm{out}}(\mathrm{Seq})$) be the set of all the edges in the inner (outer) domain of $\tilde{Z}$. Then, we may construct a new graph, which is called the *P–C graph* of Seq and denoted by $\tilde{G}_{\mathrm{pc}}(\mathrm{Seq}) = (V_{\mathrm{pc}}, E_{\mathrm{pc}})$ in the following way:

$$V_{\mathrm{pc}} = \{f | f \in F_{\mathrm{in}}(\mathrm{Seq})\};\ E_{\mathrm{pc}} = E_{\mathrm{pc}}^{(1)} + E_{\mathrm{pc}}^{(2)}, \tag{7.2.2}$$

where

$$\begin{cases} E_{\mathrm{pc}}^{(1)} = \{(f_1, f_2) | \forall f_1, f_2 \in V_{\mathrm{pc}},\ f_1 \operatorname{adj} f_2\}; \\ E_{\mathrm{pc}}^{(2)} = \{(f_1, f_2) | \exists (u, v) \in E_{\mathrm{out}}(\mathrm{Seq}), \\ \qquad\qquad (u \operatorname{ind} f_1) \wedge (v \operatorname{ind} f_2)\}. \end{cases}$$

Because $\tilde{G}_{\mathrm{pc}}(\mathrm{Seq})$ can be embedded in the plane such that $f \in V_{\mathrm{pc}}$ is a point in the face $f \in F_{\mathrm{in}}(\mathrm{Seq})$, $(f_1, f_2) \in E_{\mathrm{pc}}^{(1)}$ is a curve crossing the common boundary of f_1 and f_2, and $(f_1, f_2) \in E_{\mathrm{pc}}^{(2)}$ is a curve passing through the edge in $E_{\mathrm{out}}(\mathrm{Seq})$. $E_{\mathrm{pc}}^{(1)}$ and $E_{\mathrm{pc}}^{(2)}$ are said to be the *crossing* and *passing* subsets of E_{pc}, respectively. Moreover, we may also construct another graph which is called the *C–P graph* of Seq and denoted by $\tilde{G}_{\mathrm{cp}}(\mathrm{Seq})$ in the way of substituting $F_{\mathrm{in}}(\mathrm{Seq}), E_{\mathrm{out}}(\mathrm{Seq})$ for $F_{\mathrm{out}}(\mathrm{Seq}), E_{\mathrm{in}}(\mathrm{Seq})$, respectively in (7.2.2). Obviously, $\tilde{G}_{\mathrm{cp}}(\mathrm{Seq}) = \tilde{G}^{*}_{\mathrm{pc}}(\mathrm{Seq})$, the planar dual of $\tilde{G}_{\mathrm{pc}}(\mathrm{Seq})$, and if Seq is a crossing sequence, then

$$\begin{aligned} \tilde{G}_{\mathrm{pc}}(\text{or } \tilde{G}_{\mathrm{cp}})(\mathrm{Seq}) &= G_S; \\ \tilde{G}_{\mathrm{cp}}(\text{ or } \tilde{G}_{\mathrm{pc}})(\mathrm{Seq}) &= G_S^{*}. \end{aligned} \tag{7.2.3}$$

If the set *TTr* for $\tilde{G}_{\mathrm{pc}}(\mathrm{Seq})$ consists of a single travel in agreement with the order of Seq by the T.T. rule in Section 3.4, then $\tilde{G}_{\mathrm{pc}}(\mathrm{Seq})$ is said to be *T.T. traceable*. Of course, the traceability implies without bicycle in $\tilde{G}_{\mathrm{pc}}(\mathrm{Seq})$ (or $\tilde{G}_{\mathrm{cp}}(\mathrm{Seq})$ as well) from Theorem 4.4.6.

Lemma 7.2.5. *For a planar splitting graph* $\tilde{G}(\mathrm{Seq})$ *of a polyhedral sequence* Seq, *(if any)* $\tilde{G}(\mathrm{Seq})$ *is traceable if, and only if,* $\tilde{G}_{\mathrm{pc}}(\mathrm{Seq})$ *is T. T. traceable.*

Proof. Because the resultant closed curve of producing all the crossing points on $\tilde{G}(\mathrm{Seq})$ is just seen as the single travel obtained by the T.T. rule in Section 3.4 on $\tilde{G}_{\mathrm{pc}}(\mathrm{Seq})$ and vice versa, the lemma follows. □

Lemma 7.2.6. *For a polyhedral sequence* Seq, *a splitting graph* $\tilde{G}(\mathrm{Seq})$ *is planar if, and only if, the splitting sequence* $\widetilde{\mathrm{Seq}}$ *can be partitioned into two Dyck words.*

Proof. From Lemma 7.2.1, $\tilde{G}(\mathrm{Seq})$ is cubic. Because $\tilde{G}(\mathrm{Seq})$ is planar, the edges not on the Hamiltonian circuit $\tilde{Z}$ corresponding to $\widetilde{\mathrm{Seq}}$ are partitioned into two parts: one consists of all the edges in the inner domain of $\tilde{Z}$; the other consists of all the edges in the outer domain of $\tilde{Z}$. From the cubicness, any pair of them are not adjacent. Then, the letters which label the ends of all the edges in each part form a Dyck word. Conversely, by all the edges incident with the same Dyck word being put in the same domain of $\tilde{Z}$, a planar embedding of $\tilde{G}(\mathrm{Seq})$ can be found. Therefore, $\tilde{G}(\mathrm{Seq})$ is planar. □

Theorem 7.2.3. *A polyhedral sequence Seq is a crossing sequence if, and only if, its splitting graph* $\tilde{G}(\mathrm{Seq})$ *is planar and its P–C graph* $\tilde{G}_{\mathrm{pc}}(\mathrm{Seq})$ *is T.T. traceable.*

Proof. From Lemmas 7.2.5 and 7.2.6 and Theorem 7.2.1, the theorem is derived. □

7.3 Algebraic principles

A single-face polyhedron Σ is said to be *singular* if for a letter a whose two occurrences are with the same power, all the letters, each of which appears exactly once between the two occurrences of a in one direction, form the principal cycle $\gamma(a)$ with $(a, \gamma(a)) = 1$

and for a with different powers, all the letters of only one occurrence between the two occurrences of a in one direction form the principal cocycle $\omega(a)$ with $(a, \omega(a)) = 1$. If a polyhedral sequence Seq has a partition of two parts of its letters: M and N such that it becomes a singular polyhedron whenever the two occurrences of each letter in M are with the same power and those of each letter in N are with different powers, then Seq is said to be *singular* as well.

Lemma 7.3.1. *A graph G without bicycle is planar if, and only if, it has a single-face embedding in a surface which is singular.*

Proof. Because G is planar without bicycle, from Theorem 4.4.6, only one travel q is obtained by the T.T. rule there. It is easily by the discussion in Section 4.4 checked that q is a singular polyhedron. The necessity is proved.

Conversely, suppose *Sin* is the singular sequence determined by the singular polyhedron of G. Let Σ be such an embedding of G on a surface that the rotation system at vertices is in agreement with that for the single travel obtained by the T.T. rule in Section 4.4. Then, to show that Σ is a polyhedron (orientable, of course) with $\epsilon - \nu + 2$ faces that are obtained by the Travel rule described in Section 2.1 of Liu [34]. By induction on the size ϵ of G. When ϵ is small, it is easy to check. In general, two cases have to be discussed.

Case 1. $\exists a \in M,\ \gamma(a) = \mathrm{Int}\, a, (a, \gamma(a)) = 1$. Let $G' = G - a$. Suppose Sin $= aAaB$, then Sin$' = AB^{-1}$ is just the corresponding singular sequence for G' as shown in Figure 7.3.1. Because G is without bicycle, so is G' from Lemma 4.4.2. By hypothesis, let Σ' be the polyhedron orientable with $(\epsilon - 1) - \nu + 2 = \epsilon - \nu + 1$ faces. Because the rotation system of G' is the restriction of that of G, the two ends of a are on the boundary of the same face which is assumed to have the form (CD). Thus,

$$\Sigma = \Sigma' - \{CD\} + \{Ca, a^{-1}D\} \tag{7.3.1}$$

is a polyhedron of G orientable with $\epsilon - \nu + 2$ faces.

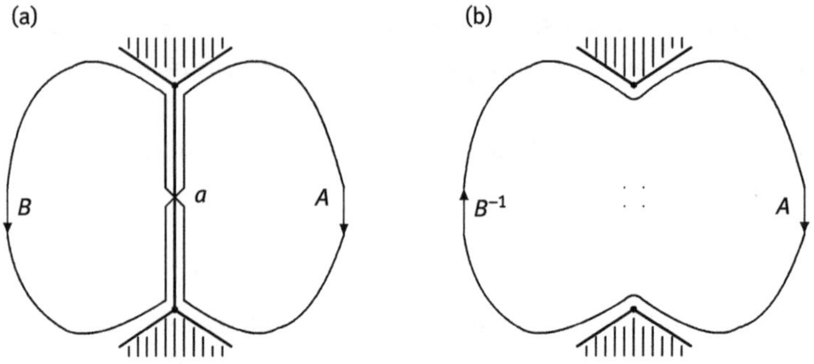

Figure 7.3.1: (a) G and (b) $G - a$.

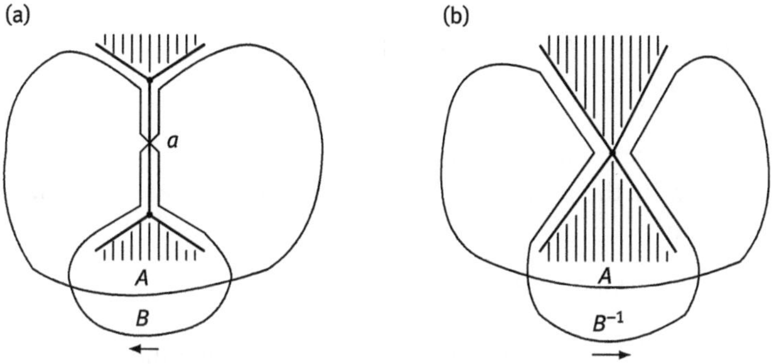

Figure 7.3.2: (a) G and (b) $G \bullet a$.

Case 2. $\exists a \in N, \omega(a) = \text{Int}\, a, (a, \omega(a)) = 1$. Let $G'' = G \bullet a$ be the resultant graph of contracting a on G. Suppose $\text{Sin} = aAa^{-1}B$. Because G is without bicycle, so is G'' from Lemma 4.4.2. Moreover, it is also easily seen that $\text{Sin}'' = AB^{-1}$ as shown in Figure 7.3.2 is the corresponding singular sequence for G''. By hypothesis, G'' has an embedding Σ'' orientable with $(\epsilon - 1) - (\nu - 1) + 2 = \epsilon - \nu + 2$ faces on a surface. Let

$$\Sigma'' = (\Sigma\langle a\rangle, C, D), \tag{7.3.2}$$

where C and D are the faces incident to the vertex into which a is contracted. By the similar reason as in case 1, we find

$$\Sigma = (\Sigma\langle a\rangle, Ca, a^{-1}D) \tag{7.3.3}$$

to be the polyhedron of G orientable with $\epsilon - \nu + 2$ faces.

In consequence, from Corollary 4.2.1, G is planar. The theorem is proved. □

Lemma 7.3.2. *A polyhedral sequence* Seq *is a crossing sequence if, and only if, it is singular.*

Proof. From Lemma 7.1.3, G_S is without bicycle and hence from Lemma 4.4.3 the crossing sequence Seq has all its letters partitioned into M and N. Moreover, from Theorem 7.2.3, G_S is planar. By Lemma 7.3.1, Seq is singular.

Conversely, because Seq is singular, from Lemma 7.3.1, the underlying graph which is isomorphic to G_S of Seq is planar without bicycle. However, the travel obtained by the T.T. rule in Section 4.4 has just Seq as a crossing sequence. □

Theorem 7.3.1. *A polyhedral sequence* Seq *is a crossing sequence if, and only if, its interlaced graph G(I) has the following properties*:

(i) *G(I) is Eulerian;*
(ii) $\forall a, b \in V(\text{Int})((a,b) \notin E(\text{Int}))$, $|Int\ a \cap \text{Int}\ b| = 0 (\text{mod}\ 2)$;
(iii) $U(\text{Int})$ *is a* 1-*coboundary.*

Proof. From Lemmas 7.1.1, 7.1.2 and 7.1.4, the necessity is obtained.

To prove the sufficiency. By the property (iii), $V(\text{Int}) = \Xi(\text{Seq})$ is partitioned into two parts: M and N such that

$$U(\text{Int}) = \{(a,b) | \forall a \in M, \forall b \in N, (a,b) \in E(\text{Int})\}. \tag{7.3.4}$$

Let G_S be the underlying graph of the polyhedron obtained by setting the two occurrences of a letter in M with the same power and those of a letter in N with different powers on Seq. If we define that

$$\gamma_S(a) = \begin{cases} \text{Int}\ a \bigcup \{a\}, & \text{if } a \in M; \\ \text{Int}\ a, & \text{if } a \in N, \end{cases} \tag{7.3.5}$$

and

$$\omega_S(a) = \begin{cases} \text{Int}\ a, & a \in M; \\ \text{Int}\ a \bigcup \{a\}, & a \in N, \end{cases} \tag{7.3.6}$$

then the following claims are satisfied. □

Claim 1. $\forall a, b \in E(G_S) = \Xi(\text{Seq})$, $(\gamma_S(a), \omega_S(b)) = 0$.

Proof. If otherwise $(\gamma_S(a), \omega_S(b)) = 1$ for some $a, b \in E(G_S)$, then in view of (7.3.5 and 7.3.6),

$$(\gamma_S(a), \omega_S(b)) = \begin{cases} |\text{Int}\ a| = 0 (\text{mod}\ 2), \\ \qquad \text{from (i), if } a = b; \\ |\text{Int}\ a \bigcap \text{Int}\ b| = 0 (\text{mod}\ 2), \\ \qquad \text{from (ii), if } a \neq b \\ \qquad \text{and } (a,b) \notin E(Int). \end{cases}$$

Now, the only possibility is that for $a \neq b$ and $(a,b) \in E(\text{Int})$. If a and b are in the same part, then

$$(\gamma_S(a), \omega_S(b)) = |(\text{Int}\ a \bigcup \{a\}) \bigcap \text{Int}\ b| (\text{mod}\ 2).$$

But, $|\text{Int}\ a \bigcap \text{Int}\ b| = 1\ (\text{mod}\ 2)$ from (iii). That implies $(\gamma_S(a), \omega_S(b)) = 0$, a contradiction to the condition. If a and b are in different parts, then

$$(\gamma_S(a), \omega_S(b)) = |\text{Int}\ a \bigcap \text{Int}\ b| = 0 (\text{mod}\ 2)$$

from (iii), a contradiction to the condition as well. Therefore, the claim is true. □

Claim 2. G_S is without bicycle.

Proof. First, because $\gamma_S(a)$ is a travel in G_S, we have $\gamma_S(a) \in Ker\ \partial_1 = \mathscr{C}(G_S)$, the cycle space. From Claim 1, $\omega_S(a) \in \mathscr{C}^{\perp}(G_S)$, the cocycle space. Then, from eqs (7.3.5) and (7.3.6),

$$\forall a \in E(G_S),\ a = \gamma_S(a) + \omega_S(a). \tag{7.3.7}$$

Therefore, from eqs (3.4.1) to (3.4.3), G_S has no bicycle. □

Claim 3. G_S is planar.

Proof. From Claims 1–2, we have known that for any $a \in E(G_S)$, $\gamma_S(a)$ and $\omega_S(a)$ are its principal cycle and cocycle, respectively. Furthermore, from the definition of G_S the polyhedron whose underlying graph is G_S determines a single-face embedding of G_S in a surface, which is singular. From Lemma 7.3.1, the claim is derived.

Because of Claim 3, the face of the single-face embedding of G_S is in agreement with the order of Seq, and hence Seq is singular. In consequence, by Lemma 7.3.2 the theorem follows. □

Since Gauss found the first two necessary conditions in Theorem 7.3.1, we call a polyhedral sequence with the conditions: (i) and (ii) in the theorem a *Gaussian sequence.*

7.4 Gauss crossing problem

By using the similar procedure of proving Theorem 5.4.2, we may further from Theorem 7.3.1 obtain the following theorem.

Theorem 7.4.1. *Let* Seq *be a Gaussian sequence and G(I), its interlaced graph with the weights on edges as* $\forall e = (a, b) \in E(\mathrm{Int})$,

$$w(e) = \begin{cases} 1, & e \in U(\mathrm{Int}); \\ 0, & \textit{otherwise}. \end{cases} \tag{7.4.1}$$

Then, the following statements are equivalent:
(1) *Seq is a crossing sequence;*
(2) *G(I) has no odd-weight circuit;*
(3) *G(I) has no odd-weight fundamental circuit;*
(4) *U(Int) is a 1-coboundary of G(I);*
(5) *G(I) is balanced.*

Now, we may see that any statement among the last four in the theorem is just what was conjectured by Gauss because it is only dependent on the interlaced graph $G(I)$ of the sequence Seq itself.

What has to be specially mentioned here is that if Seq is a crossing sequence, then from Lemma 7.3.2, the underlying graph G_S of the corresponding polyhedron is isomorphic to G_{pc} or its dual G^*_{pc} which is introduced in Section 7.2. Moreover, that is independent of the choice of the order for the splitting operations. However, if Seq is not a crossing sequence, although G_{pc} or G^*_{pc} can be determined sometime, it is never isomorphic to any underlying graph G_S of the polyhedra traceable corresponding to Seq. Let Seq = *acdbadcb*. Its interlaced graph $G(I)$ as shown in Figure 7.4.1(a), which is not Eulerian. Thus, Seq is not a crossing sequence. However, the splitting sequence

$$\widetilde{\text{Seq}} = \lambda_d\lambda_c\lambda_b\lambda_a\text{Seq} = a'b'c'a'd'c'd'b'$$

has G_{pc} as shown in Figure 7.4.1(b) which is without bicycle but not traceable.

If a polyhedral sequence Seq has the form Seq = CD such that both C and D would be polyhedral as well when they were in a cyclic order, then it is said to be *separable*. Further, if C or D as a polyhedral sequence is not separable any more, then it is called a *component* of Seq. Of course, a polyhedral sequence which is not separable is treated as a component itself.

Lemma 7.4.1. *A crossing sequence* Seq *is a component if, and only if, its P–C graph* $G_{\text{pc}}(\text{Seq})$ *is a block.*

Proof. Because *Seq* is a crossing sequence, $G_{\text{pc}}(\text{Seq})$ is isomorphic to G_S as defined in Theorem 7.3.1. For the necessity, suppose $G_S = G_1 \cup G_2$ provided $G_1 \cap G_2 = \{v\}$, where v is a vertex of G_S. Since G_S is planar without bicycle from Lemma 7.2.5, the single travel obtained by the T.T. rule in Section 3.4 on G_S has to have the form

$$q = AB = \text{Seq},$$

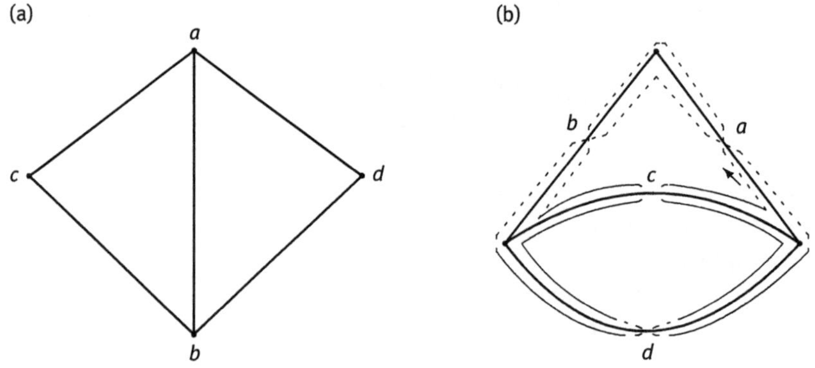

Figure 7.4.1: Interlaced graphs of two sequences.

such that letters in A consist of those in one of G_i, $i = 1, 2$, and letters in B of those in the other. Thus, the sequence Seq is separable. And, the converse procedure is for the sufficiency. □

Lemma 7.4.2. *For a crossing sequence* Seq*, its P–C graph* $G_{pc}(\text{Seq})$ *is a block if, and only if, its interlaced graph* $G(I)$ *is connected.*

Proof. Because $G(I)$ is connected if, and only if, its corresponding sequence Seq is a component, the lemma is soon found from Lemma 7.4.1. □

Theorem 7.4.2. *A crossing sequence* Seq *has* r (≥ 1) *components if, and only if, there are* 2^{r-1} *topologically non-equivalent closed curves in the plane, each of which has* Seq *as its crossing sequence.*

Proof. Since Seq is a crossing sequence with r (≥ 1) components, from Lemmas 7.3.1 to 7.3.2, the theorem is obtained when $r = 1$. By induction on r, for $r > 1$, suppose $\text{Seq} = AB$ such that $\text{Seq}_2 = B$ is a component and hence $\text{Seq}_1 = A$ is with $r - 1$ components. From Theorem 7.3.1, we see that a crossing sequence is uniquely represented by a closed curve if, and only if, it is a component itself because the partition of the letters into M and N is uniquely derived from $U(\text{Int})$. However, for a realization of Seq by a closed curve, if M and N form the partition of all the letters of Seq, suppose M_i and N_i form the partition of Seq_i which is restricted on M and N for $i = 1, 2$, respectively, then for M_1 and N_1 fixed, we may find that both $\{M_1 \cup M_2, N_1 \cup N_2\}$ and $\{M_1 \cup N_2, N_1 \cup M_2\}$ are partitions of Seq and hence Seq have two topologically non-equivalent realizations. By hypothesis, Seq_1 has 2^{r-2} different realizations. Therefore, there are $2 \bullet 2^{r-2} = 2^{r-1}$ closed curves in the plane, of which Seq is the crossing sequence.

Conversely, the sufficiency is obtained by a similar discussion in the proof of Theorem 5.3.4. □

7.5 Notes

7.5.1 The problem considered in this chapter was first proposed and studied by Gauss [96]. He conjectured that the characterization of a crossing sequence is only dependent on the interlaced graph and found the first two conditions in Theorem 7.3.1. However, he could not solve it at that time. The first solver of the problem is Dehn in Dehn [68]. But the Dehn's solution as shown in Theorem 7.2.1 is in a way different from that conjectured by Gauss. The first characterization as what Gauss conjectured was found by Treybig in Treybig [329]. However, it looks complicated and unsuitable for computation. Further discussions related to the problem can be seen in Francis [90], Marx [268], Rosenstiehl [297], Rosenstiehl and Read [298], Read and Rosenstiehl [281], Titus [328] and Whitney [394].

7.5.2 On account of algorithms, the Dehn's solution described in Theorem 7.2.1 is a rather simple one, which leads to an $O(n^2)$ algorithm, although his paper Dehn [68] is rather long and seems to be largely overlooked as noticed by Grunbaum in Grunbaum [109]. On the other hand, the algorithm based on the Gauss conjecture described in Theorem 7.3.1, from Rosenstiehl and Read [298], it is seen as $O(n^3)$. However, based on Theorem 7.2.3, an $O(n)$ algorithm can be designed by using the result on testing the planarity of a Hamiltonian cubic graph, which will be discussed in Section 8.1. Of course, a number of $O(n)$ algorithms for several kinds of generalized problems in Rosenstiehl [297] and Rosenstiehl and Tarjan [300] can also be found by similar discussions in Liu [206].

7.5.3 The construction of a 4-valent planar graph as $G(Z)$, having a pair of the associated graphs $G_S(Z)$ and $G_S^*(Z)$, which are mutually planar dual, was used for enumerating planar quadrangulations, via which the number of rooted c-nets is determined, in Mullin and Schellenberg [272].

7.5.4 It looks interesting that the interlaced graphs of sequences here and the planarity auxiliary graphs exploited in Liu [192, 193, 202, 205] are completely different in approach, but equally satisfactory in results, on solving the corresponding problems. Planarity auxiliary graphs will appear in Chapter 8, and then in Chapter 10 further.

7.5.5 The theory here plays an important role in the study of the knot problem in topology. This will be seen in Chapter 15.

7.5.6 This chapter enables us to investigate the classification of knots, or links, in certain types via 4-regular networks with binary weights on edges. One might think of the relationship between embeddings of the network on surfaces and knots, or links. It looks a new way to access the famous knot classification problem in topology.

8 Cohomology on Graphs

8.1 Immersions

An *immersion* of a graph $G = (V, E)$ in the plane, or equivalently in the sphere P_0 is such a mapping $\mu : G \to P_0$ that $\mu(v)$ is a point for $v \in V$ and $\mu(e)$, a segment of a Jordan curve connecting $\mu(u)$ and $\mu(v)$, which is topologically equivalent to a straight segment, for $e = (u, v) \in E$ with the following properties:

Im.1. $\mu_V : V \to P_0$ is injective.

Im.2. $\forall e = (u, w) \in E,\ \neg v \in V(v \neq u,\ v \neq w),\ \mu(v) \in \mu(e)$.

By geometric observations, it is allowed to consider simply crossing at a finite number of points of $\mu(a)$ and $\mu(b)$ for $a, b \in E$. By *simply crossing*, we shall mean that there are points of one of $\mu(a)$ and $\mu(b)$ on the both sides of the other in any small neighbourhood at the crossing point of $\mu(a)$ and $\mu(b)$. For any pair of $\alpha, \beta \in E$, α is not adjacent to β or simply write as $\alpha\ \overline{\text{adj}}\ \beta$, define

$$I_\mu(\alpha, \beta) = \begin{cases} 1, & \text{if } |\mu(\alpha) \bigcap \mu(\beta)| = 1 (\text{mod } 2); \\ 0, & \text{otherwise,} \end{cases} \tag{8.1.1}$$

which is called the *crossing index* of α and β for the immersion μ.

Because the crossing of $\mu(a)$ and $\mu(b)$ for a adj b is not substantial, we are allowed only to observe the crossness of $\mu(\alpha)$ and $\mu(\beta)$, for α not adjacent to β, i.e. $\alpha\ \overline{\text{adj}}\ \beta$, $\alpha, \beta \in E$ in an immersion μ. Let $\mathscr{D} = \{(\alpha, \beta) |\ \forall \alpha, \beta \in E,\ \alpha\ \overline{\text{adj}}\ \beta\}$ and $\mathscr{N} = \{(a, b) |\ \forall a, b, a \text{ adj } b\}$. For two immersions μ_1 and μ_2, define

$$\mu_1 \sim_{\text{im}} \mu_2 \Leftrightarrow \forall (\alpha, \beta) \in \mathscr{D},\ I_{\mu_1}(\alpha, \beta) = I_{\mu_2}(\alpha, \beta). \tag{8.1.2}$$

Of course, $\sim_{\text{im}}$ is an equivalence on the set of all immersions by inspection of the reflective, symmetry and transitive laws in Section 1.1. In what follows, we do not distinguish an immersion and its equivalent class.

Theorem 8.1.1. *An immersion μ of a connected graph G determines a planar embedding if, and only if,*

$$\forall (\alpha, \beta) \in \mathscr{D},\ I_\mu(\alpha, \beta) = 0. \tag{8.1.3}$$

Proof. The necessity is obvious because the edges and vertices of a planar polyhedron whose underlying graph is G determines an immersion of G which satisfies the condition (8.1.3).

Conversely, for any immersion μ of G with eq. (8.1.3), it can be seen that any pair of $\mu(\alpha)$ and $\mu(\beta)$, $\alpha\ \overline{\text{adj}}\ \beta$, $\alpha, \beta \in E$ can always be done without a common point from

DOI 10.1515/9783110479492-008

Jordan theorem mentioned in Section 5.2. Because all the connected components of the complement of $\mu(G)$ in P_0 form the set of faces of a planar polyhedron Σ whose underlying graph is G. Therefore, $\mu(G) = G(\Sigma)$ is a planar embedding of G. □

Let $\mathscr{I}$ be the set of all immersions of G. In order to investigate the relationship between immersions and planar embeddings of a graph G, one might think of treating G as a system

$$\Pi = \Pi(G) = \left([\alpha, \beta], [v, a], [u, v] \mid \forall u, v \in V, \forall \alpha, \beta, a \in E\right),$$

where $[\alpha, \beta] = [\beta, \alpha]$, $[v, a] = [a, v]$ and $[u, v] = [v, u]$ are called a 2-*cell*, 1-*cell* and 0-*cell*, respectively, and introducing the space $\mathscr{L}_2 = \langle [\alpha, \beta] \mid \forall (\alpha, \beta) \in \mathscr{D}\rangle$ over GF(2). $\mathscr{L}_2$ is called a 2-*space* of G, whose elements are called 2-*vectors*. For an immersion μ of G, let

$$\psi = \psi_\mu = \sum_{(\alpha,\beta)\in\mathscr{D}} I_\mu(\alpha, \beta)[\alpha, \beta], \tag{8.1.4}$$

which is a vector in $\mathscr{L}_2$. Similarly to the discussion in Chapter 4, let $\mathscr{L}_1$ and $\mathscr{L}_0$ be the 1-*space* and 0-*space*, whose elements are called 1-*vectors* and 0-*vectors*, generated by $\langle [v, e] \mid \forall v \in V, \forall e \in E, v\, \overline{\text{ind}}\, e\rangle$ and $\langle [u, v] \mid \forall u, v \in V, u\, \overline{\text{adj}}\, v\rangle$ over GF(2), respectively. Here, $\overline{\text{ind}}$ denotes that v is not incident to e or that e is not incident to v. The boundary mappings ∂_i, $\mathscr{L}_i \to \mathscr{L}_{i-1}, i = 2, 1, 0$, $\mathscr{L}_{-1} = 0$, are extended by those defined on the basis of a space as

$$\begin{cases} \partial_2([\alpha, \beta]) & = [u(\alpha), \beta] + [v(\alpha), \beta] \\ & \quad +[\alpha, u(\beta)] + [\alpha, v(\beta)]; \\ \partial_1([v, \beta]) & = [v, u(\beta)] + [v, v(\beta)]; \\ \partial_0([u, v]) & = 0, \end{cases} \tag{8.1.5}$$

where $u(e), v(e)$ are the two ends of $e \in E$. Dually, the coboundary mappings $\delta_i : \mathscr{L}_i \to \mathscr{L}_{i+1}$, $i = 0, 1, 2$, $\mathscr{L}_3 = 0$ are extended by those defined on the basis of a space as

$$\begin{cases} \delta_0([v, u]) = \sum_{e\in E_v} [e, u] + \sum_{e\in E_u} [v, e]; \\ \delta_1([v, a]) = \sum_{e\in E_v} [e, a]; \\ \delta_2([\alpha, \beta]) = 0, \end{cases} \tag{8.1.6}$$

where the following relations have to be defined beforehand for the convenience of treatment:

$$\begin{cases} \forall a, b \in E,\ [a, b] = 0,\ \text{if } (a, b) \notin \mathscr{D}; \\ \forall v \in V, a \in E, [v, a] = 0,\ \text{if } v \text{ ind } a; \\ \forall u, v \in V, [u, v] = 0,\ \text{if } u \text{ adj } v \text{ or } u = v. \end{cases} \tag{8.1.7}$$

Further, let $\Omega = \{(a,b) | \forall a, b \in E,\ a \neq b\}$. If we define

$$\begin{aligned}\sum_{(\alpha,\beta)\in\Omega} A(\alpha,\beta)[\alpha,\beta] &= \sum_{\alpha\in E}\left[\alpha, \sum_{\beta\in E} A(\alpha,\beta)\beta\right] \\ &= \sum_{\beta\in E}\left[\sum_{\alpha\in E} A(\alpha,\beta)\alpha, \beta\right],\end{aligned} \tag{8.1.8}$$

then by recalling the boundary and coboundary mappings among $\mathscr{G}_0$, $\mathscr{G}_1$ and $\mathscr{G}_2$ in Chapter 4, we have that $\forall u, v \in V,\ \alpha, \beta, a \in E$,

$$\begin{cases}\partial_2([\alpha,\beta]) = [\partial_1\alpha, \beta] + [\alpha, \partial_1\beta]; \\ \partial_1([v,a]) = [v, \partial_1 a]; \\ \partial_0([u,v]) = 0\end{cases} \tag{8.1.9}$$

and that $\forall u, v \in V,\ \alpha, \beta, a \in E$,

$$\begin{cases}\delta_0([u,v]) = [\delta_0 u, v] + [u, \delta_0 v]; \\ \delta_1([v,a]) = [\delta_0 v, a]; \\ \delta_2([\alpha,\beta]) = 0.\end{cases} \tag{8.1.10}$$

Moreover, it is also known that

$$\partial_i \in \mathrm{Hom}(\mathscr{L}_i, \mathscr{L}_{i-1}) \text{ and } \delta_i \in \mathrm{Hom}(\mathscr{L}_i, \mathscr{L}_{i+1}),$$

$i = 0, 1, 2$. In other words, we have the diagram commutative as follows:

$$\begin{array}{ccccccccc} & & \mathscr{L}_0 & \xrightarrow{\delta_0} & \mathscr{L}_1 & \xrightarrow{\delta_1} & \mathscr{L}_2 & \xrightarrow{\delta_2} & 0 \\ & & \downarrow & & \downarrow & & \downarrow & & \\ 0 & \xleftarrow{\partial_0} & \mathscr{L}_0 & \xleftarrow{\partial_1} & \mathscr{L}_1 & \xleftarrow{\partial_1} & \mathscr{L}_2, & & \end{array}$$

where $\mathscr{L}_i = \mathscr{L}_i(\Pi, \mathrm{GF}(2))$, $i = 0, 1, 2$.

Now, we turn to observe immersions of a graph G. Obviously, any immersion μ of G is a 2-cocycle because $\mu \in \mathscr{L}_2$ which is defined to be the 2-*cocycle space* of Π in its own right from Ker $\delta_2 = \mathscr{L}_2$. Here, we also call the subspace Im δ_1 of $\mathscr{L}_2$ the 2-*coboundary space* of Π.

Theorem 8.1.2. *If G is planar, then for any $0 \neq [a,v] \in \mathscr{L}_1$, $a \in E$, $v \in V$ on Π there is an immersion μ such that $\psi_\mu = \delta_1([a,v])$.*

Proof. Because G is planar, G has an immersion μ which is a planar embedding, notably with the condition:$\forall(\alpha,\beta) \in \Omega, I_\mu(\alpha,\beta) = 0$. Then, we prove that for any $[v,a]$,

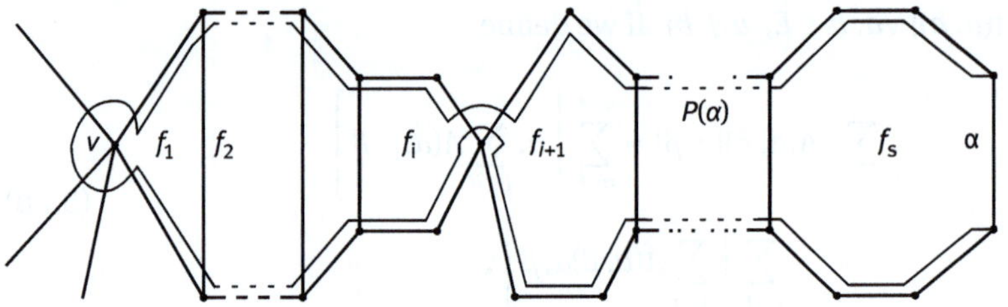

Figure 8.1.1: From one immersion to another.

there exists an immersion μ_1 such that $\psi_{\mu_1} = \delta_1([v, a])$. In fact, we know that on the planar polyhedron determined by μ, from the connectedness of G, there is a sequence of faces f_1 ind $v, f_2, \ldots, f_s$ ind a, $s > 0$, such that each common boundary of f_i and f_{i+1}, $i = 1, 2, \ldots, s-1$ has at least one point. Then, we may find a curve $P(a)$ from one end to the other of a such that

$$P(a) \bigcap \mu(b) = \begin{cases} 0, & \text{if } b \notin E_v; \\ 1, & \text{if } b \in E_v \end{cases}$$

as shown in Figure 8.1.1. That enables us to construct an immersion μ_1 as

$$\mu_1(b) = \begin{cases} \mu(b), & \text{if } b \neq a; \\ P(a), & \text{if } b = a. \end{cases}$$

It is easily seen that $\psi_{\mu_1} = \delta_1([a, v])$. The theorem is obtained. □

In fact, if G is planar, then all the 2-vectors determined by immersions coincide with the whole 2-coboundary space of Π. However, if G is not planar, then for any $[a, v] \in \mathscr{L}_1$, there is no immersion μ such that $\psi_\mu = \delta_1([a, v])$. They will be shown in Section 8.2.

8.2 Realization of planarity

Let $\mathscr{B}_2 = \operatorname{Im} \delta_1$ be the 2-coboundary space of $\Pi(G)$ and $\mathscr{H}_{[2]} = \mathscr{H}_{[2]}(\Pi, \mathrm{GF}(2)) = \operatorname{Ker} \delta_2 / \operatorname{Im} \delta_1 = \mathscr{L}_2/\mathscr{B}_2$, which is called the 2-*cohomology space* of Π as well. Further, let

$$\Psi = \Psi(G) = \{\psi_\mu | \ \forall \mu \in \mathscr{I}\}. \tag{8.2.1}$$

An *elementary operator* denoted by $\pi_{(\gamma,v)}$ on an immersion $\mu \in \mathscr{I}$ for $\gamma \in E, v \in V$, $\gamma \ \overline{\text{ind}}\ v$ is defined to be such a transformation $\pi_{(\gamma,v)}\mu = \mu_1 \in \mathscr{I}$ as

$$\mu_1(e) = \begin{cases} \mu(e), & \text{if } e \neq \gamma; \\ \varphi(\gamma, v), & \text{otherwise,} \end{cases} \tag{8.2.2}$$

where $\varphi(\gamma, v)$ is such a segment of Jordan curve that there is only one vertex which is v in the inner domain of the Jordan closed curve $\mu(\gamma) \cup \varphi(\gamma, v)$. It is easily seen that

$$I_{\mu_1}(\gamma, a) = \begin{cases} I_\mu(\gamma, a), \text{ if } a \notin E_v; \\ I_\mu(\gamma, a) + 1, \text{ if } a \in E_v \end{cases} \tag{8.2.3}$$

from the Jordan curve theorem discussed in Section 5.2.

Lemma 8.2.1. $\forall \mu_1, \mu_2 \in \mathscr{I}(\mu_1 \not\sim_{\text{im}} \mu_2)$, $\exists(\gamma_i, v_i)(\gamma_i \in E, v_i \in V, \gamma_i \overline{\text{ind}}\, v_i)$, $1 \le i \le s$, $\ni$

$$\mu_1 \sim_{\text{im}} \prod_{1 \le i \le s} \pi_{(\gamma_i, v_i)} \mu_2, \tag{8.2.4}$$

where

$$\prod_{1 \le i \le s} \pi_{(\gamma_i, v_i)} \mu = \prod_{1 \le i \le s-1} \pi_{(\gamma_i, v_i)} (\pi_{(\gamma_s, v_s)} \mu)$$

for $\mu = \mu_2$.

Proof. For any $a \in E$, let $n_{\mu_1,\mu_2}(a)$ be the number of vertices in the inner domain of the Jordan closed curve $\mu_1(a) \cup \mu_2(a)$ and $n_{\mu_1,\mu_2} = \sum_{a \in E} n_{\mu_1,\mu_2}(a)$. By induction on the number $n = n_{\mu_1,\mu_2}$. When $n = 1$, suppose $\gamma \in E$ has the property that v is the vertex in the inner domain of $\mu_1(\gamma) \cup \mu_2(\gamma)$, then from eq. (8.2.3), $\mu_1 \sim_{\text{im}} \pi_{(\gamma,v)} \mu_2$. In general, suppose there is a vertex v in the inner domain of $\mu_1(\gamma) \cup \mu_2(\gamma)$, $\gamma \in E$. We may find $\mu_3 = \pi_{(\gamma,v)} \mu_2$. Then by Jordan curve theorem, $n_{\mu_1,\mu_3}(\gamma) = n_{\mu_1,\mu_2}(\gamma) - 1$. However, for any $\beta \in E$, $\beta \neq \gamma$, $n_{\mu_1,\mu_3}(\beta) = n_{\mu_1,\mu_2}(\beta)$. We have that $n_{\mu_1,\mu_3} = n - 1$. By the hypothesis, there exists $\gamma_i \in E$, $v_i \in V$, $\gamma_i \overline{\text{ind}}\, v_i$, $1 \le i \le l$ such that

$$\begin{aligned} \mu_1 &\sim_{\text{im}} \prod_{1 \le i \le l} \pi_{(\gamma_i, v_i)} \mu_3 \\ &\sim_{\text{im}} \prod_{1 \le i \le l} \pi_{(\gamma_i, v_i)} (\pi_{(\gamma,v)} \mu_2). \end{aligned}$$

This is just eq. (8.2.4) with $\gamma_{l+1} = \gamma_s = \gamma$, $v_{l+1} = v_s = v$. □

Lemma 8.2.2. $\Psi \in \widetilde{\mathscr{H}}_{[2]}$.

Proof. First, to show that for any $\psi_1 = \psi_{\mu_1}$, $\psi_2 = \psi_{\mu_2} \in \Psi$, $\mu_1, \mu_2 \in \mathscr{I}$, $\psi_1 + \psi_2 \in \mathscr{B}_2$. In fact, from eq. (8.2.3), it is known that for any $[v, \gamma] \in \mathscr{L}_1$, $v \in V$, $\gamma \in E$, $v \,\overline{\text{ind}}\, \gamma$, we have $\psi_{\pi_{(\gamma,v)}\mu_2} = \psi_{\mu_2} + \delta_1([v, \gamma])$. From Lemma 8.2.1, $\mu_1 \sim_{\text{im}} \prod_{1 \le i \le s} \pi_{(\gamma_i, v_i)} \mu_2$. Therefore,

$$\psi_1 + \psi_2 = \sum_{1 \le i \le s} \delta_1([\gamma_i, v_i]) \in \mathscr{B}_2. \tag{8.2.5}$$

Then, to show that for any $\psi \in \mathscr{L}_2$, if there is $\psi_1 \in \Psi$ such that $\psi + \psi_1 \in \mathscr{B}_2$ then $\psi \in \Psi$. Because $\mathscr{B}_2$ has a basis in the set of $\delta_1([\gamma, v])$ for all $[\gamma, v] \in \mathscr{L}_1$, from $\psi + \psi_1 \in \mathscr{B}_2$ there is a sequence of $[\gamma_i, v_i]$, $1 \le i \le k$ such that

$$\psi + \psi_1 = \sum_{i=1}^{k} \delta_1([\gamma_i, v_i]).$$

Therefore, an immersion

$$\mu \sim_{\text{im}} \prod_{1 \le i \le k} \pi_{(\gamma_i, v_i)} \mu_1$$

is found such that $\psi = \psi_\mu$ and hence $\psi \in \Psi$. □

Theorem 8.2.1. *A connected graph G is planar if, and only if, $\Psi = 0 \in \widetilde{\mathscr{H}}_{[2]}$.*

Proof. According to the planarity of G, G has a planar embedding $\mu \in \mathscr{I}$, $\psi_\mu = 0 \in \mathscr{B}_2$. From Lemma 8.2.2, $\Psi = \mathscr{B}_2$, i.e. $\Psi = 0 \in \widetilde{\mathscr{H}}_{[2]}$.

Conversely, from $\Psi = 0 \in \widetilde{\mathscr{H}}_{[2]}$, $\Psi = \mathscr{B}_2$. Because $\psi_\mu = 0 \in \mathscr{B}_2$, there is a planar embedding $\mu \in \mathscr{I}$. This is the theorem. □

On the basis of the theorem, we are allowed to introduce variables

$$x_{b,v} = x_{v,b} \in \text{GF}(2) \tag{8.2.6}$$

for any $b \in E$, $v \in V$. For an immersion μ of G, we observe the following equation system:

$$I_\mu(\alpha, \beta) = \sum x_{b,v} \ (\text{mod } 2), \ (\alpha, \beta) \in \mathscr{D} \tag{8.2.7}$$

where the summation runs over all (b, v) such that $b = \alpha$ and $\beta \in E_v$, or $b = \beta$ and $\alpha \in E_v$.

Lemma 8.2.3. *For a graph G, if there is an immersion $\mu_0 \in \mathscr{I}$ such that the equation system (8.2.7) has a solution, then for any $\mu \in \mathscr{I}$, it has a solution as well.*

Proof. Suppose the equation system (8.2.7) has a solution which is denoted by $x_{b,v}^{(0)}$ for an immersion μ_0 of G. Then,

$$\begin{aligned}\psi_{\mu_0} &= \sum_{(\alpha,\beta)\in\mathscr{D}} I_{\mu_0}(\alpha,\beta)[\alpha,\beta]\\ &= \sum_{(\alpha,\beta)\in\mathscr{D}} \left(\sum_{b\in E, v\in V} x_{b,v}^{(0)}\right)[\alpha,\beta]\\ &= \sum_{(\alpha,\beta)\in\mathscr{D}} \left(\sum_{\beta\in E_v} x_{\alpha,v}^{(0)} + \sum_{\alpha\in E_v} x_{\beta,v}^{(0)}\right)[\alpha,\beta]\end{aligned}$$

by eq. (8.1.7),

$$= \sum_{v\in V}\left(\sum_{\alpha\in E} x_{\alpha,v}^{(0)}[\alpha,\delta_0 v] + \sum_{\beta\in E} x_{\beta,v}^{(0)}[\delta_0 v,\beta]\right)$$

and by eq. (8.1.10),

$$\begin{aligned}&= \sum_{v\in V}\left(\sum_{\alpha\in E} x_{\alpha,v}^{(0)}\delta_1([\alpha,v]) + \sum_{\beta\in E} x_{\beta,v}^{(0)}\delta_1([v,\beta])\right)\\ &\in \operatorname{Im}\delta_1 = \mathscr{B}_2.\end{aligned}$$

Therefore, from Lemma 8.2.2, $\Psi = \mathscr{B}_2$. That implies $\psi_\mu \in \mathscr{B}_2$ for any $\mu \in \mathscr{I}$ and hence eq. (8.2.7) for μ has a solution as well. □

Theorem 8.2.2. *A connected graph G is planar if, and only if, for an immersion μ of G, eq. (8.2.7) has a solution.*

Proof. Because G is planar, G has an immersion which is a planar embedding $\mu_0 \in \mathscr{I}$. However, for a planar embedding μ_0, $\forall(\alpha,\beta) \in \mathscr{D}, I_{\mu_0}(\alpha,\beta) = 0$. Therefore, eq. (8.2.7) has the solution of all $x_{b,v} = 0$, for $b \in E, v \in 8$. From Lemma 8.2.3, eq. (8.2.7) also has a solution for μ. Conversely, from Lemma 8.2.3, $\Psi = \mathscr{B}_2 = 0 \in \widetilde{\mathscr{H}}_{[2]}$. By Theorem 8.2.1, the theorem is obtained. □

8.3 Reductions

Although there are too many variables and equations in eq. (8.2.7), the theorem has a very important advantage of choosing a suitable immersion of G in order to reduce the variables and equations. Suppose T is a tree on G. We are allowed to consider an immersion μ such that the following properties are satisfied.

Im.3. Any two edges on T do not cross.

Im.4. Any two edges that have a common end do not cross.

Im.5. Any cotree edge does not cross with any tree edge.

We call such an immersion a *T-immersion* of G. Obviously, from the Jordan curve theorem and Theorem 2.1.1 of Liu [217], for any tree T on a connected graph G, a T-immersion does always exist.

Lemma 8.3.1. *For a tree T given on G, let T^* be the cotree determined by T, then*

$$\{[C_a^*, C_b^*] \mid \forall a, b \in T\} \bigcup \{[C_a^*, \alpha] \mid \forall a \in T, \forall \alpha \in T^*\} \tag{8.3.1}$$

contains a basis of the space $\mathscr{B}_2$.

Proof. From Lemmas 4.3.1 and eq. (4.3.1), $\forall a \in T$,

$$\delta_1([v, a]) = \left[\sum_{e \in E_v \cap T} C_e^*, C_a^* + \sum_{e \in C_a^* \cap T^*} e \right]. \tag{8.3.2}$$

Thus, we have

$$\delta_1([v, \alpha]) = \sum_{e \in E_v \cap T} [C_e^*, \alpha],\ \alpha \in T^*;$$

$$\delta_1([v, a]) = \sum_{e \in E_v \cap T} [C_e^*, C_a^*] + \sum_{e \in E_v \cap T} \sum_{\alpha \in C_a^* \cap T^*} [C_e^*, \alpha],\ a \in T.$$

Because $\mathscr{B}_2$ has a basis which consists of $\delta_1([v, a])$ for $[v, a] \in \mathscr{L}_1$, the lemma follows. □

Theorem 8.3.1. *A connected graph G is planar if, and only if, for a T-immersion μ, the equation system*

$$I_\mu(\alpha, \beta) = \sum_{\substack{a \in C_\alpha,\, b \in C_\beta \\ (a, b) \in \mathscr{N} \\ a, b \in T}} x_{a,b} + \sum_{\substack{b \in C_\beta \\ (b, \alpha) \in \mathscr{N} \\ b \in T}} x_{b,\alpha} + \sum_{\substack{a \in C_\alpha \\ (a, \beta) \in \mathscr{N}\, a \in T}} x_{a,\beta} \tag{8.3.3}$$

for $(\alpha, \beta) \in \mathscr{D}$, $\alpha, \beta \in T^*$ has a solution.

Proof. From Theorem 8.2.1 and Lemma 8.3.1, G is planar if, and only if, for a T-immersion μ, there exist $x_{a,b} \in GF(2)$ and $y_{a,\gamma} \in GF(2)$ for $a, b \in T$, $\gamma \in T^*$ such that the following relation holds:

$$\psi_\mu = \sum_{a,b \in T} x_{a,b}[C_a^*, C_b^*] + \sum_{a \in T, \gamma \in T^*} y_{a,\gamma}[C_a^*, \gamma]. \tag{8.3.4}$$

Moreover, from Lemmas 2.1.3 to 2.1.4, it is known that $\forall a, b \in T$, $\forall \alpha, \beta, \gamma \in T^*$,

$$\begin{cases} ([\alpha,\beta],[C_a^*,C_b^*]) = 1 \Leftrightarrow (a \in C_\alpha) \wedge (b \in C_\beta); \\ ([\alpha,\beta],[C_a^*,\gamma]) = 1 \Leftrightarrow (\gamma = \alpha, a \in C_\beta) \\ \qquad\qquad\qquad\qquad\qquad \vee(\gamma = \beta, a \in C_\alpha). \end{cases} \tag{8.3.5}$$

By eqs (8.2.4) and (8.2.5), G is planar if, and only if, the following equation system

$$I_\mu(\alpha,\beta) = \sum_{a\in C_\alpha, b\in C_\beta} x_{a,b} + \sum_{b\in C_\beta} y_{b,\alpha} + \sum_{a\in C_\alpha} y_{a,\beta} \tag{8.3.6}$$

for $(\alpha,\beta) \in \mathscr{D}$ has a solution.

Because μ is a T-immersion, we have

$$\forall (a,b) \in \mathscr{D}(a,b \in T),\ x_{a,b} = 0. \tag{8.3.7}$$

By Lemmas 2.1.3 and 2.1.4 of Liu [217], we also know that $\forall (t,\alpha) \in \mathscr{D}(t \in T, \alpha \in T^*)$,

$$([t,\alpha],\ [C_a^*, C_b^*]) = 1 \Leftrightarrow (t = a, b \in C_\alpha) \vee (t = b, a \in C_\alpha)$$

and hence that

$$y_{t,\alpha} + \sum_{b\in C_\alpha} x_{t,b} = 0. \tag{8.3.8}$$

Let $x_{t,\alpha} = y_{t,\alpha} + \sum_{b\in C_\alpha} x_{t,b}$ be a new variable instead of $y_{t,\alpha}$ for $t \in T, \alpha \in T^*$, then

$$\forall (t,\alpha) \in \mathscr{D}\ (t \in T, \alpha \in T^*),\ x_{t,\alpha} = 0. \tag{8.3.9}$$

By substituting eqs (8.3.7) (8.3.8) and (8.3.9) into eq. (8.3.6), we have

$$\begin{aligned} I_\mu(\alpha,\beta) &= \sum_{\substack{a\in C_\alpha,\, b\in C_\beta \\ (a,b)\in\mathscr{N} \\ a,b\in T}} x_{a,b} + \sum_{\substack{b\in C_\beta \\ (b,\alpha)\in\mathscr{N} \\ b\in T}} \left(\sum_{\substack{a\in C_\alpha \\ a\in T}} x_{b,a} + x_{b,\alpha} \right) \\ &\quad + \sum_{\substack{a\in C_\alpha \\ (a,\beta)\in\mathscr{N} \\ a\in T}} \left(\sum_{\substack{b\in C_\beta \\ b\in T}} x_{a,b} + x_{a,\beta} \right) \\ &= \sum_{\substack{a\in C_\alpha,\, b\in C_\beta \\ (a,b)\in\mathscr{N} \\ a,b\in T}} x_{a,b} + \sum_{\substack{b\in C_\beta \\ (b,\alpha)\in\mathscr{N} \\ b\in T}} x_{b,\alpha} + \sum_{\substack{a\in C_\beta \\ (a,\beta)\in\mathscr{N} \\ a\in T}} x_{a,\beta} \end{aligned}$$

for $\alpha, \beta \in T^*$, $(\alpha,\beta) \in \mathscr{D}$. This completes the proof. □

8.4 Planarity auxiliary graphs

In this section, the purpose is to introduce a kind of auxiliary graph from a graph $G = (V, E)$ given for testing the planarity.

First, we have to reduce the number of variables in the equations by which the planarity of a graph can be justified.

Lemma 8.4.1. *Each equation in eq. (8.2.10) has the form: for $(\alpha, \beta) \in \mathscr{D}$, $\alpha, \beta \in T^*$,*

$$\sum (x_{s,t} + x_{s,r} + x_{r,t}) = I_\mu(\alpha, \beta). \tag{8.4.1}$$

Proof. We may choose a vertex as the root of T. All the edges in T are oriented in agreement with the direction from the root along T passing through them.

When $C_\alpha \cap C_\beta = P(u, v)$, a path from u to v on T, we have that eq. (8.2.10) becomes

$$I_\mu(\alpha, \beta) = \sum_u + \sum_v,$$

where $\sum_w$ represents the summation at vertex $w = u$ or v over $s, t, l \in (C_\alpha \cup C_\beta) \cap E_w$, and

$$\sum_w = \begin{cases} x_{s,t} + x_{s,l} + x_{t,l}, & \text{if } r \in C_\alpha \bigcup C_\beta \\ & \text{and suppose } r = l; \\ (x_{s,t} + x_{r,s} + x_{r,t}) + (x_{s,l} + x_{r,l} + x_{r,s}) & \\ \quad +(x_{t,l} + x_{r,t} + x_{r,l}), & \text{if } r \notin C_\alpha \bigcup C_\beta, \end{cases} \tag{8.4.2}$$

where r is the tree edge coming to w.

When $C_\alpha \cap C_\beta = \{v\}$, then eq. (8.2.10) is as

$$I_\mu(\alpha, \beta) = \sum_v,$$

where for $s, t \in C_\alpha \cap E_v$, and $p, q \in C_\beta \cap E_v$,

$$\sum_v = \begin{cases} (x_{s,p} + x_{s,t} + x_{p,t}) + (x_{s,q} + x_{s,t} + x_{q,t}), \\ \quad \text{if } r \in C_\alpha \bigcup C_\beta \text{ and suppose } r = t; \\ (x_{t,p} + x_{t,r} + x_{p,r}) + (x_{t,q} + x_{t,r} + x_{q,r}) \\ \quad +(x_{s,p} + x_{s,r} + x_{p,r}) + (x_{s,q} + x_{s,t} + x_{q,r}), \\ \quad \text{if } r \notin C_\alpha \bigcup C_\beta. \end{cases} \tag{8.4.3}$$

Therefore, the lemma holds. □

Theorem 8.4.1. *A graph G (connected, of course) is planar if, and only if, for a T-immersion μ of G, the equation system*

$$\sum_{\substack{t\in C_\alpha, s\in C_\beta\\ (s,t)\in\mathcal{N}\\ s,t\neq r}} x_{t,s} = I_\mu(\alpha,\beta) \tag{8.4.4}$$

for $(\alpha,\beta)\in\mathcal{D}$ has a solution. Here, r is the tree edge whose direction is coming to v, while $r,s,t\in E_v$, for $v\in C_\alpha\cap C_\beta$.

Proof. According to Lemma 8.4.1, we are allowed to introduce the substitution

$$\widetilde{x}_{s,t} = x_{r,s} + x_{r,t} + x_{s,t}$$

and to use the symbol $x_{s,t}$ as well instead of $\widetilde{x}_{s,t}$ for the sake of simplicity. Then, eq. (8.4.1) becomes eq. (8.4.4). From Theorem 8.2.3, the theorem is obtained. □

Now, one might think that eq. (8.4.4) has much fewer variables than eq. (8.2.7) even eq. (8.2.10) does. However, since each equation is allowed to have six variables (but not more) involved, it seems that the only efficient way to solve the equation system is employing the Gaussian elimination method which is still of $O(v^6)$ in computing complexity.

If μ is a T-immersion of G for T being an OD-tree on G, then μ is called an *OD-immersion* of G. Whenever an OD-immersion μ is given for a graph G, it always implies that G has been oriented in the way of all the tree edges having the directions in agreement with those from the starting vertex, the root, to their ends along the tree according to the OD-procedure described in Section 2.2 of Liu [217]. Of course, all cotree edges in G have the directions in agreement with those of the fundamental dicircuits on which they are.

Lemma 8.4.2. *Suppose Tod is an OD-tree on G. Let C_α be the fundamental circuit (precisely, dicircuit) formed by Tod with a cotree edge α as before. Then, for $(\alpha,\beta)\in\mathcal{D}$, $\alpha,\beta\in Tod^*$, the corresponding cotree of Tod on G, there are exactly two variables involved in eq. (8.4.4) on the subgraph $C_\alpha\cup C_\beta$ of G: one is formed by two adjacent edges, at least one of them is a tree edge, going away from the common end and the other is formed by two adjacent edges with different directions at the common end.*

Proof. Of course, if $C_\alpha\cap C_\beta=\emptyset$, i.e. no common vertex, then no variable is involved with $C_\alpha\cup C_\beta$.

Case 1. $C_\alpha\cap C_\beta=\{v\}$, $v\in 8$. Because both C_α and C_β are dicircuits, from $(\alpha,\beta)\in\mathcal{D}$, we know that v is not the root which has no tree edge coming to and that the tree edge r whose direction is coming to v has to be on one of C_α and C_β. By the symmetry between

α and β, we are allowed to assume r is on C_α. The other edge of C_α at v is denoted by s which is allowed to be a cotree edge. Moreover, the cotree edge β on C_β has to be incident with v. The other edge of C_β at v is denoted by t which is only allowed to be a tree edge. Thus, we have exactly two variables: $x_{s,t}$ and $x_{s,\beta}$ with the properties of a lemma.

Case 2. $C_\alpha \cap C_\beta = P(u,v)$, a path from u to v on G. From Lemma 2.1.3 of Liu [217], $P(u,v)$ is on the tree *Tod*. The tree edge on $P(u,v)$ at u is denoted by l whose direction is going away from u. Because both C_α and C_β are dicircuits, each of C_α and C_β has an edge coming to u. From $(\alpha,\beta) \in \mathscr{D}$, and by the symmetry between α and β, we are allowed to assume C_α has the tree edge coming to u. Thus, the edge other than l of C_β at u is only allowed to be the cotree edge β. Hence, there is only one variable $x_{l,\beta}$ at u. Similarly, we know that there is one edge going away from v on each of C_α and C_β. Let s and t be the edges on C_α and C_β respectively. Then, one of s and t has to be a tree edge. Hence, there is only one variable $x_{s,t}$ as well. Moreover, $x_{s,t}$ and $x_{l,\beta}$ have the properties required. The lemma is proved. □

According to the lemma above, we are allowed to call the variables corresponding to a pair of adjacent edges, both of which have the direction of going away from the common end *tree variables* denoted by $x_{s,t}$ for $(s,t) \in \mathscr{N}$. Of course, one of s and t has to be a tree edge. The variables corresponding to pairs of adjacent edges with different directions at the common ends are called *cotree variables* denoted by $y_{l,\gamma}$ for $(l,\gamma) \in \mathscr{N}$, $l \in Tod$ and $\gamma \in Tod^*$. Let $\mathscr{X}_G(T), \mathscr{Y}_G(T^*)$ be the set of all tree and cotree variables of G, respectively. For two variables $x_{s,t} \in \mathscr{X}_G(T)$ and $y_{l,\gamma} \in \mathscr{Y}_G(T^*)$, if there exists a pair of cotree edges α, β, $(\alpha,\beta) \in \mathscr{D}$ in G such that $x_{s,t}$ and $y_{l,\gamma}, \gamma = \alpha$, or β are the two variables determined by Lemma 8.4.2 for $C_\alpha \cup C_\beta$ associated with the OD-tree *Tod*, then they are said to be *co-variables* denoted by $x(T;\alpha,\beta)$ and $y(T^*;\alpha,\beta)$.

Theorem 8.4.2. *For a T-immersion μ of G, eq. (8.4.4) has a solution if, and only if, for an OD-immersion μ_0 of G, the equation system*

$$x(T;\alpha,\beta) + y(T^*;\alpha,\beta) = I_{\mu_0}(\alpha,\beta) \tag{8.4.5}$$

for $(\alpha,\beta) \in \mathscr{D}$, $\alpha,\beta \in Tod^*$ has a solution.

Proof. Because for a T-immersion μ, eq. (8.4.4) has a solution, from Theorem 8.4.1, eq. (8.4.1) has a solution. Then, from Theorem 8.2.3, $\psi_\mu \in \mathscr{B}_2$. In consequence, from Lemma 8.2.2, $\Psi = \mathscr{B}_2$. That implies for an OD-immersion μ_0, we have $\psi_{\mu_0} \in \mathscr{B}_2$ as well. Since for μ_0 eq. (8.2.10) has a solution, from Theorem 8.4.1, eq. (8.4.4) for μ_0 has a solution. Hence, by Lemma 8.4.2, eq. (8.4.5) has a solution. Conversely, from eq. (8.4.5) having a solution for μ_0, $\psi_{\mu_0} \in \mathscr{B}_2$ can be derived. Then, by the similar procedure, we may find that eq. (8.4.4) has a solution for a T-immersion $\mu \in \Psi$. □

For an OD-immersion μ_0 of G given, eq. (8.4.5) determines a graph denoted by $\mathrm{Aux}_0(G) = (V^0_{\mathrm{au}}, E^0_{\mathrm{au}})$, or simply written as Aux_0 and called a *planarity 0-auxiliary graph* of G where

$$\begin{aligned} V^0_{\mathrm{au}} =& V^0_{\mathrm{au}}(G) = \{x \mid \forall x \in \mathscr{X}_G(T)\} \\ &+ \{y \mid \forall y \in \mathscr{Y}_G(T^*)\}; \end{aligned} \tag{8.4.6a}$$

$$\begin{aligned} E^0_{\mathrm{au}} =& E^0_{\mathrm{au}}(G) = \Big\{(x, y) \mid \exists(\alpha, \beta) \\ &\in \mathscr{D}(\alpha, \beta \in Tod^*) \\ &\ni \Big(x = x(T; \alpha, \beta) \wedge \\ & y = y(T^*; \alpha, \beta)\Big)\Big\}. \end{aligned} \tag{8.4.6b}$$

Moreover, each edge e of $\mathrm{Aux}_0(G)$ has a weight $w(e) = I_{\mu_0}(\alpha, \beta) \in \mathrm{GF}(2)$. Define the weight of a subgraph of $\mathrm{Aux}_0(G)$ to be the sum of all the weights of their edges over GF(2). If the weight is 1, then the subgraph is said to be of *odd weight;* otherwise, of *even weight.*

Lemma 8.4.3. *For an OD-immersion μ_0, eq.* (8.4.5) *has a solution if, and only if, there is no circuit of odd weight on $\mathrm{Aux}_0(G)$.*

Proof. By the 1-boundary mapping ∂_1 on the space $\mathscr{G}_1(\mathrm{Aux}_0)$ as defined in 4.1, eq. (8.4.5) becomes $\forall e \in E^0_{\mathrm{au}}$,

$$\partial_1 e = w(e). \tag{8.4.7}$$

Necessity. If not, suppose C is an odd-weight circuit on $\mathrm{Aux}_0(G)$, i.e.

$$\sum_{e \in C} w(e) = 1.$$

However, from eq. (8.4.7), we have

$$0 = \sum_{e \in C} \partial_1 e = \sum_{e \in C} w(e) = 1.$$

This is a contradiction.

Sufficiency. Because of no odd-weight circuit on $\mathrm{Aux}_0(G)$, eq. (8.4.7) is equivalent to those in eq. (8.4.7) only for $e \in T(\mathrm{Aux}_0)$ where $T(\mathrm{Aux}_0)$ is a tree (in general, a forest but it is allowed to assume to be a tree without loss of generality) on $\mathrm{Aux}_0(G)$. However, for $T(\mathrm{Aux}_0)$, a solution of eq. (8.4.7) is easily determined only by choosing a vertex as the root and setting the variable corresponding to the root to be 0 or 1. □

Theorem 8.4.3. *For an OD-immersion μ_0 of G, eq. (8.4.5) has a solution if, and only if, there is no fundamental circuit of odd weight for a tree $T(Aux_0)$ on $Aux_0(G)$.*

Proof. Let $\mathscr{C}(\mathrm{Aux}_0)$ be the cycle space of $\mathrm{Aux}_0(G)$. It is easily shown that the set $\mathscr{C}_0(\mathrm{Aux}_0)$ of all even-weight cycles is a subspace of $\mathscr{C}(\mathrm{Aux}_0)$, which is generated by even-weight circuits. From Lemma 8.4.3, Eq. (8.4.5) has a solution if, and only if, $\mathscr{C}(\mathrm{Aux}_0)/\mathscr{C}_0(\mathrm{Aux}_0) = 0$, i.e. $\mathscr{C}(\mathrm{Aux}_0) = \mathscr{C}_0(\mathrm{Aux}_0)$. Therefore, from Theorem 4.2.1 the theorem follows. □

8.5 Basic conclusions

A graph $G = (V, E)$ is said to be *balanced* for a weighting on E by elements in GF(2) if there exists a kind of labelling vertices on V by $l(v)$ = "+" or "−" such that $\forall e = (a, v) \in E$,

$$\begin{cases} l(u) = l(v), \text{ if } w(e) = 0; \\ l(u) \neq l(v), \text{ if } w(e) = 1. \end{cases} \tag{8.5.1}$$

Lemma 8.5.1. *For an OD-immersion μ_0 of G, the planarity 0-auxiliary graph $Aux_0(G)$ is balanced if, and only if, $Aux_0(G)$ has no circuit of odd weight.*

Proof. Because $\mathrm{Aux}_0(G)$ is balanced, we have that $\forall z \in V^0_{\mathrm{au}}$,

$$z = \begin{cases} 1, \text{ if } l(z) = \text{" + "}; \\ 0, \text{ if } l(z) = \text{" − "} \end{cases}$$

is a solution of eq. (8.4.7) i.e. eq. (8.4.5). By Lemma 8.4.3, the necessity is obtained.

Conversely, from Lemma 8.4.3, eq. (8.4.5) has a solution and hence eq. (8.4.7). Suppose $\forall z \in V^0_{\mathrm{au}}$, $z = a$ is a solution of eq. (8.4.7). Then, $\forall z \in V^0_{\mathrm{au}}$,

$$l(z) = \begin{cases} \text{" + "}, \text{ if } a = 0; \\ \text{" − "}, \text{ if } a = 1 \end{cases} \tag{8.5.2}$$

has the property in Theorem (8.5.1). That means $\mathrm{Aux}_0(G)$ is balanced. The sufficiency is found. □

Lemma 8.5.2. *For an OD-immersion μ_0 of G, $Aux_0(G)$ has no circuit of odd weight if, and only if, the set $U_1 = \{e | \forall e \in E^0_{\mathrm{au}}, w(e) = 1\}$ is a 1-coboundary (of course, a cocycle) of $Aux_0(G)$.*

Proof. From Lemma 8.5.1, V^0_{au} is partitioned into two parts:

$$V^+_{au} = \{z \mid \forall z \in V^0_{au},\ l(z) = \text{“}+\text{"}\}$$

and

$$V^-_{au} = \{z \mid \forall z \in V^0_{au},\ l(z) = \text{“}-\text{"}\}.$$

Then, $\delta_0 V^+_{au} = \delta_0 V^-_{au} = U_1$. This is the necessity. Conversely, because U_1 is a 1-coboundary, from Lemma 4.1.1 any circuit has even number of edges in common with U_1 and hence is of even weight. This is the sufficiency. □

Theorem 8.5.1. *The following statements are equivalent:*
(1) *G is planar.*
(2) *For an OD-immersion μ_0 of G, $Aux_0(G)$ has no odd-weight circuit.*
(3) *For an OD-immersion μ_0 of G, $Aux_0(G)$ has no fundamental circuit of odd weight.*
(4) *For an OD-immersion μ_0 of G, the set of edges of odd weights forms a 1-coboundary of $Aux_0(G)$.*
(5) *For an OD-immersion, $Aux_0(G)$ is balanced.*

Proof. From Theorems 8.5.1 and 8.5.2 and Lemma 8.4.3, it is known that (1) ⇔ (2). From Lemma 8.4.3 and Theorem 8.4.2, (2) ⇔ (3). From Lemma 8.5.2, (3) ⇔ (4) is obtained. Further, from Lemmas 8.5.1–2, we have that (4) ⇔ (5). The theorem is proved. □

Let $E_1(X, Y)$, $E_2(X)$ be two equation systems on GF(2) and $\mathscr{E}_1(X, Y)$, $\mathscr{E}_2(X)$ be the sets of all the solutions of $E_1(X, Y)$, $E_2(X)$, respectively. And, let

$$\mathscr{E}_1(X, Y)/Y = \bigcup_Y \{X \mid \forall (X, Y) \in \mathscr{E}_1(X, Y)\}. \tag{8.5.3}$$

If $\mathscr{E}_1(X, Y)/Y = \mathscr{E}_2(X)$, then $E_1(X, Y)$ and $E_2(X)$ are said to be *co-consistent*.

Lemma 8.5.3. *Equation systems $E_1(X, Y; y) = \{x_i + y_i = w_i, i = 1, 2, \dots, n, x_j + y = c$ for a fixed $j, 1 \le j \le n\}$ and $E_2(X, Y) = \{x_i + y_i = w_i,\ i = 1, 2, \dots, n\}$ are co-consistent.*

Proof. Obviously, $\mathscr{E}_1(X, Y; y)/y \subseteq \mathscr{E}_2(X, Y)$. Moreover, for any $(X, Y) \in \mathscr{E}_2(X, Y)$, we have $(X, Y; x_j + c) \in \mathscr{E}_1(X, Y; y)$ and hence

$$(X, Y) \in \mathscr{E}_1(X, Y; y)/y.$$

Therefore, $\mathscr{E}_2(X, Y) \subseteq \mathscr{E}_1(X, Y; y)/y$. □

Lemma 8.5.4. *Let* $\mathscr{D}_0 = \{(\alpha,\beta) \mid \forall(\alpha,\beta) \in \mathscr{D}(\alpha,\beta \in Tod^*), C_\alpha \cap C_\beta = P(u,v),\ u \neq v\}$. *Then eq. (8.4.5) for* $(\alpha,\beta) \in \mathscr{D}$ *has a solution if, and only if, eq. (8.4.5) for* $(\alpha,\beta) \in \mathscr{D}_0$ *has a solution.*

Proof. Because for any $(\alpha,\beta) \in \mathscr{D}$, all the equations involving the cotree variable $y = y(T^*;\alpha,\beta)$ associated with $C_\alpha \cap C_\beta = \{v\}$ in eq. (8.4.5) are the same, by Lemma 8.5.3 the equation can be cancelled. The lemma follows. □

Lemma 8.5.5. *Equation systems* $E_1(Y;x) = \{x + y_i = w_i,\ i = 1,2,\ldots,n\}$ *and* $E_2(Y) = \{y_i + y_{i+1} = w_i + w_{i+1},\ i = 1,2,\ldots,n-1\}$ *are co-consistent.*

Proof. Since $E_1(Y;x) = \{y_i + y_{i+1} = w_i + w_{i+1}, i = 1,2,\ldots,n-1, y_1 + x = w_1\}$, from Lemma 8.5.3, the assertion follows. □

For a tree variable $x_{s,t} \in \mathscr{X}_G(T)$, let $y_{\gamma_1},\ldots,y_{\gamma_k}$ be all the cotree variables; each of which appears in an equation with $x_{s,t}$ of eq. (8.4.5) in the order of the occurrences

$$\gamma_1 = \gamma_1(x_{s,t}),\ldots,\gamma_k = \gamma_k(x_{s,t}), k = k(x_{s,t}),$$

when moving from the root of *Tod*, along the tree edges in the direction determined by the OD-procedure in section 2.2 of Liu [217], and at each vertex, along the rotation given, for instance, in the clockwise direction, with the coming-in tree edge as the starting one, to the common end of s and t on G.

Lemma 8.5.6. *For an OD-immersion* μ_0 *of* G, *eq. (8.4.5) has a solution if, and only if,* $\forall x_{s,t} \in \mathscr{X}_G(T)$,

$$y_{\gamma_i} + y_{\gamma_{i+1}} = w_i + w_{i+1}, i = 1,2,\ldots,k(x_{s,t}) \tag{8.5.4}$$

has a solution, where w_i *is the constant term of the equation involving both* $x_{s,t}$ *and* y_{γ_i} *in eq. (8.4.5).*

Proof. By using Lemma 8.5.5 on each tree variable, from the finiteness of tree variables the lemma follows. □

Let $(V,\preceq)$ be the poset defined by an OD-tree on G. Of course, the root of the tree is the minimum. For any cotree edge $\gamma = \langle u,v\rangle$, let $h(\gamma) = v$ and $t(\gamma) = u$, which are said to be the *head* and the *tail* of γ, respectively. Naturally, we always have $h(\gamma) \prec t(\gamma)$.

It is easily seen that for any $u,v \in V, u \wedge v = g.l.b(u,v)$ is well defined. For any $\alpha,\beta \in Tod^*$, let $\langle\alpha,\beta\rangle = \langle t(\alpha),t(\beta)\rangle = t(\alpha) \wedge t(\beta)$ and let

$$V_\gamma = \{v \mid \forall v \in V,\ u_0(\gamma) \preceq v\}, \tag{8.5.5}$$

where $u_0(\gamma) = \min\{u| \forall u \in V, \langle\alpha,\beta\rangle \prec u \preceq t(\gamma)\}$, $\gamma = \alpha$, or β.

For an OD-immersion μ_0 of G given, suppose the rotation at $v \in V$ is $\varrho(v) = (abc\ldots d)$, where a is the tree edge coming to v. Then, define $a \circ\prec b \circ\prec c \circ\prec \ldots \circ\prec d$. Further, we define a partial order denoted by $*\prec$ on the set of all cotree edges as $\forall\alpha,\beta \in Tod^*$,

$$\alpha *\prec \beta \Leftrightarrow \begin{cases} h(\alpha) \prec h(\beta), \text{ if } h(\alpha) \neq h(\beta); \\ \alpha \circ\prec \beta, \text{ if } h(\alpha) = h(\beta). \end{cases} \tag{8.5.6}$$

Let $\Lambda(\alpha,\beta) = \{\gamma| \forall\gamma \in Tod^*, t(\gamma) \in V_\alpha \cup V_\beta,\ h(\gamma) \prec \langle\alpha,\beta\rangle\}$ for $\alpha,\beta \in Tod^*$. Then, two cotree edges α,β (suppose $\alpha *\prec \beta$ without loss of generality) are said to be *successive* if

$$\neg\gamma \in \Lambda(\alpha,\beta),\ \alpha *\prec \gamma *\prec \beta,$$

that is, to say α is covered by β.

For $\alpha,\beta \in Tod^*$ successive, if they are only allowed on the same side, or different sides of Tod in the OD-immersion in order to avoid crossing with others, then y_α and y_β are said to be *adjacent*.

Lemma 8.5.7. *For an OD-immersion of G, two cotree variables y_α and y_β are adjacent if, and only if, α and β are successive and occur in one of the following three typical configurations: Type A, Type B and Type C as*

Type A $\exists\gamma \in Tod^*$,
(i) $h(\gamma) \prec h(\alpha) \prec h(\beta)$;
(ii) $h(\beta) \prec \langle t(\alpha),\Delta\rangle \prec \Delta = \langle t(\gamma), t(\beta)\rangle$.

Type B $\exists\gamma,\delta \in Tod^*$,
(i) $\max(h(\gamma), h(\delta)) \prec h(\alpha) = h(\beta)$;
(ii) $\langle t(\gamma), t(\alpha)\rangle = \Delta_1 \succ \langle\Delta_1,\Delta_2\rangle \prec \Delta_2 = \langle t(\delta), t(\beta)\rangle$.

Type C $\exists\gamma \in Tod^*$,
(i) $h(\gamma) \prec h(\alpha) \preceq h(\beta)$;
(ii) $h(\beta) \prec \langle t(\gamma),\Delta\rangle \prec \Delta = \langle t(\alpha), t(\beta)\rangle$.

Proof. By the properties of an OD-tree and an OD-immersion and by the procedure of finding eq. (8.5.4), from exhaustive discussions the lemma can be found. Details are left to the reader. □

A *left (right)-most immersion* denoted by $\mu_{lm}(\mu_{rm})$ of G is such an OD-immersion that any cotree edge $\gamma = \prec u, v \succ$ is on the left (right)-hand side of the path from v to u on the OD-tree Tod. A left-or right-most immersion is also called a *standard immersion* of G.

Lemma 8.5.8. *Let $\mathscr{A}(G; Tod^*)$ be the set of all the pairs of cotree edges which correspond to adjacent pairs of cotree variables for an OD-tree Tod on G. Then, for the OD-immersion, eq. (8.4.4) has a solution if, and only if, the equation system*

$$y_\alpha + y_\beta = \lambda(\alpha, \beta) \tag{8.5.7}$$

for $(\alpha, \beta) \in \mathscr{A}(G; Tod^*)$ has a solution, where

$$\lambda(\alpha, \beta) = \begin{cases} 1, \text{ if } \alpha, \beta \text{ are in a type A, or type B;} \\ 0, \text{ if } \alpha, \beta \text{ are in a type C.} \end{cases} \tag{8.5.8}$$

Proof. First, we prove that eq. (8.5.4) becomes eq. (8.5.7) when the OD-immersion is replaced by a standard immersion. In fact, it is easy to check by following the procedure of finding eq. (8.5.4). Then, for the consistency of eq. (8.5.4) is independent of the choice of an OD-immersion, the lemma is obtained. □

From Lemma 8.5.8, it is allowed to construct a graph denoted by $\mathrm{Aux}_1(G) = (V^1_{\mathrm{au}}, E^1_{\mathrm{au}})$ and called a *planarity* 1-*auxiliary graph* of G as

$$\begin{cases} V^1_{\mathrm{au}} = V^1_{\mathrm{au}}(G) = \{\gamma \mid \forall \gamma \in Tod^*,\ \exists \alpha \in Tod^*, \\ \qquad\qquad\qquad (\gamma, \alpha) \in \mathscr{A}(G; Tod^*)\}; \\ E^1_{\mathrm{au}} = E^1_{\mathrm{au}}(G) = \mathscr{A}(G; Tod^*). \end{cases} \tag{8.5.9}$$

Let $\mathscr{A}_0(G; Tod^*) = \{(\alpha, \beta) \mid \forall (\alpha, \beta) \in \mathscr{A}(G; Tod^*), \lambda(\alpha, \beta) = 0\}$ and

$$\begin{aligned} \mathscr{A}_1(G; Tod^*) &= \mathscr{A}(G; Tod^*) - \mathscr{A}_0(G; Tod^*) \\ &= \{(\alpha, \beta) \mid \forall (\alpha, \beta) \in \mathscr{A}(G; Tod^*), \\ &\qquad \lambda(\alpha, \beta) = 1\}. \end{aligned}$$

Then, $\forall e \in E^1_{\mathrm{au}}$,

$$\lambda(e) = \begin{cases} 1, & \text{if } e \in \mathscr{A}_1(G; Tod^*); \\ 0, & \text{if } e \in \mathscr{A}_0(G; Tod^*) \end{cases} \tag{8.5.10}$$

is treated as the weight.

Theorem 8.5.2. *The following statements are equivalent:*
(1) *G is planar;*
(2) *$Aux_1(G)$ has no odd-weight circuit;*
(3) *$Aux_1(G)$ has no fundamental circuit of odd weight;*
(4) *$\mathscr{A}_1(G; Tod^*)$ is a 1-coboundary of $Aux_1(G)$;*
(5) *$Aux_1(G)$ is balanced.*

Proof. Similar to the procedure of the proof of Theorem 8.5.1. □

Now, one might pay attention to the main difference that the order of $\text{Aux}_1(G)$ is bounded by a linear function of the order of G while $\text{Aux}_0(G)$ by a quadratic one.

8.6 Notes

8.6.1 The foundation of Theorem 8.2.1 was first established by Wu in Wu [402–406] via chain groups over GF(2), and independently by Tutte in Tutte [346] via chain groups over the real field. However, Theorem 8.2.1 carries out their idea in terms of vector spaces over GF(2), and was called Wu–Tutte theorem in Liu [216, 217].

8.6.2 On characterizing the planarity of a graph, the first paper was published by Kuratowski in 1930 in Kuratowski [166]. A few years later, Whitney [391, 392] and then MacLane [261, 262] also published their planarity characterizations of graphs in different ways. Wu found his theorem, the equivalent form of Theorem 8.2.1, by investigating a more general problem on the realization of complexes in the Euclidean space in Wu [402] of the last 1950s. Moreover, Tutte provided an equivalent form of Theorem 8.2.1 based on a theory of chain groups over the real fields in the beginning of 1970s in Tutte [346]. In fact, Tutte introduced this kind theory of chain groups in the 1950s as Tutte [335]. In Liu [205], Liu showed that the results obtained independently by Wu and Tutte are in fact the same, in view of the space theory over GF(2). Liu also discussed the problem by quadratic Boolean functions in an independent way as Liu [206–208]. These may form different kinds of theoretical foundations on the related topics.

8.6.3 The planarity theory in terms of an auxiliary graph was done in 1978 from Liu [192]. Although the auxiliary graph depends on the choice of a spanning tree, the results such as Theorem 8.4.3 is independent of the tree. This enables us to make the theorem with efficiency. The theta graph in Archdeacon and Siran [14], more than 20 years later, is also in the class of auxiliary graphs. Although theta graphs are independent of a tree, their theorem can never be made efficiency shown in Wei [364].

8.6.4 Since the beginning of 1960s when Auslander and Parter published their paper on embedding graphs on the sphere in Auslander and Parter [16], a large number of papers have appeared on discussing algorithms such as Goldstein [103], Fisher [87], Fisher and Wing [88], Demoucron [69], Weinberg [378], Lin [184], Hotz [142, 143], Dambit [65], Mondshein [271], Hopcroft and Tarjan[136, 140], Deo [70], etc. Notably, in Hopcroft and Tarjan [140], a linear-time algorithm was found. Some corrections of this paper were pointed out in Deo [70]. Most papers in this aspect are based on a principal idea of finding a way to embed circuits of a graph step by step into the plane. In Tutte [339], Tutte provided an algorithm in a completely different way of determining the coordinates of vertices at a time by solving a linear equation system. Wu further developed his theoretical results for designing algorithms by solving equation systems on GF(2) in Wu [403, 405]. Theorems 8.2.2 and 8.3.1 reflect his substantial ideas.

8.6.5 The ideas of transforming the planarity problem into justifying if a graph has an odd-weight circuit, or if a subset of edges is a 1-coboundary, or if a graph is balanced are from Liu [192, 193, 202, 205], then also in Fraysseix and Rosenstiehl [92, 93] and Rosenstiehl [296].

8.6.6 Based on Theorem 5.4.1, an algorithm of $O(n^2)$, n is the order of the graph, can be easily found in Liu [192]. However, the computational results showed that it is much more efficient than Hopcroft and Tarjan's, especially within ten thousands of vertices of graphs, in Sun [323, 324], Xu [407]. In Xu [407], an $O(n \log n)$ algorithm was presented via some improvement on data structure.

8.6.7 Although Theorem 8.5.2 shows that the order of $\mathrm{Aux}_1(G)$ is bounded by a linear function of the order of G, the size of $\mathrm{Aux}_1(G)$ is still by a quadratic function of the order of G. In Chapter 10, theorems shall be explained in order to reach at the linearity of the computing complexity. The original discussions can be seen in Liu [202, 205, 208].

8.6.8 The auxiliary graphs discussed in Sections 8.3–8.5 depend on the choice of an OD-tree. They are in a type of generalization of interlaced graphs in Chapter 7. However, how to characterize the auxiliary graphs of a graph is still open from Liu [206].

8.6.9 Theorem 8.5.2 is in fact independent of the choice of an immersion, because the constants $\lambda(\alpha, \beta)$ are not determined by an immersion anymore. Although the planarity 1-auxiliary graph still depends on the partial order determined by an OD-tree, the planarity of a graph in Theorem 8.5.2 is independent of whatever an OD-tree chosen as well.

8.6.10 The dual space $\mathscr{H}_{[2]}$ of $\overline{\mathscr{H}_{[2]}}$ in Section 8.2 can be imagined from the duality for polyhedra between skeletons and supports shown in Liu [227]. Then, another pairs of theorems are done for exacting a new criterion for the embeddability of a graph on a surface, orientable and non-orientable, of genus given. By the orientable case of genus 0, the three important theorems of, separately, Lefschetz, MacLane and Whitney can be deduced directly as well. However, the construction of $\mathscr{H}_{[2]}$ on a graph has not yet been declared up to now.

9 Embeddability on Surfaces

9.1 Joint tree model

An *embedding* (i.e. *cellular embedding* in early references particularly in topology and geometry) of a graph is such a polyhedron whose skeleton is the graph.

The distinction of embeddings are the same as polyhedra. Precisely speaking, two distinct embeddings on a 2-dimensional manifold are not equivalent topologically in 1-dimensional sense.

According to Chapter 2, all embeddings always imply to be classic.

For a graph $G = (V, E)$, the Heffter-Edmonds model of an embedding of G by rotation system at vertices, in fact, only for orientable case in Heffter [133] and Edmonds [83].

Let $\rho = \{\rho_v | v \in V\}$ be the rotation system on G where σ_v is the cyclic order of semi-edges at $v \in V$. Then, by the following procedure to find an embedding of G:

Procedure 9.1.1. First, put different vertices in different positions marked by a hole circle or a bold point on the plane. Draw lines for edges such that no interior point passes through a vertex and ρ_v is in clockwise when v is a hole circle; in anticlockwise, otherwise.

Then, by travelling along an edge in the rule: passing through on the same side when the two ends of the edges are in same type; crossing to the other side, otherwise. Find all cycles such that each edge occurs just twice. The set of cycles is denoted by P_G.

Lemma 9.1.1. *P_G is a polyhedron.*

Proof. On the basis of Procedure 9.1.1, it is easily checked from the definition of a polyhedron. □

Lemma 9.1.2. *P_G is orientable.*

Proof. On the basis of Theorem 2.2.1, because of the consistency of eq. (2.2.2) on the support of P_G, the lemma is true. □

Theorem 9.1.1. *The dual of P_G is an orientable embedding of G on a surface.*

Proof. Because the support of the dual of P_G is G itself, the theorem is deduced. □

However, P_G in general is not classic except that all vertices are of the same type.

Theorem 9.1.2. *For a given rotation system ρ of a graph G, let $P_G(\rho; 0)$ be the polyhedron obtained by the procedure above for all vertices of same type, then $P_G(\rho; 0)$ is unique.*

DOI 10.1515/9783110479492-009

Proof. From the uniqueness of classic polyhedron in this case, the theorem is done. □

On the basis of Theorem 9.1.2, it suffices only to make all vertices with the same type, e.g. in clockwise. Further, in order to extend to non-orientable case, on account of Lemma 2.5.1, edges in a set not containing cocycle are marked for crossing one side to the other in the Heffter-Edmonds model. The marked edges are called *twist*. This model as well as the Procedure 9.1.1 here is called an *expansion*.

Theorem 9.1.3. *The dual of what is obtained in an expansion is a unique non-orientable embedding of G for twist edges fixed.*

Proof. Because one obtained in an expansion is a classic polyhedron, from the uniqueness of the dual of a polyhedron, the theorem deduced. □

Theorem 9.1.4. *All embeddings of a graph G obtained by expansions for all possible rotation system and twist edges in a subset of the cotree $\bar{T}$ of a given spanning tree T on G are distinct.*

Proof. As a result of Theorem 9.1.3. □

This theorem enables us to choose a spanning tree T on a graph G for discussing all embeddings of G on surfaces.

Let T_1 and T_2 be two spanning trees of a graph G. The sets of all embeddings of G as shown in Theorem 9.1.4 for T_1 and T_2 are, respectively, denoted by $\mathscr{E}_1$ and $\mathscr{E}_2$.

Theorem 9.1.5. *Let $\mathscr{E}_1^g$ and $\mathscr{E}_2^g$ be, respectively, the subsets of $\mathscr{E}_1$ and $\mathscr{E}_2$ on surfaces of genus g (orientable $g = p \geq 0$, or non-orientable $g = q \geq 1$). Then $\mathscr{E}_1^g = \mathscr{E}_2^g$.*

Proof. Because of Theorem 9.1.4, it suffices only to discuss expansions for T_1 and T_2. Since $|\bar{T}_1| = |\bar{T}_2|$, Theorems 9.1.2 and 9.1.3 imply the theorem. □

For an embedding $P \in \mathscr{E}_1^g$, if $P \notin \mathscr{E}_2^g$, then there exits a twist edge e in T_2. By doing a switch with the fundamental cocircuit containing e for T_2, an embedding P' in the same distinct class with P is found. If no twist edge is in $\bar{T}_2$, then P' is the classic embedding in $\mathscr{E}_2^g$ corresponding to P. Otherwise, by the finite recursion, a classic embedding $Q \in \mathscr{E}_2^g$ in the same distinct class with P is finally found. In this way, the 1-to-1 correspondence between $\mathscr{E}_1^g$ and $\mathscr{E}_2^g$ is established.

The last two theorems form the foundation of the joint tree model shown in Liu [195, 196]. Related topics are referred to Stahl [318, 319].

9.2 Associate polyhegons

Given a graph $G = (V, E)$ and a spanning tree T, the edge set E is partitioned into E_T(tree edge) and $\bar{E}_T$(cotree edge), i.e. $E = E_T + \bar{E}_T$. Let $\bar{E}_T = \{i|i = 1, 2, \ldots, \beta\}$, $\beta = \beta(G)$ be the Betti number(or corank) of G. If $i = (u[i], v[i])$, then i_u and i_v, respectively, meant the semi-edges of i incident with $u[i]$ and $v[i]$. Write $G' = (V + V_1, E_T + E_1)$, where $V_1 = \{v_i, \bar{v}_i|1 \le i \le \beta\}$ and $E_1 = \{(u[i], v_i), (v[i], \bar{v}_i)|1 \le i \le \beta\}$. Because G' is a tree itself, G' is called an *expanded tree* of T on G, and denoted by $\hat{T}_G$, or $\hat{T}$ in general case without specification for G from Liu [218, 219].

Let $\lambda = (\lambda_1, \lambda_2, \ldots, \lambda_\beta)$ be a binary vector, or as a binary number of β digits. Denoted by $\hat{T}^\lambda$ that $\hat{T}$, edges $(u[i], v_i)$ and $(v[i], \bar{v}_i)$ are labelled by i with indices: +(always omitted) or −, $1 \le i \le \beta$, where $\lambda_i = 1$ means that the two indices are the same; otherwise, i.e., $\lambda = 0$, different. Then, λ is called an *assignment* of indices on $\hat{T}$.

For $v \in V$, let ρ_v be a rotation at v and $\rho_G = \{\rho_v|\forall v \in V\}$, the rotation of G, then $\hat{T}_\rho$ determine an embedding of $\hat{T}$ on the plane.

Theorem 9.2.1. *For any ρ as a rotation and λ as an assignment of indices, $\hat{T}^\lambda_\rho$ determines a joint tree.*

Proof. By the definition of a joint tree, it is soon seen. □

According to the theory described in Theorem 3.5, the orientability and genus of surface are naturally defined to be those of a joint tree.

Lemma 9.2.1. *Joint tree $\hat{T}^\lambda_\rho$ is orientable if, and only if, $\lambda = 0$.*

Proof. Because $\lambda = 0$ implies each label with its two occurrences of different indices, the lemma is true. □

On a joint tree $\hat{T}^\lambda_\rho$, the surface determined by the boundary of the infinite face on the planar embedding of $\hat{T}_\rho$ with λ on label indices is said to be an *associate surface*.

Lemma 9.2.2. *The genus of a joint tree $\hat{T}^\lambda_\rho$ is that of its associate surface.*

Proof. Only from the definition of orientability of a joint tree. □

Two associate surfaces are the *same* means that they have the same assignment with the same rotation system. Otherwise, *distinct*. Let $\mathcal{F}(\beta)$ be the set of distinct surfaces on $I_\beta = \{1, 2, \ldots, \beta\}$.

For a surface $F \in \mathcal{F}(\beta)$ and a tree T on a graph G, if there exists a joint tree $\hat{T}^\lambda_\rho$ such that F is its associate surface, then F is said to be *admissible*. Let $\mathcal{F}_T(\beta)$ be the set of all distinct associate surfaces.

Given two integers $p, p \geq 0$, and $q, q \geq 1$, let $\mathscr{F}_T(\beta; p)$(or $\mathscr{F}_T(\beta; q), q \geq 1), p \geq 0$, be all distinct admissible surfaces of orientable genus p (or non-orientable genus q).

Theorem 9.2.2. *For any integer $p \geq 0$ (or $q \geq 1$), the cardinality $|\mathscr{F}_T(\beta; p)|$(or $|\mathscr{F}_T(\beta; q)|$) is independent of the choice of tree T on G. Further, it is the number of distinct embeddings of G on a surface of orientable genus p (or non-orientable genus q).*

Proof. According to Theorem 9.1.5, a 1-to-1 correspondence between two sets of embeddings generated by two distinct spanning trees can be found such that the same embeddings are in correspondence. This implies the theorem. □

Because of

$$|\mathscr{F}_T(\beta)| = \sum_{p\geq 0} |\mathscr{F}_T(\beta; p)| + \sum_{q\geq 1} |\mathscr{F}_T(\beta; q)|,$$

the following conclusion is found from the theorem.

Corollary 9.2.1. *The cardinality $|\mathscr{F}_T(\beta)|$ is independent of the choice of tree T on G. Further, it is the number of distinct embeddings of G.*

From Lemma 9.2.1, the non-orientability of an associate surface can be easily justified by only checking if it has a label i with the same index, i.e. $\lambda(i) = 1$.

Theorem 9.2.3. *There is a 1-to-1 correspondence between associate surfaces and embeddings of a graph.*

Proof. First, we can easily see that each embedding determines an associate surface. Then, we show that each associate surface is determined by an embedding. Because of Theorem 9.2.2, this statement is derived. □

From what is mentioned above, it is soon seen that the problem of determining the genus distribution of all embeddings for a graph is transformed into that of finding the number of all distinct admissible associate surfaces in each elementary equivalent class and the problem on minimum and maximum genus of a graph is that among all admissible associate surfaces of the graph. All of them are done on a polyhegon.

9.3 A transformation

Given a surface $S = (A)$, it is divided into segments layer by layer as in the following.

The 0th layer contains only one segment, i.e. $A(= A_0)$.

The 1st layer is obtained by dividing the segment A_0 into l_1 segments, i.e. $S = (A_1, A_2, \ldots, A_{l_1})$, where $A_1, A_2, \ldots, A_{l_1}$ are called the 1*st layer segments*.

Suppose that on $k-1$st layer, the $k-1$st layer segments are $A_{\underline{n}_{(k-1)}}$, where $\underline{n}_{(k-1)}$ is an integral $k-1$-vector satisfied by

$$\underline{1}_{(k-1)} \leq (n_1, n_2, \ldots, n_{k-1}) \leq \underline{N}_{(k-1)},$$

with $\underline{1}_{(k-1)} = (1, 1, \ldots, 1)$,

$$\underline{N}_{(k-1)} = (N_1, N_2, \ldots, N_{k-1}),$$

$N_1 = l_1 = N_{(1)}$, $N_2 = l_{A_{\underline{N}_{(1)}}}$, $N_3 = l_{A_{\underline{N}_{(2)}}}$, …, $N_{k-1} = l_{A_{\underline{N}_{(k-2)}}}$, then the kth layer segments are obtained by dividing each $k-1$st layer segment as

$$A_{\underline{n}_{(k-1)},1}, A_{\underline{n}_{(k-1)},2}, \ldots, A_{\underline{n}_{(k-1)},l_{A_{\underline{n}_{(k-1)}}}}, \tag{9.3.1}$$

where

$$\underline{1}_{(k)} = (\underline{1}_{(k-1)}, 1) \leq (\underline{n}_{(k-1)}, i) \leq \underline{N}_{(k)} = (\underline{N}_{(k-1)}, N_k)$$

and $N_k = l_{A_{\underline{N}_{(k-1)}}}$, $1 \leq i \leq N_k$. Segments in eq. (9.3.1) are called *sons* of $A_{\underline{n}_{(k-1)}}$. Conversely, $A_{\underline{n}_{(k-1)}}$ is the *father* of any one in eq. (9.3.1).

A layer segment that has only one element is called an *end segment*, and others are *principle segments*.

For example, let

$$S = (1, -7, 2, -5, 3, -1, 4, -6, 5, -2, 6, 7, -3, -4).$$

Figure. 9.3.1 shows a layer division of S and Table 9.3.1, the principle segments in each layer.

For a layer division of a surface, if principle segments are dealt with vertices and edges are with the relationship between father and son, then what is obtained is a tree denoted by T. On T, by adding cotree edges as end segments, a graph $G = (V, E)$ is induced. For example, the graph induced from the layer division shown in Figure 9.3.1 is as

$$V = \{A, B, C, D, E, F, G, H, I\} \tag{9.3.2}$$

and

$$E = \{a, b, c, d, e, f, g, h, 1, 2, 3, 4, 5, 6, 7\}, \tag{9.3.3}$$

where

$$a = (A, B), b = (A, C), c = (A, D), d = (B, E),$$
$$e = (C, F), f = (C, G), g = (D, H), h = (D, I),$$

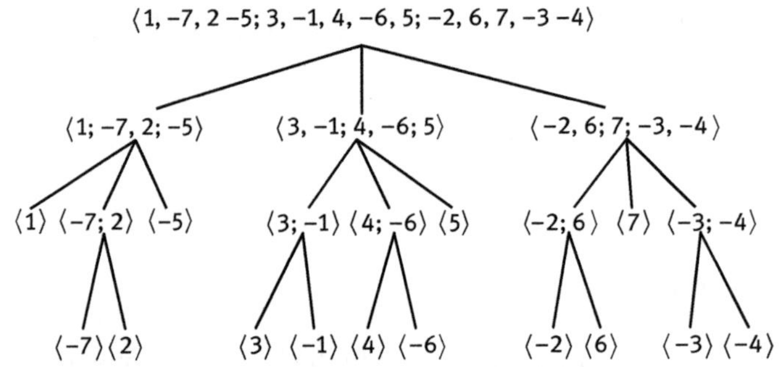

Figure 9.3.1: A layer division of S.

Table 9.1: Layers and principle segments.

Layers	Principle segments
0th layer	$A = \langle 1, -7, 2 - 5; 3, -1, 4, -6, 5; -2, 6, 7, -3 - 4\rangle$
1st layer	$B = \langle 1; -7, 2; -5\rangle, C = \lambda 3, -1; 4, -6; 5\rangle,$ $D = \langle -2, 6; 7; -3, -4\rangle$
2nd layer	$E = \langle -7; 2\rangle, F = \langle 3; -1\rangle, G = \langle 4; -6\rangle,$ $H = \langle -2; 6\rangle, I = \langle -3; -4\rangle$

and

$$1 = (B, F), 2 = (E, H), 3 = (F, I), 4 = (G, I),$$
$$5 = (B, C), 6 = (G, H), 7 = (D, E).$$

By considering $E_T = \{a, b, c, d, e, f, g, h\}$, $\bar{E}_T = \{1, 2, 3, 4, 5, 6, 7\}$, $\delta_i = 0, i = 1, 2, \ldots, 7$, and the rotation ρ implied in the layer division, a joint tree $\hat{T}^{\delta}_{\rho}$ is produced.

Theorem 9.3.1. *A layer division of a surface determines a joint tree. Conversely, a joint tree determines a layer division of its associate surface.*

Proof. From the procedure of constructing a layer division, a joint tree is determined. Conversely, it is natural. □

Then, an operation on a layer division is discussed for transforming an associate surface into another in order to visit all associate surfaces without repetition.

A layer segment with all its successors is called a *branch* in the layer division. The operation of interchanging the positions of two layer segments with the same father in a layer division is called an *exchanger*.

Lemma 9.3.1. *A layer division of an associate surface of a graph under an exchanger is still a layer division of another associate surface. Conversely, the latter under the same exchanger becomes the former.*

Proof. From the correspondence between layer divisions and associate surfaces, the lemma can be obtained. □

On the basis of this lemma, exchanger can be seen as an operation on the set of all associate surfaces of a graph.

Lemma 9.3.2. *The exchanger is closed in the set of all associate surfaces of a graph.*

Proof. From the correspondence between joint trees and layer divisions, the conclusion of the lemma is seen. □

Lemma 9.3.3. *Let $\mathscr{A}(G)$ be the set of all associate surfaces of a graph G, then for any $S_1, S_2 \in \mathscr{A}(G)$, there exists a sequence of exchangers on the set such that S_1 can be transformed into S_2.*

Proof. By considering the joint trees and layer divisions, the lemma is right. □

If $\mathscr{A}(G)$ is dealt as the vertex set and an edge as an exchanger, then what is obtained is called the *associate surface graph* of G, and denoted by $\mathscr{H}(G)$. From Theorem 9.2.3, it is also called the *surface embedding graph* of G.

Theorem 9.3.2. *In $\mathscr{H}(G)$, there is a Hamilton path. Further, for any two vertices, $\mathscr{H}(G)$ has a Hamilton path with the two vertices as ends.*

Proof. By arranging an order, a Hamiltonian path can be extracted based on the procedure of the layer division.

First, starting from a surface in $\mathscr{A}(G)$, by doing exchangers at each principle segment in one layer to another, a Hamilton path can always be found in considering Theorem 9.2.3. This implies the first statement.

Further, for chosen $S_1, S_2 \in \mathscr{A}(G) = V(\mathscr{H}(G))$ adjective, starting from S_1, by doing exchangers avoid S_2 except in the final step, on the basis of the strongly finite recursion principle, a Hamilton path between S_1 and S_2 can be obtained. This implies that $\mathscr{H}(G)$ has a Hamilton circuit and hence the last statement. □

This theorem tells us that the problem of determining the minimum or maximum genus of graph G has an algorithm in time linear on $\mathscr{H}(G)$.

9.4 Criteria of embeddability

Let $G = (V, E)$ be a graph. This section is concerned with the problem of determining if it has an embedding on a surface of genus given beforehand. In general, this problem asks us to find a structural condition of G such that G has an embedding on the given surface. Because of the difficulty for a surface of genus not zero, only one approach is presented for solving the problem in a general constructive manner instead of considering the interior structure of the graph itself based on the theory of joint trees.

Although for the sphere, or equivalently the plane, this problem has been solved in certain ways via the interior structure of a graph, e.g. Chapters 8 and 10 both via planarity auxiliary networks which are determined by the interior structure of the graph considered a criterion in general genus is further provided in what follows.

Let $\mathscr{R}_G$ and Λ_G be, respectively, the sets of all rotation systems and binary numbers on cotree for a spanning three T chosen on G.

Lemma 9.4.1. *An embedding of a graph G is orientable if, and only if, the associate polyhegon for $\rho \in \mathscr{R}_G$ and $\lambda \in \Lambda_G$ has all letters with the two occurrences each of different powers, i.e. $\lambda = 0$.*

Proof. First from Theorem 9.2.3, an embedding of a graph G is orientable if, and only if, the associate polyhegon is orientable. Then from Theorem 3.1.3, an associate polyhegon is orientable if, and only if, each letter is with different powers, i.e. $\lambda_G = 0$. □

Theorem 9.4.1. *A graph $G = (V, E)$ with $\rho \in \mathscr{R}_G$ and $\Lambda_G = 0$ is embeddable on the sphere, or on the plane if, and only if, its associate polyhegon has no pair of letters corresponding cotree edges interlaced.*

Proof. First from Theorem 9.2.3, an embedding of graph G is on the sphere if, and only if, the associate polyhegon is on the sphere. Then from Lemmas 9.4.1 and 3.1.2, the conclusion of the lemma is deduced. □

For a graph embeddable on an orientable surface of $p \neq 0$, the following theorem provides a characterization.

Theorem 9.4.2. *A graph G with $\rho \in \mathscr{R}_G$ and $\Lambda_G = 0$ is embeddable on an orientable surface of genus $p \geq 1$ if, and only if, its associate polyhegon is irreducible for length $4p$.*

Proof. On the basis of Lemma 3.2.2 and the results mentioned above, we have seen that an orientable polyhegon is of genus $p \neq 0$ if, and only if, its irreducible polyhegon is of length $4p$. This is the lemma. □

In what follows, the non-orientability of an embedding of a graph is characterized in terms of a polyhegon.

Lemma 9.4.2. *An embedding of graph G is non-orientable if, and only if, there exists a letter of its two occurrences with the same power,* i.e. $\Lambda_G \neq 0$.

Proof. First from Theorem 9.2.3, an embedding of graph G is non-orientable if, and only if, the associate polyhegon is non-orientable. Then, from Theorem 3.1.3, the conclusion of the lemma is deduced. □

The embeddability of a graph with given rotation system and an assignment on a non-orientable surface is characterized in the following theorem.

Theorem 9.4.3. A graph G with $\rho \in \mathscr{R}_G$ and $\Lambda_G \neq 0$ is embeddable on an non-orientable surface of genus $q \geq 1$ if, and only if, its associate polyhegon is non-orientably irreducible for length $2q$.

Proof. From Lemma 9.4.2 and Lemmas 3.2.3–3.2.4, by considering that the pregenus is the genus in non-orientable case, the theorem is obtained. □

9.5 Notes

9.5.1 In Liu [196], the idea of joint tree was initiated. Particularly, one might see the theoretical basis of Section 9.1.

9.5.2 Theorem 9.1.5 shows that all embeddings with any topological invariant given, not necessary to be genus, can be extracted as well in a manner, independent of the choice of the spanning tree, on the basis of joint trees.

9.5.3 By employing the joint trees and associate polyhegons, a number of new results, on genus distributions, for graphs more general, have been obtained such as in Wan and Liu [355–357], and for some specific graphs in Zhao and Liu [423, 425], Li and Liu [183], Zhu and Liu [426]. More extensions of applications for joint trees can be seen in Liu [227]. For directed graphs and directed embeddings, see Hao and Liu [118, 119] and Hao et al. [121].

9.5.4 The crucial step in employing joint trees is to find the way, for certain classification of associate polyhegons, on the basis of Section 9.2 and Section 9.4. For algorithms and implement on a computer, see Liu [219] and Wang and Liu [363].

9.5.5 Although embeddings of a graph on a surface were investigated in the monographs Liu [216, 217], only constructions of embeddings on surfaces with maximum and minimum genera were mentioned for, respectively, a general graph and a graph with higher symmetry, by which the *cover space* method from topology, or in other words, via *quotient embedding*, can be employed.

9.5.6 Three approaches other than what is mentioned in Section 9.4 for testing the embeddability of a graph on a surface, orientable or non-orientable, of genus arbitrarily given can be seen from Liu [229, 231, 232].

9.5.7 Joint trees are also employed to determine the genus (minimum!) of a graph, not necessary with symmetry, shown in Shao Liu [313, 315] and to estimate a lower bound of the number of genus embeddings as in Shao and Liu [312, 316] and Shao et al. [317].

9.5.8 Notably, for a large number of types of graphs or digraphs, their genus distributions, or genus (handle or crosscap) polynomials have been determined by employing the joint trees as in Chen [51], Hao and Liu [118, 119], Hao et al. [121], Shao and Liu [311, 314], Wan and Liu [355–359], Wan et al. [360], Wan et al. [361], Zeng and Liu [420, 421], Zeng et al. [422], etc.

9.5.9 Some better bounds of average genus of a graph have been derived by employing joint trees, whenever the genus distribution, or genus polynomial, is not yet known, shown as in Chen and Liu [47], Chen et al. [50], etc.

9.5.10 Joint trees are employed to determine the number of embeddings of a graph on a surface of genus given shown as in Yang and Liu [409–414], etc.

9.5.11 Joint trees are applied for proving some equality and inequality involved with genus of polyhedra and graphs, shown in Zeng and Liu [419].

10 Embeddings on Sphere

10.1 Left and right determinations

First, we observe the problems on the planarity and planar embeddings for cubic, or say 3-valent, graphs with an OD-tree on a Hamiltonian circuit. These kinds of graphs are especially important on the Gauss crossing problem as shown in Chapter 7 because the splitting graphs of a crossing sequence have the property.

Lemma 10.1.1. *For a cubic graph G with an OD-tree Tod being a path (in particular on a Hamiltonian circuit), each cotree edge $\gamma \in Tod^*$ is associated with at most two variables: one of which is tree variable; the other, cotree variable in* $\mathrm{Aux}_0(G)$.

Proof. Because of the cubicness, any two fundamental circuits C_α and C_β that have a vertex in common are with at least one common edge. Suppose $h(\alpha) \prec h(\beta)$ without loss of generality on the OD-tree *Tod* which determines a poset $(V; \preceq)$ as described in Section 8.4. Then, the two variables in $C_\alpha \cup C_\beta$, from Lemma 8.4.2, are as follows: the cotree variable denoted by y_β is formed by β and the tree edge going away from $h(\beta)$ and the tree variable denoted by x_α; or x_β, is respectively formed by α and the tree edge going away from $t(\alpha)$ if $t(\alpha) \prec t(\beta)$; or by β and the tree edge going away from $t(\beta)$, otherwise because of *Tod* as a path. Since y_β and x_α are the unique cotree variable and tree variable associated with β and α, respectively, the lemma follows. □

From this lemma, we may see that the order of $\mathrm{Aux}_0(G)$ for G in question is bounded by a linear function of the order of G.

However, the size of $\mathrm{Aux}_0(G)$ is still by a quadratic one. In order to reduce the size of the planarity auxiliary graphs, it can be imagined to do the procedure in two stages: one is for testing the planarity to find a subgraph of $\mathrm{Aux}_0(G)$ as small as possible such that $\mathrm{Aux}_0(G)$ has a circuit of odd weight if, and only if, so does the subgraph; the other is for embedding a planar graph in the plane such that the subgraph has the same number of components as $\mathrm{Aux}_0(G)$ does.

Since *Tod* is a path, the poset $(V, \preceq)$ determines a linear order. Let $\mathrm{Seq}(G)$ be the sequence of variables at corresponding vertices along the linear order.

Moreover, the least and the greatest elements of $\mathrm{Seq}(G)$ are defined to be y and x, respectively, whence (y, x) is the cotree edge on the Hamiltonian circuit of G. A subsequence

$$\mathrm{Sub} = \tilde{y}_\beta y_\alpha y_\gamma x_\beta x_\alpha \tilde{x}_\gamma (\text{or } y_\beta \tilde{y}_\alpha y_\gamma x_\beta \tilde{x}_\alpha x_\gamma)$$

of $\mathrm{Seq}(G)$ for $y \preceq \tilde{y}_\beta \prec y_\alpha$ and $x_\alpha \prec \tilde{x}_\gamma \preceq x$ (or for $y \preceq y_\beta \prec \tilde{y}_\alpha$ and $\tilde{x}_\alpha \prec x_\gamma \preceq x$) is said to be *forbidden* if there are $l, k \geq 1, l = k (\bmod 2)$ such that $\mathrm{Seq}(G)$ has the following two subsequences:

DOI 10.1515/9783110479492-010

$$(\tilde{y}_\beta, x_\beta) = \begin{cases} y_k y_\alpha \left(\prod_{\substack{i=1 \\ y_1 = y_\beta}}^{k-1} y_i x_{i+1} \right) y_\gamma x_\beta, \; k \geq 2; \\ y_\beta y_\alpha y_\gamma x_\beta, \; k = 1, \end{cases} \tag{10.1.1}$$

where $\tilde{y}_\beta = y_k, k \geq 2;\ y_\beta, k = 1$ and

$$(y_\gamma, \tilde{x}_\gamma) = \begin{cases} y_\gamma x_\beta \left(\prod_{\substack{i=1 \\ x'_1 = x_\gamma}}^{l-1} y'_{i+1} x'_i \right) x_\alpha x'_l, \; l \geq 2; \\ y_\gamma x_\beta x_\alpha x_\gamma, \; l = 1 \end{cases} \tag{10.1.2}$$

where $\tilde{x}_\gamma = x'_l, l \geq 2; x_\gamma, l = 1$ (or if there is a $k = 1(\bmod 2),\ k \geq 3$, such that Seq($G$) has the following subsequence:

$$(\tilde{y}_\alpha, \tilde{x}_\alpha) = y_k y_\gamma \left(\prod_{i=1}^{k-1} y_i x_{i+1} \right) x_\beta x_1$$

where $\tilde{y}_\alpha = y_k$ and $\tilde{x}_\alpha = x_1$).

Lemma 10.1.2. *For a cubic graph G with Tod being a path on a Hamiltonian circuit, G is planar if, and only if, Seq(G) has no forbidden subsequence.*

Proof. Suppose Seq(G) has forbidden subsequence

$$\mathrm{Sub} = \tilde{y}_\beta y_\alpha y_\gamma x_\beta x_\alpha \tilde{x}_\gamma.$$

Then, it is easily shown that G is not planar by using Theorem 8.5.1 because we may always find a circuit of odd weight in $\mathrm{Aux}_0(G)$ as explained in Figure 10.1.1 for $k = 1, l = 3$ of the form (10.1.1) and (10.1.2). Here, the bold lines represent the edges of *Tod* in (a) and those with weight 1 in (b); others, the edges of *Tod** in (a) and those of weight 0 in (b). The arrows show the procedure of finding a circuit of odd weight along the linear order of Sub.

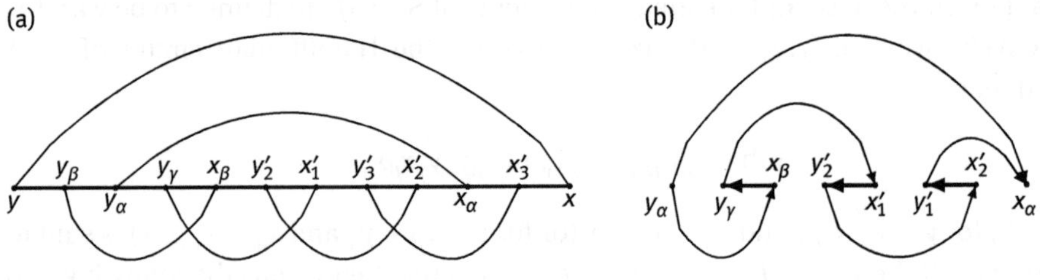

Figure 10.1.1: (a) Forbidden subsequence and (b) in $\mathrm{Aux}_0(G)$.

Conversely, from Lemma 7.2.6, for cubic Hamiltonian graphs, we may see that any non-planar graph G in question has a forbidden subsequence by the exhaustive discussion. □

Whenever G is determined to be planar by means of the procedure as shown in the proof of Lemma 10.1.2, we are allowed to find a spanning forest of $\mathrm{Aux}_0(G)$ which has the same number of components as $\mathrm{Aux}_0(G)$ does along the linear order defined by *Tod* on G. The forest is called a *planarity c-auxiliary graph* denoted by $\mathrm{Aux}_c(G)$ for G being cubic with a Hamiltonian circuit.

In order to make an immersion well defined, we always immerse the Hamiltonian circuit as a Jordan closed curve and all the edges not on the circuit in the same domain (inner, or outer) of the curve. In this case, eq. (5.3.5) is divided into two types:

$$\begin{cases} y_\alpha + x_\alpha = 0, \ \forall \alpha \in Tod^*; \\ y_\alpha + x_\beta = 1, \ \forall \alpha, \beta \in Tod^* \\ \qquad\qquad (y_\alpha \prec x_\beta \prec x_\alpha \prec x_\beta). \end{cases} \tag{10.1.3}$$

From the first type of eq. (10.1.3), any solution has to be with the property that $y_\alpha = x_\alpha$, $\alpha \in Tod^*$. It enables us to write

$$\begin{cases} \mathscr{A}_0 = \{\alpha | \ \forall \alpha \in Tod^*, y_\alpha = x_\alpha = 0\}; \\ \mathscr{A}_1 = \{\alpha | \ \forall \alpha \in Tod^*, y_\alpha = x_\alpha = 1\}. \end{cases} \tag{10.1.4}$$

Attention 10.1.1. The planar representation of a cubic Hamiltonian graph in the following theorem presents a different way from the known in that its determination of cotree edges is derived from a solution of an equation system over GF(2) shown in eq. (10.1.3).

Theorem 10.1.1. *For a cubic planar graph G with Tod being a path on a Hamiltonian circuit, G has such a planar embedding that all edges in $\mathscr{A}_0$ are on the right-hand side and all the edges in $\mathscr{A}_1$ on the left-hand side of Tod or vice versa.*

Proof. From Theorem 8.5.1, any solution of eq. (10.1.3) on $\mathrm{Aux}_c(G)$ can be extended to that on $\mathrm{Aux}_0(G)$. By virtue of the second type of eq. (10.1.3), $\alpha \in \mathscr{A}_0$ and $\beta \in \mathscr{A}_1$ or vice versa if α and β are crossing in the immersion, the theorem is obtained. □

On the other hand, one might like to consider $\mathrm{Aux}_1(G)$ for G which is cubic with an OD-tree *Tod* being a path not necessary on a Hamiltonian circuit. Because *Tod* is a path the partial order $* \prec$ defined by eq. (5.4.6) on Tod^*, or on $V^1_{\mathrm{au}}(G)$ correspondingly, becomes a linear order determined by that $\alpha * \prec \beta \Leftrightarrow h(\alpha) \prec h(\beta)$. Furthermore, it is also allowed to define a linear order denoted by $\prec *$ on Tod^* (or on $\mathscr{X}_G(T)$ correspondingly) as $\forall \alpha, \beta \in Tod^* (x_\alpha, x_\beta \in \mathscr{X}_G(T))$,

$$\alpha \prec * \beta (x_\alpha \prec * x_\beta), \Leftrightarrow t(\alpha) \prec t(\beta). \tag{10.1.5}$$

From Lemma 10.1.1, we have

$$|\mathscr{X}_G(T)| \le \nu(G) \tag{10.1.6}$$

for eq. (5.4.3) there.

For a pair of cotree variables (y_α, y_β), if α covers, or is covered by β under the linear order $* \prec$ or $\preceq$ as in the cubic case, then y_α and y_β are said to be *successive*. Let $\mathscr{A}_{0s}(G; Tod^*)$ be the set of all the adjacent cotree variable pairs that are successive in $\mathscr{A}_0(G; Tod^*)$. Of course, $\mathscr{A}_{0s}(G; Tod^*) \subseteq \mathscr{A}_0(G; Tod^*) \subset E^1_{\text{au}}(G)$ as defined in eq. (8.5.9) and

$$|\mathscr{A}_{0s}(G; Tod^*)| \le \nu(G). \tag{10.1.7}$$

Let $\mathscr{A}_{0n}(G; Tod^*) = \mathscr{A}_0(G; Tod^*) - \mathscr{A}_{0s}(G; Tod^*)$. Since for any

$$(y_\alpha, y_\beta) \in \mathscr{A}_{0n}(G; Tod^*),$$

from eqs. (8.5.8) and (8.5.10), there is $\xi \in Tod^*$ in Type C as shown in Lemma 8.5.7 such that $h(\alpha) \prec h(\xi) \prec h(\beta)$ and $t(\xi) \preceq \langle t(\alpha), t(\gamma)\rangle$. Let $t(\xi_0) = \max t(\xi)$, over all ξ with the property. Define a mapping $\tau : \mathscr{A}_{0n}(G; Tod^*) \to \mathscr{X}_T(G)$ by

$$\tau(y_\alpha, y_\beta) = x_{\xi_0}. \tag{10.1.8}$$

Lemma 10.1.3. *The mapping τ is injective.*

Proof. By contradiction, suppose $(y_\alpha, y_\beta), (y_\gamma, y_\delta) \in \mathscr{A}_{0n}(G; Tod^*)$ with $\tau(y_\alpha, y_\beta) = \tau(y_\gamma, y_\delta) = x_\xi$, $\alpha \prec \beta$, $\gamma \prec \delta$ and $(\alpha, \beta) \neq (\gamma, \delta)$ without loss of generality. Because $(h(\alpha), h(\beta))$ and $(h(\gamma), h(\delta))$ are not allowed to be interlaced by the definition of x_ξ, from the symmetry between (α, β) and (γ, δ), only two possibilities may happen.

Case 1. $h(\alpha) \preceq h(\gamma) \prec h(\xi) \prec h(\delta) \preceq h(\beta)$ $(h(\alpha) \prec h(\gamma) \vee h(\delta) \prec h(\beta))$. Because $(y_\gamma, y_\delta) \in \mathscr{A}_0(G; Tod^*)$, i.e. both γ and δ are in Type C, we have $\max\{t(\delta), t(\gamma)\} \prec \min\{t(\alpha), t(\beta)\}$. However, since $t(\xi) \prec \min\{t(\gamma), t(\delta)\}$, we have $\tau(y_\alpha, y_\beta) \neq x_\xi$. That contradicts to $\tau(y_\alpha, y_\beta) = x_\xi$.

Case 2. $h(\gamma) \prec h(\alpha) \prec h(\xi) \prec h(\beta) \prec h(\delta)$. Because $(y_\alpha, y_\beta) \in \mathscr{A}_0(G; Tod^*)$, we have $t(\xi) \prec \min\{t(\alpha), t(\beta)\}$. This is again a contradiction to $\tau(y_\gamma, y_\delta) = x_\xi$.

Therefore, the lemma holds. □

From (10.1.6) and Lemma 10.1.3, we have

$$|\mathscr{A}_{0n}(G; Tod^*)| \le \nu(G). \tag{10.1.9}$$

Furthermore, from Lemma 8.5.6, for each tree variable, there is at most one pair of cotree variables in Type A or Type B; in this case, we have

$$|\mathscr{A}_1(G; Tod^*)| \leq \nu(G). \tag{10.1.10}$$

Theorem 10.1.2. *For any cubic graph G with an OD-tree Tod being a path (not necessary on a Hamiltonian circuit), we have*

$$|E^1_{\mathrm{au}}(G)| \leq 3\nu(G). \tag{10.1.11}$$

Proof. From eq. (8.5.9) and what discussed earlier, we have

$$E^1_{\mathrm{au}}(G) = \mathscr{A}_{0s}(G; Tod^*) + \mathscr{A}_{0n}(G; Tod^*) + \mathscr{A}_1(G; Tod^*).$$

Then, by eqs. (10.1.7), (10.1.9) and (10.1.10) the theorem is obtained. □

By the similar discussion, we may find that for any graph G not necessary to be cubic with *Tod* being a path, its planarity 1-auxiliary graph $\mathrm{Aux}_1(G)$ satisfies eq. (10.1.11) as well because Lemma 10.1.1 and eqs. (10.1.5)–(10.1.10) can be extended for this case. The only thing as shown in [323] has to be specially mentioned is that ξ_0 in eq. (10.1.8) is taken to be ξ_0' such that $h(\xi_0') = \min\{h(\xi) | h(\alpha) \prec h(\xi) \prec h(\beta), t(\xi) = t(\xi_0)\}$, whereas the linear order $\prec *$ on $\mathscr{X}_G(T)$ is extended for the general case because there are more cotrees going away from a vertex here than only one in the cubic case. However, we may also allow to transform the more general case to the cubic one. Details are given in the next section.

Moreover, by partitioning an OD-tree *Tod* into paths and decomposing the problem on G into those for paths, the result similar to Theorem 10.1.2 for general graph can also be found whenever the decomposition is determined through a number of treatments.

10.2 Forbidden configurations

Let $(V, \preceq)$ be the poset determined by an OD-tree *Tod* on a graph $G = (V, E)$. The purpose of this section and the next is to find a complete set of *forbidden configurations*, some kind of minimal subgraphs, whose occurrence causes G to be non-planar, described by the partial order $\preceq$ for the planarity of G. From the discussion similar to that in Section 7.4, we are allowed to observe 3-connected graphs only for the purpose here because any minimal non-planar subgraph of a graph has to be 3-connected. Such a forbidden configuration for the planarity is usually called a *planarity obstacle* of G as well.

First, we investigate general Hamiltonian graphs in each of which an OD-tree *Tod* on the Hamiltonian circuit is chosen beforehand. Let $G = (V, E)$ be such a graph with *Tod* being a Hamiltonian path. For $v \in V$, we may write

$$\begin{cases} B(v) = \{\alpha | \forall \alpha \in Tod^*(\alpha \neq \alpha_0),\ t(\alpha) = v\}; \\ H(v) = \{\alpha | \forall \alpha \in Tod^*(\alpha \neq \alpha_0),\ h(\alpha) = v\}, \end{cases} \tag{10.2.1}$$

where α_0 is the cotree edge on the Hamiltonian circuit. Because *Tod* is a path on a Hamiltonian circuit, we have

$$\forall v \in V,\ \rho(v) = |B(v)| + |H(v)| + 2. \tag{10.2.2}$$

From that G is without multi-edge, we may write

$$\begin{cases} B(v) = \{\alpha'_i | h(\alpha'_1) \prec h(\alpha'_2) \prec \cdots \prec h(\alpha'_t)\}; \\ H(v) = \{\alpha_i | t(\alpha_1) \prec t(\alpha_2) \prec \cdots \prec t(\alpha_h)\}, \end{cases} \tag{10.2.3}$$

where $t = |B(v)|$ and $h = |H(v)|$.

For any $v \in V,\ \rho(v) > 3$, we substitute for v the $\rho(v) - 2$ vertices which are denoted by $v'_1 \prec v'_2 \prec \cdots \prec v'_t \prec v_1 \prec v_2 \prec \cdots \prec v_h, v'_t$ is in the position of v, on *Tod* such that

$$\begin{cases} t(\alpha'_{t+1-i}) = v'_i, & i = 1, 2, \ldots, t; \\ h(\alpha_j) = v_{h+1-j}, & j = 1, 2, \ldots, h. \end{cases} \tag{10.2.4}$$

Then, the resultant graph denoted by $\tilde{G}$ is a cubic one with $\widetilde{Tod}$ being a path on a Hamiltonian circuit as shown in Figure 10.2.1. We call one step or the whole procedure as described earlier from G to $\tilde{G}$ *lengthening*. The inverse operation of lengthening is called *shortening*.

Lemma 10.2.1. *For a graph $G = (V, E)$ with Tod on a Hamiltonian circuit, G is planar if, and only if, $\tilde{G} = (\tilde{V}, \tilde{E})$ as defined earlier.*

Proof. In fact, if the linear order $* \prec$ on Tod^* is chosen to be the one which is in agreement with $(\tilde{V}, \preceq)$ defined by $\widetilde{Tod}$ on $\tilde{G}$. From *Tod* being a path, only Type A and

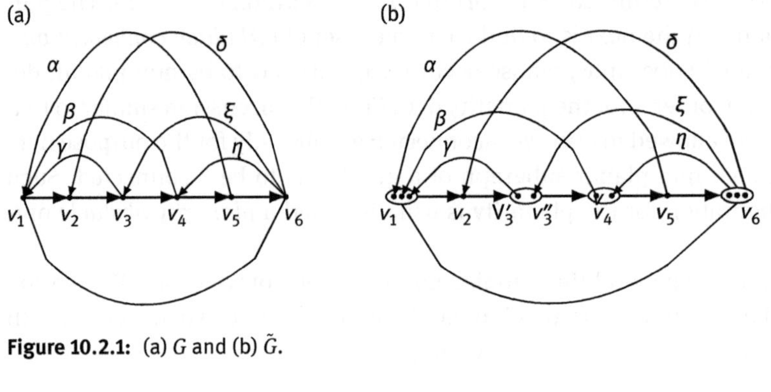

Figure 10.2.1: (a) G and (b) $\tilde{G}$.

Type C are allowed to occur in G. Further, it is easily shown that $\mathrm{Aux}_1(G) \cong \mathrm{Aux}_1(\tilde{G})$ in accordance with the natural correspondence between cotree variables in G and $\tilde{G}$ and hence between $V^1_{\mathrm{au}}(G)$ and $V^1_{\mathrm{au}}(\tilde{G})$. Therefore, from Theorem 8.5.2, the lemma follows. □

Based on this lemma, we are allowed to establish an estimate of the size of $\mathrm{Aux}_1(G)$ for G under the given condition.

Theorem 10.2.1. *For any graph $G = (V, E)$ with Tod on a Hamiltonian circuit, we have*

$$\epsilon(\mathrm{Aux}_1(G)) \leq 12\nu(G) - 36. \tag{10.2.5}$$

Proof. From Theorem 10.1.2, we see that $\epsilon(\mathrm{Aux}_1(\tilde{G})) \leq 3\nu(\tilde{G})$. Because it is known that $\epsilon(G) = |E(G)| \leq 3\nu(G) - 6$ according to the discussion in Section 5.1, there are at most $2(3\nu(G) - 6 - \nu(G) + 1 - 1) - \nu(G) = 3\nu(G) - 12$ additional vertices in $\tilde{G}$. Moreover, according to the proof of Lemma 10.2.1, we have

$$\epsilon(\mathrm{Aux}_1(G)) = \epsilon(\mathrm{Aux}_1(\tilde{G})) \leq 3\,(\nu(G) + 3\nu(G) - 12),$$

and hence eq. (10.2.5). The theorem is obtained. □

In a graph $G = (V, E)$ with *Tod* on a Hamiltonian circuit, two cotree edges $\alpha, \beta \in Tod^*$ are called *parallel* and denoted by $\alpha // \beta$ if $h(\alpha) \preceq h(\beta) \preceq t(\beta) \preceq t(\alpha)$ is satisfied; *alternate* and denoted by $\alpha \bowtie \beta$, if $h(\alpha) \prec h(\beta) \prec t(\alpha) \prec t(\beta)$; *cascaded* and denoted by $\alpha| - |\beta$, otherwise, i.e. they are with the relation:

$$h(\alpha) \prec t(\alpha) \preceq h(\beta) \prec t(\beta).$$

Lemma 10.2.2. *For a graph $G = (V, E)$ with Tod on a Hamiltonian circuit, G is planar if, and only if, all the cotree edges in Tod* except only for the one on the Hamiltonian circuit can be partitioned into two parts: I and O such that in each of them any two elements are never alternate.*

Proof. From Lemma 10.2.1, it is only necessary to discuss $\tilde{G}$ which is cubic with $\widetilde{Tod}$ on a Hamiltonian circuit. By Lemma 7.2.6, $\tilde{G}$ has two Dyck words. However, each Dyck work determines a subset of cotree edges in which each pair of elements is never alternate. Moreover, it is, in view of eq. (10.2.4), easily seen that any pair of cotree edges in G is alternate if, and only if, so is the corresponding pair of cotree edges in $\tilde{G}$. From Lemma 10.2.1, the lemma is derived. □

In a graph $G = (V, E)$ with an OD-tree *Tod* which determines a poset $(V, \preceq)$, for two vertices u and v, they are said to be *shortenable* if the following conditions are satisfied:

(1) u is covered by v or $\forall w \in V(u \prec w \prec v)$, $\rho_G(w) = 2$;

(2) $\max\{t(\alpha)|\forall\alpha \in Tod^*,\ h(\alpha) = v\}$
$\prec \min\{t(\beta)|\forall\beta \in Tod^*,\ h(\beta) = u\}$;

(3) $\max\{h(\alpha)|\forall\alpha \in Tod^*,\ t(\alpha) = v\}$
$\prec \min\{h(\beta)|\forall\beta \in Tod^*,\ t(\beta) = u\}$.

From condition (1), we are allowed to call the path $P(u, v)$ *shortenable* as well. If *Tod* is a path on a Hamiltonian circuit, then G is planar if, and only if, so is $\prod_{e \in P(u,v)} G \cdot e$ for $P(u, v)$ shortenable. If a graph G' is obtained by contracting a series of shortenable paths on *Tod* from G, then G is said to be *shortenable to* G'.

Theorem 10.2.2. *A graph $G = (V, E)$ with Tod on a Hamiltonian circuit is planar if, and only if, G has no subgraph which is shortenable to Conf.A or Conf.B as follows:*

Conf. A: $h(\delta) = h(\dot\beta) \prec h(\alpha) \prec h(\dot\gamma) \prec t(\dot\beta) \prec t(\alpha) \prec t(\dot\gamma) = t(\delta)$ for $\alpha, \beta, \gamma, \delta \in Tod^*$, *where for* $l = k \pmod 2$,

$$\dot\beta = \begin{cases} \beta, \text{ if } l = 1; \\ h(\beta_l)\left(\prod_{i=1}^{l-1} h(\beta_i)t(\beta_{i+1})\right) t(\beta_1), \\ \qquad\qquad \text{if } l > 1 \end{cases} \tag{10.2.6}$$

while $h(\delta) = h(\beta_l) = h(\dot\beta)$, $h(\beta_{l-1}) = h(\alpha)$, $t(\beta_2) = h(\dot\gamma)$, $t(\beta_1) = t(\dot\beta)$, $t(\gamma_i) = h(\gamma_{i-2})$, $i = 3, 4, \ldots, l$, $\beta_i \in Tod^*$, $1 \le i \le l$, and

$$\dot\gamma = \begin{cases} \gamma, \text{ if } k = 1; \\ h(\gamma_1)\left(\prod_{i=2}^{k} h(\gamma_i)t(\gamma_{i-1})\right) t(\gamma_k), \\ \qquad\qquad \text{if } k > 1, \end{cases} \tag{10.2.7}$$

while $h(\gamma_1) = h(\dot\gamma)$, $h(\gamma_2) = t(\dot\beta)$, $t(\gamma_{k-1}) = t(\alpha)$, $t(\gamma_k) = t(\dot\gamma) = t(\delta)$, $t(\gamma_1) = h(\gamma_{i+2})$, $i = 1, 2, \ldots, k-2$, $\gamma_i \in Tod^*$, $1 \le i \le k$.

Conf. B: $h(\delta) = h(\beta) = h(\hat\alpha) \prec h(\gamma) \prec t(\beta) \prec t(\hat\alpha) = t(\gamma) = t(\delta)$ for $\alpha, \beta, \gamma, \delta \in Tod^*$, where for $r = 1 \pmod 2$, $r \ge 3$,

$$\hat\alpha = h(\alpha_1)h(\alpha_2)\left(\prod_{i=1}^{r-2} t(\alpha_i)h(\alpha_{i+2})\right) t(\alpha_{r-1})t(\alpha_r), \tag{10.2.8}$$

while $h(\hat\alpha) = h(\alpha_1) = h(\beta) = h(\delta)$, $h(\alpha_2) = h(\gamma)$, $t(\alpha_{r-1}) = t(\beta)$, $t(\hat\alpha) = t(\alpha_r) = t(\gamma) = t(\delta)$, $t(\alpha_i) = h(\alpha_{i+2})$, $i = 1, 2, \ldots, r-2$, $\alpha_i \in Tod^*$, $1 \le i \le r$.

Proof. From Lemma 10.2.1, we are allowed to observe the cubic graph $\tilde G = (\tilde V, \tilde E)$ with an OD-tree $\widetilde{Tod}$ on a Hamiltonian circuit obtained by lengthening on G.

According to Theorem 10.1.1, $\tilde{G}$ is planar if, and only if, $\tilde{G}$ contains no forbidden sequence

$$\mathrm{Sub}_1 = \tilde{y}_\beta y_\alpha y_\gamma x_\beta x_\alpha \tilde{x}_\gamma \text{ or } \mathrm{Sub}_2 = y_\beta \tilde{y}_\alpha y_\gamma x_\beta \tilde{x}_\alpha x_\gamma.$$

Because $\widetilde{Tod}$ is on a Hamiltonian circuit, we may always choose that δ is the cotree edge on the Hamiltonian circuit.

Moreover, Sub_1 appears in $\tilde{G}$ if, and only if, Conf.A does in G by shortening, the inverse of lengthening; and Sub_2 appears in $\tilde{G}$ if, and only if, Conf.B does in G. Therefore, from Lemma 10.2.1, the theorem is deduced. □

Of course, we may, by employing Lemma 10.2.2, prove the theorem directly in a similar way to the proof of Theorem 10.1.1. Unfortunately, this theorem cannot be extended to graphs with an OD-tree *Tod* being a path not on a Hamiltonian circuit. Characterization of planarity for general graphs by forbidden configurations will be provided in the next section.

Now, we investigate $\mathrm{Aux}_1(G; Tod^*)$ of a graph with *Tod* being a path (not necessary on a Hamiltonian circuit) for providing an estimate of its size directly.

Let $(V, \preceq)$ be the poset determined by *Tod* which is a path on $G = (V, E)$. From Lemma 10.1.1, we are still allowed to distinguish tree variables by cotree edges as $\mathscr{X}_G(T) = \{x_\alpha | \forall \alpha \in Tod^*\}$ here. Because it may happen that for two cotree edges $\alpha, \beta \in Tod^*$, $h(\alpha) = h(\beta)$ or $t(\alpha) = t(\beta)$, the poset $(V, \preceq)$ does not induce a linear order on $\mathscr{Y}_G(T)$ or on $\mathscr{X}_G(T)$, we define that $\forall \alpha, \beta \in Tod^*$,

$$y_\alpha * \prec y_\beta \Leftrightarrow \begin{cases} \alpha \prec \beta, & \text{if } h(\alpha) \prec h(\beta); \\ t(\alpha) \succ t(\beta), & \text{otherwise,} \end{cases} \tag{10.2.9}$$

and

$$x_\alpha \prec * x_\beta \Leftrightarrow \begin{cases} t(\alpha) \prec t(\beta), \text{ if } t(\alpha) \neq t(\beta); \\ h(\alpha) \succ h(\beta), \text{ otherwise.} \end{cases} \tag{10.2.10}$$

It is easily seen that $* \prec$ or $\prec *$ is a linear order on $\mathscr{Y}_G(T)$ or $\mathscr{X}_G(T)$, respectively. It seems that $* \prec$ defined by eq. (10.2.9) is only a special case of that by eq. (8.5.6) if $* \prec$ here is also treated as a linear order on Tod^*. However, there is no difference between them for our purpose here.

If two variables $y_\alpha, y_\beta \in \mathscr{Y}_G(T)$ are with that one covers the other under $* \prec$, then they are called successive as well and denoted by $y_\alpha \mathrm{suc}_{*\prec}\, y_\beta$ or vice versa. Let

$$\begin{aligned} \mathscr{A}_{0s}(G; Tod^*) = \{(y_\alpha, y_\beta) | \forall (y_\alpha, y_\beta) \in \\ \mathscr{A}_0(G; Tod^*), y_\alpha \,\mathrm{suc}_{*\prec}\, y_\beta\} \end{aligned} \tag{10.2.11}$$

and then $\mathscr{A}_{0n}(G; Tod^*) = \mathscr{A}_0(G; Tod^*) - \mathscr{A}_{0s}(G; Tod^*)$. Thus, we have

$$E^1_{au}(G) = \mathscr{A}_{0s}(G; Tod^*) + \mathscr{A}_{0n}(G; Tod^*) + \mathscr{A}_1(G; Tod^*), \tag{10.2.12}$$

the edges set of the planarity 1-auxiliary graph defined by eq. (8.5.9) in this case.

Lemma 10.2.3. *For a graph $G = (V, E)$ with Tod being a path, we have*

$$|\mathscr{A}_{0n}(G; Tod^*)| \le \nu(G). \tag{10.2.13}$$

Proof. First, we may construct a mapping $\tau : \mathscr{A}_{0n}(G; Tod^*) \to \mathscr{X}_G(T)$ such that for any $(y_\alpha, y_\beta) \in \mathscr{A}_{0n}(G; Tod^*)$, $\tau(y_\alpha, y_\beta) = x_\xi$, where

$$x_\xi = \max_{\prec *}\{x_\gamma | \forall \gamma \in Tod^*, y_\alpha * \prec y_\gamma * \prec y_\beta\} \tag{10.2.14}$$

Then, we show that τ is injective. If otherwise, we may suppose that $(y_{\alpha_1}, y_{\beta_1})$, $(y_{\alpha_2}, y_{\beta_2}) \in \mathscr{A}_{0n}(G; Tod^*)$,

$$\tau(y_{\alpha_1}, y_{\beta_1}) = \tau(y_{\alpha_2}, y_{\beta_2}) = x_\xi,$$

without loss of generality $y_{\alpha_1} * \prec y_{\beta_1}$ and $y_{\alpha_2} * \prec y_{\beta_2}$, but $(y_{\alpha_1}, y_{\beta_1}) \neq (y_{\alpha_2}, y_{\beta_2})$. Because of the adjacentness, only two possible cases have to be discussed.

Case 3. $y_{\alpha_2} * \prec y_{\alpha_1} * \prec y_{\beta_1} * \prec y_{\beta_2}$ and at most one of $y_{\alpha_2} = y_{\alpha_1}$ and $y_{\beta_1} = y_{\beta_2}$ is allowed. However, if the case happens, then we have that

$$t(\xi) \prec \min\{t(\alpha_1), t(\beta_1)\} \preceq \min\{t(\alpha_2), t(\beta_2)\}.$$

Suppose α_1, without loss of generality, satisfies that $y_{\alpha_2} * \prec y_{\alpha_1} * \prec y_{\beta_2}$ from the definition of Type C in Section 8.5. Because $t(\alpha_1) \succ t(\xi)$, a contradiction to $\tau(y_{\alpha_2}, y_{\beta_2}) = x_\xi$ occurs.

Case 4. $y_{\alpha_1} * \prec y_{\alpha_2} * \prec y_{\beta_2} * \prec y_{\beta_1}$. From the adjacentness and the definition of Type C in Section 5.4, it is only possible that $t(\xi) \prec \max\{t(\alpha_1), t(\beta_1)\} \prec \min\{t(\alpha_2), t(\beta_2)\}$. However, this leads to a contradiction to $\tau(y_{\alpha_1}, y_{\beta_1}) = x_\xi$.

In consequence, we have that

$$|\mathscr{A}_{0n}(G; Tod^*)| \le |\mathscr{X}_G(T)| \le 3\nu(G) - 6.$$

This is the lemma. □

Theorem 10.2.3. *For a graph $G = (V, E)$ with Tod being a path, we have*

$$|E^1_{au}(G)| \le 9\nu(G) - 19. \tag{10.2.15}$$

Proof. Because for a graph with the given condition, each edge in $\mathscr{A}_1(G; Tod^*) \subseteq E^1_{\mathrm{au}}(G)$ is only allowed to associate with Type A described in Section 5.4 for any $x_\alpha \in \mathscr{X}_G(T)$, at most one adjacent pair of cotree variables appears in $\mathscr{A}_1(G, Tod^*)$ from eq. (5.4.4). That implies

$$|\mathscr{A}_1(G; Tod^*)| \leq |\mathscr{X}_G(T)| \leq 3\nu(G) - 6. \tag{10.2.16}$$

From eqs. (10.2.12), (10.2.15)–(10.2.16) and that it is obvious that

$$|\mathscr{A}_{0s}(G; Tod^*)| \leq |\mathscr{Y}_G(T)| - 1 \leq 3\nu(G) - 7,$$

we have $|E^1_{\mathrm{au}}(G)| \leq 3(3\nu(G) - 6) - 1$.

The theorem is obtained. □

10.3 Basic order characterization

Let $G = (V, E)$ be a general graph, connected of course with an OD-tree *Tod*. The poset $(V, \preceq)$ determined by *Tod* produces an order, which is called the *basic order* of G and denoted by $* \prec$ as shown in eq. (5.4.6) as an example, on Tod^* as the set of all cotree edges. Based on the basic order, the planarity 1-auxiliary graphs denoted by $\mathrm{Aux}_1(G)$ are defined in Section 8.5.

Lemma 10.3.1. *For a graph $G = (V, E)$ with an OD-tree Tod, G is planar if, and only if, all the cotree edges on Tod^* can be partitioned into two parts: $\mathscr{P}_1$ and $\mathscr{P}_2$ such that any pair of cotree edges in the same part $\mathscr{P}_1$, or $\mathscr{P}_2$ does not appear in Type A or Type B defined in Lemma* 8.5.7 *and that any pair of cotree edges in different parts: $\mathscr{P}_1$ and $\mathscr{P}_2$ do not appear in Type C defined in Lemma* 8.5.7 *as well.*

Proof. If G is planar, then from Theorem 8.5.2(5), $\mathrm{Aux}_1(G)$ is balanced. Let $\mathscr{P}_1$ and $\mathscr{P}_2$ be the sets of cotree edges of G which correspond the vertices of the same label "+" and "−" in $\mathrm{Aux}_1(G)$, respectively. By the definition of $\mathrm{Aux}_1(G)$ and from Theorem 8.5.2(4), any two edges in the same set $\mathscr{P}_1$ or $\mathscr{P}_2$ do never appear in Type A or in Type B and those in different sets: $\mathscr{P}_1$ and $\mathscr{P}_2$ do never appear in Type C. This is the necessity.

Conversely, according to the properties that $\mathscr{P}_1$ and $\mathscr{P}_2$ satisfy, all the edges with weight 1 in $\mathrm{Aux}_1(G)$, each of which has its two ends corresponding to two cotree edges in different sets: $\mathscr{P}_1$ and $\mathscr{P}_2$, form a coboundary of $\mathrm{Aux}_1(G)$. From Theorem 8.5.2(4), the sufficiency is derived. □

For two cotree edges $\alpha, \beta \in Tod^*$, let us write $\alpha\ \mathrm{Sam}_0\ \beta$ when α and β appear in Type C and $\alpha\ \mathrm{Dif}_0\ \beta$ when α and β are in Type A or in Type B. Of course, both Dif_0 and Sam_0 are with the symmetry. Then, the lemma suggests us to extend the two binary relations: Dif_0 and Sam_0 into Dif and Sam, respectively, for any pair of cotree edges that have not been defined with one of Dif and Sam yet by using the following rules as far as possible:

Rule 1. $(\alpha \operatorname{Dif} \beta) \wedge (\beta \operatorname{Dif} \gamma) \Rightarrow \alpha \operatorname{Sam} \gamma$;

Rule 2. $(\alpha \operatorname{Sam} \beta) \wedge (\beta \operatorname{Sam} \gamma) \Rightarrow \alpha \operatorname{Sam} \gamma$;

Rule 3. $(\alpha \operatorname{Dif} \beta) \wedge (\beta \operatorname{Sam} \gamma) \Rightarrow \alpha \operatorname{Dif} \gamma$.

It may happen that for two cotree edges $\alpha, \beta \in Tod^*$, $\alpha \operatorname{Dif} \beta$ is defined in one way and $\alpha \operatorname{Sam} \beta$ in another. If it does never happen for any pair of cotree edges in Tod^*, then the relations are said to be *well defined* on Tod^*.

Lemma 10.3.2. *For a graph $G = (V, E)$ with an OD-tree Tod, all the edges in Tod^* can be partitioned into two parts: $\mathscr{P}_1$ and $\mathscr{P}_2$ with the properties described in Lemma* 10.3.1, *if, and only if, the relations* Dif *and* Sam *are well defined on Tod^*.*

Proof. From Lemma 10.3.1 and Theorem 8.5.2, it suffices to prove that $\operatorname{Aux}_1(G)$ has no odd weight circuit if, and only if, so does the enlarged graph $\widetilde{\operatorname{Aux}_1}(G)$ obtained by adding edges on $\operatorname{Aux}_1(G)$ according to Rules 1–3 above such that Dif and Sam correspond to the respective values 1 and 0 of the weight. In fact, the rules do never produce any odd weight circuit if $\operatorname{Aux}_1(G)$ does not have an odd weight circuit and do never reduce the number of odd weight circuits whenever $\operatorname{Aux}_1(G)$ has an odd weight circuit. Therefore, the lemma is soon obtained. □

Theorem 10.3.1. *For a graph $G = (V, E)$ with a basic order $* \prec$ on Tod^* determined by an OD-tree Tod, G is planar if, and only if, there is no one of the configurations denoted by Obs.1–5 on G as follows:*

Ob.1 $\exists \alpha, \beta \in Tod^* (\alpha * \prec \beta)$,

$$(\alpha \operatorname{Dif}_O \beta) \wedge (\alpha \operatorname{Sim}_O \beta);$$

Ob.2 $\exists \alpha, \beta, \gamma \in Tod^* (\alpha * \prec \beta * \prec \gamma)$,

$$(\alpha \operatorname{Dif} \beta) \wedge (\beta \operatorname{Dif} \gamma) \wedge (\alpha \operatorname{Dif} \gamma);$$

Ob.3 $\exists \alpha, \beta, \gamma \in Tod^* (\alpha * \prec \beta * \prec \gamma)$,

$$(\alpha \operatorname{Sam} \beta) \wedge (\beta \operatorname{Sam} \gamma) \wedge (\alpha \operatorname{Dif} \gamma);$$

Ob.4 $\exists \alpha, \beta, \gamma \in Tod^* (\alpha * \prec \beta * \prec \gamma)$,

$$(\alpha \operatorname{Dif} \beta) \wedge (\beta \operatorname{Sam} \gamma) \wedge (\alpha \operatorname{Sam} \gamma);$$

Ob.5 $\exists\alpha,\beta,\gamma \in Tod^*(\alpha * \prec \beta * \prec \gamma)$,

$$(\alpha \operatorname{Sam} \beta) \wedge (\beta \operatorname{Dif} \gamma) \wedge (\alpha \operatorname{Sam} \gamma).$$

Proof. From Lemmas 10.3.1 and 10.3.2, if G is planar, then Dif and Sam defined earlier are well defined. However, if one of Obs.1–5 appears, then Dif and Sam are not well defined because there are two cotree edges, which have to satisfy both Sam and Dif according to the rules, in the configurations described in Obs.1–5. The necessity is proved. Conversely, both Dif and Sam have to be well defined. If otherwise, one of the two possibilities, Case 1 and Case 2, will happen.

Case 1. $\exists\alpha,\beta \in Tod^*$, $(\alpha \operatorname{Dif}_0 \beta) \wedge (\alpha \operatorname{Sam}_0 \beta)$. This is just Ob.1 that contradicts to the given condition.

Case 2. $\exists\alpha,\beta \in Tod^*$, $(\alpha \operatorname{Dif} \beta) \wedge (\alpha \operatorname{Sam} \beta)$. By the rules described earlier, there exists $\gamma \in Tod^*, \gamma * \prec \alpha * \prec \beta$, $\alpha * \prec \gamma * \prec \beta$, or $\alpha * \prec \beta * \prec \gamma$(suppose $\alpha * \prec \beta$ without loss of generality) such that one of Obs.2–5 appears. This is again a contradiction to the condition.

Therefore, from Lemma 10.3.2 and Lemma 10.3.1, the sufficiency is obtained. □

Although it is allowed to design a linear time algorithm in its own right by employing the theorem without considering the planarity auxiliary graphs anyhow, one might still like to know if there is a kind of planarity auxiliary graphs whose size is bounded by a linear function of the order of the original graph G. In what follows, we introduce those as planarity 2-auxiliary graphs denoted by $\mathrm{Aux}_2(G)$ and show that the size of $\mathrm{Aux}_2(G)$ is bounded by a linear function of $\nu(G)$.

Lemma 10.3.3. *Let $s = (u, v)$ and $t = (u, w)$ be two edges with the direction going away from u. And, let us write*

$$\begin{cases} E(s, Tod^*) = \{\alpha | \forall\alpha \in Tod^*, \\ \qquad (h(\alpha) \prec u) \wedge (v \preceq t(\alpha))\}; \\ E(t, Tod^*) = \{\alpha | \forall\alpha \in Tod^*, \\ \qquad (h(\alpha) \prec u) \wedge (w \preceq t(\alpha))\} \end{cases} \tag{10.3.1}$$

for $G = (V, E)$ with an OD-tree Tod which determines the poset $(V, \preceq)$. If $\min\{h(\alpha)| \forall\alpha \in E(t, Tod^*)\} \prec \min\{h(\alpha)|\forall\alpha \in E(s; Tod^*)$, then for planar graph G, we have

$$\begin{cases} \forall\alpha,\beta \in E(s, Tod^*),\ \alpha\, \overline{\operatorname{Dif}}\, \beta; \\ \forall\alpha,\beta \in E(t, Tod^*),\ \alpha \operatorname{Sam} \beta. \end{cases} \tag{10.3.2}$$

Proof. Suppose $E(s; Tod^*) = \{\alpha_i \in Tod^* | \forall i, 1 \le i \le k, h(\alpha_1) \preceq \cdots \preceq h(\alpha_k) \prec u\}$ and $E(t; Tod^*) = \{\beta_j \in Tod^* | \forall j, 1 \le j \le l,\ h(\beta_1) \preceq \cdots \preceq h(\beta_k) \prec u\}$. Then, the given condition means $\alpha_1 \prec \beta_1$ to prove the last statement of eq. (10.3.2). If $l = 1$, i.e. $E(t; Tod^*) = \{\beta_1\}$, then β_1 itself can be seen as β_1 Sam β_1. It holds. By induction on l, the cardinality is $E(t, Tod^*)$. Suppose the lemma holds for $E(t, Tod^*) = \{\beta_2, \beta_3, \ldots, \beta_l\}$.

Case 3. If $\neg\alpha \in E(t, Tod^*)$, $h(\beta_1) \prec h(\alpha) \prec h(\beta_2)$, then from $h(\alpha_1) \prec h(\beta_1)$, β_1 and β_2 are in Type C described in Lemma 8.5.7 and hence β_1 Sam_0 β_2. By the hypothesis, it is true.

Case 4. Otherwise, assume that $h(\beta_1) \prec h(\alpha_{i_1}) \preceq h(\alpha_{i_1+1}) \preceq \ldots \preceq h(\alpha_{i_1+s}) \prec h(\beta_2)$. By Lemmas 8.5.7–8.5.8, we have

$$\beta_1 \text{ Dif } \alpha_{i_1} \text{ Sam } \alpha_{i_1+1} \text{ Sam } \cdots \text{ Sam } \alpha_{i_1+s} \text{ Dif } \beta_2.$$

From Rules 1–3 above for planar graphs, we have β_1 Sam β_2. By the hypothesis, the last statement of eq. (10.3.2) is obtained.

Moreover, if $h(\alpha_1) \preceq h(\alpha_2) \cdots \preceq h(\alpha_i) \preceq h(\beta_1)$ and $h(\beta_1) \prec h(\alpha_{i+1})$, then from Lemmas 8.5.7–8.5.8, we have

$$\alpha_1 \ \overline{\text{Dif}} \ \alpha_2 \ \overline{\text{Dif}} \ \cdots \ \overline{\text{Dif}} \ \alpha_i$$

and from the last statement of eq. (10.3.2), we have

$$\alpha_{i+1} \text{ Sam } \alpha_{i+2} \text{ Sam } \cdots \text{Sam} \alpha_k.$$

Therefore, the first statement of eq. (10.3.2) is obtained. □

According to Lemma 10.3.3, if there are

$$\alpha \in E(s; Tod^*) \text{ and } \beta \in E(t; Tod^*)$$

with the relation α Dif_0 β, or α Sim_0 β, then s and t are defined to have the relation s Dif t, or s Sam t, respectively.

For a graph $G = (V, E)$, let

$$A(v; Tod) = \{e | \forall e \in E_v,\ v \to e\}, \tag{10.3.3}$$

where $v \to e$ represents that e has the direction of going away from v. If G is planar, then from Lemmas 10.3.2 to 10.3.3 for any $v \in V$ the relations: Dif and Sam are well defined on $A(v; Tod)$. Thus, for the planarity, or planar embeddings as well, it is only necessary to discuss graphs with this property which is sometimes called *planarity preferable*.

We may define a linear order on $A(v; Tod)$ arbitrarily at all vertices $v \in V$. For $a, b \in A(v; Tod)$, that a is covered by b under the linear order is denoted by $a| \prec b$.

Let $\hat{A}, \hat{B}$ and $\hat{C}$ be Types A, B and C as defined in Lemma 5.4.7, respectively, when the tree variables involved in them have their corresponding two edges, one covered by the other at the common vertices, under the linear order chosen beforehand. And, let $(\alpha, \beta) \subset \hat{A}, \hat{B}$ or $\hat{C}$ denote that α and β appear in respective $\hat{A}, \hat{B}$ or $\hat{C}$. Then for a graph $G = (V, E)$ with the poset $(V, \preceq)$ defined by an OD-tree *Tod*, we are allowed to construct what is called a *planarity 2-auxiliary graph*, denoted by $\mathrm{Aux}_2(G) = (V^2_{\mathrm{au}}, E^2_{\mathrm{au}})$ as follows:

$$V^2_{\mathrm{au}} = V^1_{\mathrm{au}};$$
$$E^2_{\mathrm{au}} = \{(y_\alpha, y_\beta) \in E^1_{\mathrm{au}} | \forall(\alpha, \beta) \subset \hat{A}, \hat{B}, \text{ or } \hat{C}\}. \tag{10.3.4}$$

Similarly, an edge (y_α, y_β) in E^2_{au} has its weight $w(y_\alpha, y_\beta) = 1$ or 0 according to $(y_\alpha, y_\beta) \subset \hat{A}, \hat{B}$ or $(y_\alpha, y_\beta) \subset \hat{C}$. And let

$$A^2_i(G; Tod^*) = \{e | \forall e \subset E^2_{\mathrm{au}},\ w(e) = i\}$$

for $i = 0, 1$.

Theorem 10.3.2. *For a graph $G = (V, E)$ (planarity preferable of course) with an OD-tree Tod, the statements below are equivalent:*
(1) *G is planar;*
(2) *$\mathrm{Aux}_2(G)$ has no odd weight circuit;*
(3) *$\mathrm{Aux}_2(G)$ has no fundamental circuit of odd weight;*
(4) *$\mathscr{A}^2_1(G; Tod^*)$ forms a 1-coboundary of $\mathrm{Aux}_2(G)$; and*
(5) *$\mathrm{Aux}_2(G)$ is balanced.*

Proof. Similar to the proof of Theorem 8.5.1. □

In general, as described above we see that $\mathrm{Aux}_2(G) \subseteq \mathrm{Aux}_1(G)$. However, if G is cubic, i.e. 3-valent, then it becomes

$$\mathrm{Aux}_2(G) = \mathrm{Aux}_1(G).$$

Lemma 10.3.4. *For a cubic graph $G = (V, E)$ with an OD-tree Tod (of course, not necessarily a path), we have*

$$\epsilon(\mathrm{Aux}_2(G)) \le \frac{3\nu(G)}{2} - 3\nu_3(Tod) + 1, \tag{10.3.5}$$

where $\nu_3(Tod) = |\{v | \forall v \in V, \rho_{Tod}(v) = 3\}|$.

Proof. Let $(V^2_{\mathrm{au}}, * \preceq)$ be the poset deduced from the basic order $* \prec$ as defined by eq. (8.5.6). Because *Tod* is not a path in general, $* \prec$ is not necessary to be a linear order here. Let $\widehat{\mathscr{A}_0}(G; Tod^*)$ and $\widehat{\mathscr{A}_1}(G; Tod^*)$ be those defined by eqs. (8.5.8) and (8.5.10) in which Type A, B or C are respectively replaced by $\hat{A}, \hat{B}$ or $\hat{C}$. We write

$$\widehat{\mathscr{A}_{0s}}(G; Tod^*) = \{(y_\alpha, y_\beta) | \forall (y_\alpha, y_\beta) \in \widehat{\mathscr{A}_0}(G; Tod^*), y_\alpha * | \prec y_\beta\} \tag{10.3.6}$$

and $\widehat{\mathscr{A}_{0n}}(G; Tod^*) = \widehat{\mathscr{A}_0}(G; Tod^*) - \widehat{A_{0s}}(G; Tod^*)$, where $*| \prec$ represents the corresponding cover relation to $* \prec$. Then, we have

$$E^2_{\mathrm{au}} = \widehat{\mathscr{A}_{0s}}(G; Tod^*) + \widehat{\mathscr{A}_{0n}}(G; Tod^*) + \widehat{\mathscr{A}_1}(G; Tod^*). \tag{10.3.7}$$

By the same reason as mentioned earlier, we know that

$$|\widehat{\mathscr{A}_{0s}}(G; Tod^*)| \leq |\mathscr{Y}_G(T)|; \; |\widehat{\mathscr{A}_1}(G; Tod^*| \leq |\mathscr{X}_G(T)|, \tag{10.3.8}$$

where $\mathscr{X}_G(T)$ and $\mathscr{Y}_G(T)$ are the sets of tree and cotree variables, respectively.

Moreover, we may also define a mapping $\tau: \widehat{\mathscr{A}_{0n}}(G; Tod^*) \to \mathscr{X}_G(T)$ by $\tau(y_\alpha, y_\beta) = x_\xi$ such that

$$t(\xi) = \max\{t(\eta) | \forall \eta \in Tod^*, \; h(\alpha) \prec h(\eta) \prec h(\beta)\}. \tag{10.3.9}$$

Because it is easily shown that τ is injective from the fact that $t(\xi) \preceq v_{\alpha,\beta}$ the vertex with which the tree variable determined by Type $\hat{A}$, $\hat{B}$ or $\hat{C}$ involving α and β associates, we have

$$|\widehat{\mathscr{A}_{0n}}(G; Tod^*)| \leq |\mathscr{X}_G(T)| - v_3(Tod). \tag{10.3.10}$$

Since $\epsilon(Tod^*) = \frac{3}{2}v(G) - v(G) + 1 = \frac{v(G)}{2} + 1$, we know that

$$|\mathscr{Y}_G(T)| \leq \frac{v(G)}{2} - 1;$$
$$|\mathscr{X}_G(T)| \leq \frac{v(G)}{2} - v_3(Tod) + 1. \tag{10.3.11}$$

In consequence, in view of eqs. (10.3.6)–(10.3.10), we have

$$\epsilon(E^2_{\mathrm{au}}) \leq |\mathscr{Y}_G(T)| + 2|\mathscr{X}_G(T)| - v_3(Tod)$$

and then from eq. (10.3.11), the lemma follows. □

Let $G = (V, E)$ be a graph planarity preferable. We introduce two kinds of operations to transform G into a cubic graph $\tilde{G}$ such that $\tilde{G}$ is planar if, and only if, so is G.

First, for $\alpha \in Tod^*(h(\alpha) = v, \; u \preceq t(\alpha), \; a = (v, u) \in E_v$ and $\rho(v) \geq 4)$, split v into $v'(= v)$ and $v''(\succeq v')$ such that a and all $\beta \in Tod^*(h(\beta) = v$ and $u \prec t(\beta))$ are incident to v'' and then lengthen v'' such that all $h(\beta)$ become 3-valent. Then, when no cotree edge is coming to a k-valent ($k \geq 4$) vertex, split v into $v_1, v_2, \ldots, v_{s-1}$, $s = |A(v; Tod)| \geq 3$ such that

$$v_i | \prec v_{i+1} \Leftrightarrow \gamma(a_s; v) \preceq \gamma(a_{i+1}; v) \preceq \gamma(a_i; v) \tag{10.3.12}$$

for $i = 1, 2, \ldots, s-2$, where

$$\gamma(a_i; v) = \min\{h(\beta) | \forall \beta \in Tod^*,\ t(\beta) \preceq h(a_i)\} \tag{10.3.13}$$

for $a_i \in A(v; Tod)$, $1 \leq i \leq s$. To do the procedure as far as possible, the resultant graph denoted by $\tilde{G} = (\tilde{V}, \tilde{E})$ is cubic and is called the *lengthened graph* of G. Of course, Tod with all the additional edges (v', v'') and (v_i, v_{i+1}) in $\tilde{G}$ forms an OD-tree $\widetilde{Tod}$ as well. This suggests that a poset $(\mathscr{X}_G(T), \preceq *)$ for a planarity preferable graph G can be defined by $(\tilde{V}, \preceq)$ as

$$\forall x_1, x_2 \in \mathscr{X}_G(T),\ x_1 \preceq * x_2 \Leftrightarrow \tilde{v}(x_1) \preceq \tilde{v}(x_2), \tag{10.3.14}$$

where $\tilde{v}(x), x \in \mathscr{X}_G(T)$, represents the vertex the tree variable x associates with, because there is an injection from tree variables to inner vertices of $\widetilde{Tod}$ from the cubicness of $\tilde{G}$. Based on $(\mathscr{X}_G(T), \preceq *)$, a linear order, denoted by $\prec *$ as well, on $A(v, Tod)$, $v \in V$, can be determined. In the following discussion, the planarity 2-auxiliary graph $\mathrm{Aux}_2(G)$ is always assumed to be related to this linear order at each vertex of G.

Lemma 10.3.5. *A graph $G = (V, E)$ (planarity preferable of course) is planar if, and only if, so is its lengthened graph $\tilde{G}$.*

Proof. Because it is easily checked that

$$\mathrm{Aux}_2(G) \subseteq \mathrm{Aux}_2(\tilde{G}), \tag{10.3.15}$$

the sufficiency is obvious.

Conversely, from Lemma 10.3.1, all the cotree edges in Tod^* can be partitioned into two parts with certain property. Moreover, this leads to the same partition for $\tilde{G}$ by considering Lemmas 10.3.2–10.3.3. By Lemma 10.3.1 again, $\tilde{G}$ is planar. The necessity is then obtained. □

Theorem 10.3.3. *For any planarity preferable graph $G = (V, E)$ with an OD-tree Tod, we have*

$$\epsilon(\mathrm{Aux}_2(G)) \leq 6\nu(G) - 3\nu_1(\widetilde{Tod}) - 11, \tag{10.3.16}$$

where $\nu_1(Tod)$ is the number of articulate vertices of Tod.

Proof. Because each vertex except for the articulate vertices and the 3-valent vertices of $\widetilde{Tod}$ in $\tilde{V}$ is incident with exactly one cotree edge, we have

$$\nu(\tilde{G}) = 2\epsilon(\widetilde{Tod^*}) + \nu_3(\widetilde{Tod}) - \nu_1(\widetilde{Tod}). \tag{10.3.17}$$

In virtue of the relations

$$v_3(\widetilde{Tod}) = v_1(\widetilde{Tod}) - 2, v_1(\widetilde{Tod}) = v_1(Tod)$$

and $\epsilon(\widetilde{Tod^*}) = \epsilon(G) - v(G) + 1$, we find

$$\begin{aligned} v(\tilde{G}) &= 2(\epsilon(G) - v(G) + 1) - 2 \\ &\leq 2(2v(G) - 6). \end{aligned} \tag{10.3.18}$$

The inequality is from the consideration of G with at most $3v(G) - 6$ edges for the planarity. By Lemma 10.3.4, we obtain

$$\begin{aligned} \epsilon(\mathrm{Aux}_2(\tilde{G})) &\leq \tfrac{3}{2}(2(2v(G) - 6)) - 3v_3(\widetilde{Tod}) + 1 \\ &= 3(2v(G) - 6) - 3v_1(Tod) + 7 \\ &= 6v(G) - 3v_1(Tod) - 11. \end{aligned}$$

In consequence, from Lemma 10.3.5 and eq. (10.3.15) the theorem is derived. □

10.4 Number of planar embeddings

Suppose $G = (V, E)$ with an OD-tree *Tod* is known to be planar in the way discussed in the last section. Let $\mathrm{Aux}_2(G) = (V^2_{\mathrm{au}}, E^2_{\mathrm{au}})$ be a planarity 2-auxiliary graph of G. We construct an equation system as

$$y_\alpha + y_\beta = \begin{cases} 0, & \text{if } (y_\alpha, y_\beta) \in \widehat{\mathscr{A}_0}(G; Tod^*); \\ 1, & \text{if } (y_\alpha, y_\beta) \in \widehat{\mathscr{A}_1}(G; Tod^*), \end{cases} \tag{10.4.1}$$

where $\widehat{\mathscr{A}_0}(G; Tod^*)$ and $\widehat{\mathscr{A}_1}(G; Tod^*)$ are as defined in eqs. (10.3.6)–(10.3.7).

From Theorems 10.3.2 and 5.3.3, we know that eq. (10.4.1) has a solution.

Lemma 10.4.1. *Equation* (10.4.1) *has at least two solutions. And, it has exactly two solutions if, and only if,* $\mathrm{Aux}_2(G)$ *is connected.*

Proof. If $Z = \{z_\alpha | \forall\alpha \in Tod^*\}$ is a solution of eq. (10.4.1), then $\bar{Z} = \{\bar{z}_\alpha | \forall\alpha \in Tod\}$ is a solution as well, where $\bar{z}_\alpha = 0$ and $\bar{z}_\alpha = 1$, if $z_\alpha = 1$, if $z_\alpha = 0$. Moreover, $Z \neq \bar{Z}$. That implies the first statement. For the second statement, the necessity is obvious because if $\mathrm{Aux}_2(G)$ is disconnected, then eq. (10.4.1) has at least four solutions. The sufficiency is derived from that if a variable y_α is fixed to be 0 or 1, then the solution of eq. (10.4.1) is well defined along a tree on $\mathrm{Aux}_2(G)$ and hence along $\mathrm{Aux}_2(G)$ in the whole. □

If $g_1(G)$ and $g_2(G)$ are two planar embeddings of a graph $G = (V, E)$, because they are both planar polyhedra, then from the duality, $g_1(G) \cong g_2(G)$ or in other words topologically equivalent if, and only if, there is a bijection $\tau : g_1(E) \to g_2(E)$ such that

$$\forall v \in V, \varrho\,(g_1(v); g_1(G)) = \varrho\,(\tau(g_1(v))\,; g_2(G)) \tag{10.4.2}$$

or

$$\forall v \in V, \varrho\,(g_1(v); g_1(G)) = \varrho^{-1}\,(\tau(g_1(v)); g_1(G))\,, \tag{10.4.3}$$

where $\varrho\,(g_1(v); g_1(G))$ is the rotation of $\{g_1(e)|\forall e \in E_v\}$ at $g_1(v)$ in $g_1(G)$ and ϱ^{-1} is the inverse of ϱ.

For $G = (V, E)$ with an OD-tree *Tod*, we may for convenience imagine the root denoted by o of *Tod* is split into (o_{-1}, o') in the corresponding OD-tree of the resultant graph such that only one cotree edge denoted by α_{-1} is incident with o_{-1}, the new root, and that $t(\alpha_{-1})$ is a maximal vertex on the poset $(V, \preceq)$. Then, all the cotree edges except only for α_{-1} are defined corresponding to variables in $\mathscr{Y}_G(T)$ and are called *active*. Let $\mathscr{A}_{\text{ct}}(G; Tod)$ be the set of all active edges. That is $\mathscr{A}_{\text{ct}}(G; Tod) = \{\alpha|\forall\alpha \in Tod^*,\ \alpha \neq \alpha_{-1}\}$. Further, for a planar embedding $g = g(G)$ of G, let $\mathscr{R}(Tod; g)$ and $\mathscr{L}(Tod; g)$ be the sets of active edges on the *right*-hand side and on the *left*-hand side, respectively, when one moves along *Tod* from the heads to the tails of them. Of course, we have

$$\mathscr{A}_{\text{ct}}(G; Tod) = \mathscr{R}(Tod; g) + \mathscr{L}(Tod; g) \tag{10.4.4}$$

for any planar embedding of G.

Theorem 10.4.1. *If g is a planar embedding of $G = (V, E)$ with an OD-tree Tod, then*

$$y_\alpha = \begin{cases} c, \ \alpha \in \mathscr{R}(Tod; g); \\ \bar{c}, \ \alpha \in \mathscr{L}(Tod; g) \end{cases} \tag{10.4.5}$$

for $\alpha \in \mathscr{A}_{ct}(G; Tod)$ is a solution of eq. (10.4.1). Conversely, if for a graph $G = (V, E)$ with an OD-tree Tod, eq. (10.4.1) has a solution

$$y_\alpha = \begin{cases} c, & \alpha \in \mathscr{B}_0(G; Tod); \\ \bar{c}, & \alpha \in \mathscr{B}_1(G; Tod). \end{cases} \tag{10.4.6}$$

If $\mathscr{A}_{ct}(G; Tod) = \mathscr{B}_0(G; Tod) + \mathscr{B}_1(G; Tod)$, then there is a planar embedding g of G such that

$$\mathscr{B}_0(G; Tod) = \mathscr{R}(Tod; g); \mathscr{B}_1(G; Tod) = \mathscr{L}(Tod; g). \tag{10.4.7}$$

Proof. Because g is an embedding of G, for $\alpha, \beta \in \mathscr{A}_{ct}(G; Tod)$ if $(y_\alpha, y_\beta) \in \widehat{\mathscr{A}_1}(G; Tod^*)$, i.e. α and β are in Type $\hat{A}$, or $\hat{B}$, then they have to be on different sides: $\mathscr{R}(Tod; g)$ and $\mathscr{L}(Tod; g)$, and hence

$$y_\alpha = c \text{ and } y_\beta = \bar{c}, \text{ or } y_\alpha = \bar{c} \text{ and } y_\beta = c$$

both of which satisfy the equation

$$y_\alpha + y_\beta = c + \bar{c} = 1;$$

otherwise, if $(y_\alpha, y_\beta) \in \widehat{\mathscr{A}_0}(G; Tod^*)$, i.e. α and β have to be on the same side: $\mathscr{R}(Tod; g)$ or $\mathscr{L}(Tod; g)$, and hence

$$y_\alpha = y_\beta = c, \text{ or } y_\alpha = y_\beta = \bar{c}$$

both of which satisfy the equation:

$$y_\alpha + y_\beta = c + c = \bar{c} + \bar{c} = 0.$$

Therefore, eq. (10.4.5) provides a solution of eq. (10.4.1).

Conversely, we are allowed to construct a planar embedding g of G satisfying eq. (10.4.7) according to eq. (10.4.6) in the following way. For $v \in V$, let t_v be the tree edge coming to v and $t_i, 1 \leq i \leq l$, $s_i, 1 \leq i \leq k$ be all the edges going away from v such that a cotree edge $\alpha = t_i$; or $h(t_i) \preceq t(\alpha)$, $\alpha \neq t_i, 1 \leq i \leq l$, $h(\alpha) \prec v$, satisfies $\alpha \in \mathscr{B}_0(G; Tod)$ and a cotree edge $\beta = s_i$, or $h(s_i) \preceq t(\beta)$, $\alpha \neq s_i$, $1 \leq i \leq k$, $h(\alpha) \prec v$, satisfies $\beta \in \mathscr{B}_1(G; Tod)$. Then,

$$(t_v t_1 \ldots t_l s_k s_{k-1} \ldots s_1) \subseteq \varrho(v; g) \tag{10.4.8}$$

is determined to be in clockwise if, and only if,

$$\begin{cases} \gamma(t_l) \preceq \gamma(t_{l-1}) \preceq \ldots \preceq \gamma(t_1); \\ \gamma(s_k) \preceq \gamma(s_{k-1}) \preceq \ldots \preceq \gamma(s_1), \end{cases} \tag{10.4.9}$$

where $\gamma(x) = h(x)$, if $x \in Tod^*$; $(h(\alpha_1), h(\alpha_2), \ldots)$, otherwise, where $\alpha_1 * \prec \alpha_2 * \prec \ldots$, for $\{\alpha_1, \alpha_2, \ldots\} = \{\alpha | \forall \alpha \in Tod^*, h(x) \preceq t(\alpha), h(\alpha) \prec v\}$, $x = t_1, t_2, \ldots, s_1, s_2, \ldots$. Further, for any $x = t_i \in Tod, 1 \leq i \leq l$, or $s_j \in Tod, 1 \leq j \leq k$, suppose $\{\xi_1, \xi_2, \ldots, \xi_s\} = \{\xi | \forall \xi \in \mathscr{B}_0(G; Tod), (h(x) \preceq t(\xi)) \wedge (h(\xi) = v)\}$, $t(\xi_1) \succeq t(\xi_2) \succeq \ldots \succeq t(\xi_s)$ and $\{\eta_1, \eta_2, \ldots, \eta_r\} = \{\eta | \forall \eta \in \mathscr{B}_1(G; Tod), (h(x) \preceq t(\eta)) \wedge (h(\eta) = v)\}$, $t(\eta_1) \preceq t(\eta_2) \preceq \ldots \preceq t(\eta_r)$, then

$$\langle \xi_1, \xi_2, \ldots, \xi_s x \eta_1 \eta_2 \ldots \eta_r \rangle \subseteq \varrho(v; g) \tag{10.4.10}$$

is determined as a segment in the rotation. Of course, this can always be done for a planar graph from Theorems 10.3.1–10.3.2. It is easily checked that g is a planar embedding of G by the Jordan curve theorem and that g satisfies eq. (10.4.7). □

Because of Lemma 10.4.1, we are allowed to call a solution and its complement a *co-pair* of the solutions of eq. (10.4.1). Let $\mathscr{S}(G; Tod)$ be the set of all co-pairs of eq. (10.4.1) for a graph G with an OD-tree Tod.

Lemma 10.4.2. *For a graph $G = (V, E)$ with an OD-tree Tod on a Hamiltonian circuit, $|\mathscr{S}(G; Tod)|$ is independent of the choice of Tod and $|\mathscr{S}(G; Tod)|$ is just the number of distinct planar embeddings of G.*

Proof. Because $\mathrm{Aux}_2(G) = \mathrm{Aux}_1(G)$ in this case, it is not necessary to assume the planarity preferability and we are allowed to discuss $\mathrm{Aux}_1(G)$ only. Moreover, from Theorem 10.4.1, each active edge has to be laid in one of the two domains: the inner and the outer domains of the closed curve which correspond to the Hamiltonian circuit. Each of the two states of an active edge corresponds to one of the two values: 0 and 1 of the corresponding variable in a solution of eq. (10.4.1) according to the definition of Type A, B or C. Therefore, from what was mentioned ealier, the number $|\mathscr{S}(G; Tod)|$ of co-pairs of solutions in eq. (10.4.1) is independent of the choice of Tod and is the number of distinct planar embeddings of G. This is the lemma. □

Theorem 10.4.2. *For a graph $G = (V, E)$ with an OD-tree Tod on a Hamiltonian circuit, the number of distinct planar embeddings of G is*

$$n_{pe}(G) = 2^{c(\mathrm{Aux}_1(G))-1}, \tag{10.4.11}$$

where $c(\mathrm{Aux}_1(G))$ is the number of connected components.

Proof. From Lemma 10.4.2, $|\mathscr{S}(G; Tod)|$ is independent of the choice of Tod in this case. From Lemma 10.4.1,

$$|\mathscr{S}(G; Tod)| = 2^{c(\mathrm{Aux}_1(G))-1}.$$

Then, from Lemma 10.4.2, we have $n_{\mathrm{pe}}(G) = |\mathscr{S}(G; Tod)|$. The theorem is obtained. □

If G is not Hamiltonian, we have to consider the occurrences of splittting pairs incident with at least three basic spliting blocks. However, if G is Hamiltonian, it is not necessary because of no multi-edge. If a splitting pair has at least three basic splitting blocks in this case, then it is only allowed to be with one of three, which is an edge not on a Hamiltonian circuit. This edge itself corresponds to an isolated vertex of $\mathrm{Aux}_1(G)$ which is of course a connected component of $\mathrm{Aux}_1(G)$. In general, from the discussion in Section 5.3, we are allowed to observe simply 2-separable planar graphs without loss of generality. Because any simply 2-separable has never a basic splitting block which consists of a single edge, $\alpha_{Bl}(G)$ defined in Section 5.3 becomes the number of all basic splitting blocks.

Lemma 10.4.3. *A graph G with an OD-tree Tod is 3-connected (not 2-separable), if, and only if, $\mathrm{Aux}_1(G)$ is connected.*

Proof. Suppose G is 3-connected and $\mathrm{Aux}_1(G)$ is disconnected. By Lemma 10.4.1, eq. (10.4.1) for $\mathrm{Aux}_1(G)$ has at least two co-pairs of the solutions. However, from Theorem 10.4.1, G has at least two distinct planar embeddings. This is a contradiction to Theorem 5.3.3. The necessity is found. Conversely, if $\mathrm{Aux}_1(G)$ is connected, then from

Lemma 10.4.1, the eq. (10.4.1) for $\mathrm{Aux}_1(G)$ has exactly one co-pair of solutions. Furthermore, by Theorem 10.4.1, only one embedding is determined in the way of eq. (10.4.7). Therefore, from Theorem 5.3.3, G is 3-connected. This is the sufficiency. □

If G is 2-separable, then it may happen that a splitting pair $\{u, v\}$ of vertices is incident with at least three splitting blocks none of which consists of only one edge. However, $\mathrm{Aux}_1(G)$ is independent of the choice of a cyclic order of the splitting blocks at a splitting pair, which is determined by a given planar embedding of G. That means that a *co-pair* of solutions of eq. (10.4.1) for $\mathrm{Aux}_1(G)$ does not uniquely produce a planar embedding of G anymore in general.

Lemma 10.4.4. *For any non-separable planar graph G (simply 2-separable of course) with an OD-tree Tod,* $\mathrm{Aux}_1(G)$ *is disconnected if, and only if, G has at least a splitting pair. All the cotree edges corresponding to the vertices on a connected component of* $\mathrm{Aux}_1(G)$ *are in the same basic splitting block of G and two cotree edges in different components of* $\mathrm{Aux}_1(G)$ *are in different basic splitting blocks of G.*

Proof. The first statement is in fact another version of Lemma 10.4.3 and hence it is true.

Suppose α and β are two cotree edges in different basic splitting blocks A and B of G, respectively. If y_α and y_β are connected in $\mathrm{Aux}_1(G)$, them from the connectedness, there is a path $P(y_\alpha, y_\beta)$ connecting y_α and y_β in $\mathrm{Aux}_1(G)$. We may without loss of generality assume that α in A and β in B are chosen to satisfy that $P(y_\alpha, y_\beta)$ is the shortest path among all possibilities of choosing A and B. Then, $P(y_\alpha, y_\beta)$ has to be an edge (y_α, y_β) in $\mathrm{Aux}_1(G)$. If α and β are in Type C, i.e. $\alpha\ \mathrm{Sam}_0\ \beta$. Because of the simply 2-separability, from Theorem 10.4.1 the planar embedding obtained by reflection of A as defined in Section 5.3 determines a solution of Eq. (10.4.1) for $\mathrm{Aux}_1(G)$ such that the value of y_α is changed and the value of y_β is unchanged. Thus, we have α Dif β which is a contradiction to $\alpha\ \mathrm{Sam}_0\ \beta$. Similarly, we may show that it is not allowed to have $\alpha\ \mathrm{Dif}_0\ \beta$ either. Therefore, y_α and y_β are disconnected in $\mathrm{Aux}_1(G)$. The second statement is true. Further, suppose $\alpha, \beta \in Tod^*$ $(y_\alpha, y_\beta \in V^1_{\mathrm{au}})$ are in the same basic splitting block A. If y_α and y_β are disconnected in $\mathrm{Aux}_1(G)$, then by Theorem 10.4.1, we may find another planar embedding such that the state of β is changed but the state of α is unchanged. From Theorem 5.3.4, A has to contain a splitting pair. This is a contradiction to that A is a basic splitting block. Therefore, y_α and y_β are connected in $\mathrm{Aux}_1(G)$. The second statement is true. □

Theorem 10.4.3. *For any non-separable planar graph G (simply 2-separable of course) with an OD-tree Tod, the number of topologically distinct planar embeddings of G is*

$$n_{\mathrm{pe}}(G) = 2^{c(\mathrm{Aux}_2(G))-1}. \tag{10.4.12}$$

Proof. Because $\mathrm{Aux}_2(G)$ is connected if, and only if, so is $\mathrm{Aux}_1(G)$, from Lemma 10.4.4 and Theorem 5.3.4, the theorem is obtained. □

Furthermore, the theorem can be extended to the case for a general non-separable planar graph G with an OD-tree *Tod*. The number of distinct planar embeddings of G is

$$n_{\mathrm{pe}}(G) = 2^{\hat{c}(\mathrm{Aux}_2(G))-1} \prod_{i\geq 1} ((s_i - 1)!)^{l_i}, \tag{10.4.13}$$

where $\hat{c}(\mathrm{Aux}_2(G))$ is the number of components excluding the isolated vertices that correspond to the cotree edges, each of which has its two ends from a splitting pair of G, and the meanings of s_i, l_i are the same as described in Section 5.3.

10.5 Notes

10.5.1 On the embeddings of planar cubic graphs, the discussion in Section 10.1 is from Liu [202]. However, formulae (10.1.1) and (10.1.2) make the configuration more precise than that appeared in the original form. Moreover, the estimation of the linearity firstly appears here. As seen in this chapter, the methods are extended to several general cases, in particular, graphs with an OD-tree *Tod* being a path not necessary on a Hamiltonian circuit [323, 407].

10.5.2 The basic order characterizations of planarity and embeddings were first studied in Liu [206–208], in which Boolean methods were used. The complete set of forbidden configurations is rather complicated there. However, the results in Section 10.3 are much simpler, and more concrete estimates of the linearity are also provided here.

10.5.3 In Section 10.4, the general result on determining the number of planar embeddings by planarity auxiliary graphs is found. In Liu [207], only the case for Hamiltonian planar graphs was presented.

10.5.4 How to characterize the planarity auxiliary graphs, even for some kind of graphs, in particular for cubic graphs, is still a open problem. Of course, the planarity 1-auxiliary graphs of a cubic graph with the OD-trees being paths are all interval graphs in Golumbic [105].

10.5.5 The five configurations, Ob.1–Ob.5 in Theorem 10.3.1, have been reduced to only one configuration as shown in Liu [238].

11 Orthogonality on Surfaces

11.1 Definitions

Suppose G is determined to be planar and a planar embedding $\mu(G)$ of G is given. Now, we ask if this embedding can be transformed into that with the following properties:

Rec.1. All vertices are on grid points in the Euclidean plane or the affine plane in general.

Rec.2. Each edge is a broken line that consists of a finite number of horizontal or vertical segments in the Euclidean plane, or of straight segments each of which is parallel to one vector or the other both of which form a basis in the affine plane.

Rec.3. The infinite face is fixed.

If it can, then $\mu(G)$ is called *rectilinearly extensible* and an embedding with the properties, Recs.1–3, a *rectilinear extension* of $\mu(G)$. Planar embeddings with the properties Rec.1 and Rec.2 are called *rectilinear embeddings* of $\mu(G)$ or G. If $\mu(G)$ has a face such that it is rectilinearly extensible for the face as the infinite face, then $\mu(G)$ is said to be *rectilinearly embeddable.* If $\mu(G)$ is always rectilinearly extensible for any face as the infinite face, then it is said to be *rectilinearly realizable*. Further, if a graph G has a planar embedding which is rectilinearly extensible, then G is said to be *rectilinearly embeddable* and if all planar embeddings of G are rectilinearly realizable, then G is said to be *rectilinearly realizable* as well. Of course, by no means any planar embedding of a graph is rectilinearly extensible. In fact, no planar embedding of a graph with a vertex of valency greater than 4 is rectilinearly extensible because only four possible ways denoted by N, W, S, E are allowed to connect an edge at a vertex.

In a rectilinear embedding, an edge may have several inner points each of which meets a horizontal and a vertical segment. Those points are called the *bends* of the edge. If a rectilinear embedding has the number of bends of each edges less than or equal to k, then it is called a *k-embedding*. Similarly, if a planar embedding of a graph can be represented by a k-embedding in correspondence with the infinite face, then it is said to be *k-extensible.* If a planar embedding is always k-extensible for any face as the infinite face, it is said to be *k-realizable.* Furthermore, if a graph has a k-embedding, then it is called *k-embeddable* and likewise, *k-realizable.*

Lemma 11.1.1. *A planar embedding of a cubic graph $G = (V, E)$ with an OD-tree Tod on a Hamiltonian circuit is 3-extensible.*

Proof. Because G is cubic, we have $\nu = |V| = 0 (\mathrm{mod}\ 2)$. Since each vertex is incident to exactly one cotree edge in Tod^* except for the two ends of Tod, each of which is incident to two cotree edges, G has $\tilde{\epsilon} = \frac{\nu}{2} + 1$ cotree edges. Suppose there are ϵ_1 cotree edges on the right-hand side and $\epsilon_2 = \tilde{\epsilon} - \epsilon_1$ cotree edges on the left-hand side of Tod

DOI 10.1515/9783110479492-011

on the planar embedding of G. From Theorem 7.1.1, cotree edges on each side can be classified by the equivalence denoted by $\sim_p$:

$$\begin{cases} \forall\alpha,\beta\in\mathscr{A}_0(\text{or}\mathscr{A}_1), \\ \alpha\sim_p\beta\Leftrightarrow(h(\alpha)\prec h(\beta)\prec t(\beta)\prec t(\alpha)) \\ \qquad\vee(h(\beta)\prec h(\alpha)\prec t(\alpha)\prec t(\beta)). \end{cases} \tag{11.1.1}$$

Let $E_1(R),\ldots,E_r(R)$ and $E_1(L),\ldots,E_l(L)$ be the classes on the right- and the left-hand sides of *Tod*, respectively. And, let

$$\begin{aligned} k(R) &= \max_{1\leq i\leq r}|E_i(R)|; \\ k(L) &= \max_{1\leq j\leq l}|E_j(L)|. \end{aligned} \tag{11.1.2}$$

The planar embedding of G can be represented in the following way.

First, the OD-tree *Tod* is arranged as a segment of $v-1$ units, each of which as an edge on a horizontal line with the direction from smaller to greater. The cotree edge incident with the two ends of *Tod* is taken to be a broken line with two bends being connected by a horizontal segment of $k(R)+1$ (or $k(L)+1$) units down (or up) to that of *Tod* when both the two ends of *Tod* are incident to edges in $\sum_{j=1}^{l}E_j(L)$ (or $\sum_{i=1}^{r}E_i(R)$), or is taken to be that with three bends: one is at the point one unit left (or right) of the root (or the other end) on the same line as that of *Tod*; the other two are connected by a horizontal segment of $k(L)+1$ (or $k(R)+1$) units up (or down) to that of *Tod* when the root (or the other end) is incident to an edge in

$$\sum_{1\leq j\leq l}E_j(L)\left(\text{or}\sum_{1\leq i\leq r}E_i(R)\right)$$

and the other end to an edge in

$$\sum_{1\leq i\leq r}E_i(R)\left(\text{or}\sum_{1\leq j\leq l}E_j(L)\right).$$

Then, embed each class of

$$E_i(R),\ 1\leq i\leq r,\ \text{and}\ E_j(L),\ 1\leq j\leq l$$

rectilinearly according to the rotation determined by the planar embedding. Assume $\tilde{E}=\{\alpha_1,\ldots,\alpha_s\}$ be a class where

$$h(\alpha_s)\prec h(\alpha_{s-1})\prec\cdots\prec h(\alpha_1).$$

All the cotree edges are represented by broken lines each of which has two bends that are connected by a horizontal segment i units away from that of *Tod* for $1 \le i \le s$. It is easily checked what obtained above is a 3-embedding. Therefore, the lemma is found. □

Lemma 11.1.2. *A planar embedding of a cubic graph $G = (V, E)$ is t-extensible, for an integer t, $0 \le t \le \lfloor \nu(G)/4 \rfloor + 2$.*

Proof. According to the rotation determined by the planar embedding of G at each vertex, we are allowed to embed rectilinearly an OD-tree *Tod* on the plane by the procedure of the two steps: **Step 1.** The longest path on *Tod* is arranged on a horizontal line; **Step 2.** At each 3-valent vertex of *Tod* on a path P which has been arranged, the longest path P_1 from the vertex passing through the edge of *Tod* which has not been arranged yet is arranged on the straight line perpendicular to P with refinement of the grid if necessary.

For a cotree edge $\alpha \in Tod^*$, let P_α be the path from $h(\alpha)$ to $t(\alpha)$ on *Tod*. If P_α has $n_b(\alpha)$ bends on the rectilinear embedding of *Tod* obtained above. From the cubicness and the properties of *Tod*, we may see that α can be rectilinearly embedded as a broken line with at most $n_b(\alpha) + 2$ bends. The worst case is as shown in Fig. 11.1.1 in which the heavy line denotes *Tod* and the thin line, the cotree edge.

Moreover, it can be shown that

$$n_{\mathrm{br}}(G) = \max_{\alpha \in Tod^*} n_{\mathrm{b}}(\alpha) \tag{11.1.3}$$

is such an integer n that

$$\sum_{i=1}^{n-1} (3+i) < \nu(G) - 3 \le \sum_{i=1}^{n} (3+i). \tag{11.1.4}$$

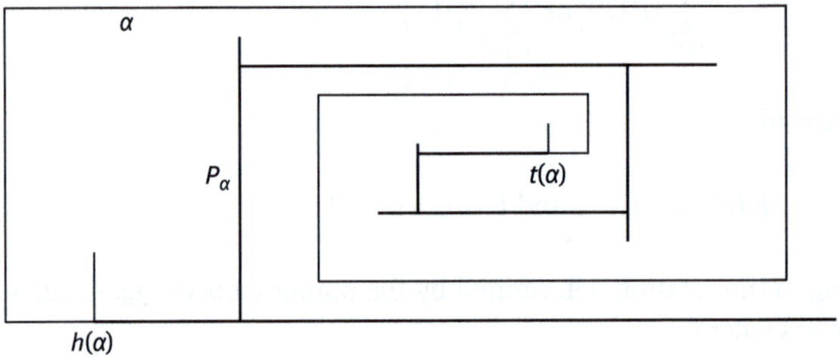

Figure 11.1.1: A rectilinear embedding of an OD-tree.

Of course, we have

$$n_{br}(G) \leq \lfloor \nu(G)/4 \rfloor. \qquad (11.1.5)$$

(The proofs of eqs. (11.1.4) and (11.1.5) are left to the reader!)

Therefore, in the similar manner of the proof of Lemma 11.1.1, the lemma can be proved. □

Theorem 11.1.1. *A planar graph $G = (V, E)$ is k-realizable ($k \leq \lfloor \nu(G)/4 \rfloor + 4$) if, and only if, the maximum of valencies of vertices in V is not greater than four.*

Proof. The necessity is obvious from the discussion above. Because any planar embedding of a general graph G can be transformed from that of a cubic graph, its lengthened graph $\tilde{G}$ according to Lemma 10.3.5 by the operations as shown in Figure 11.1.2 and 11.1.3. Whenever noticing that the operations may cause two more bends increased for an edge, from Lemma 11.1.2 and the arbitrariness of choices of a planar embedding the theorem is obtained. □

In a rectilinear embedding, if any horizontal or vertical line never has two or more segments at least one of which is finite in common with the inner domain of a face then the face is said to be *convex* (rectilinearly).

Further, if all the faces are convex, then it is called a *rectilinearly convex embedding*, and similarly, *convex k-embedding* if it is a k-embedding.

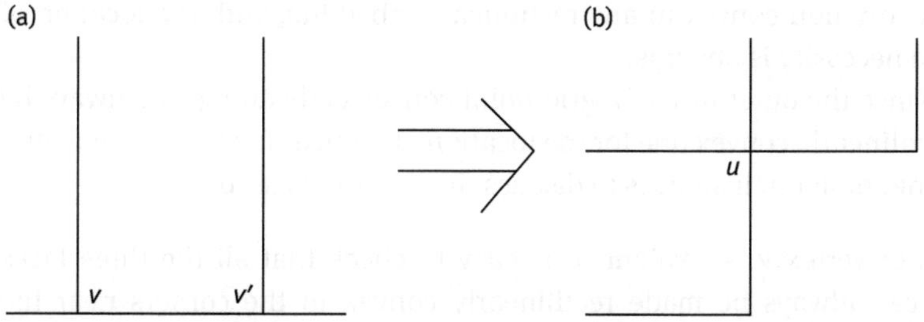

Figure 11.1.2: An operation on a rectilinear embedding.

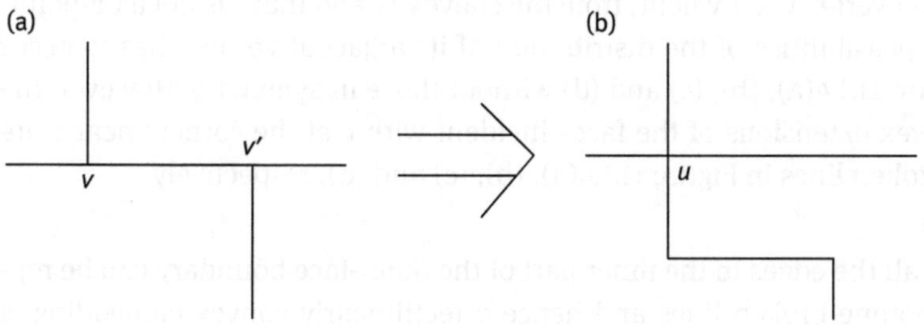

Figure 11.1.3: An operation on a rectilinear embidding.

A planar embedding of a graph which can be represented by a rectilinearly convex embedding with the infinite face fixed is said to be *rectilinearly convex extensible*.

Moreover, if a planar embedding is always rectilinearly convex extensible for any face as the infinite face, then it is said to be *rectilinearly convex realizable*. Similarly, we may see what the meaning of *convex k-extensible* or *convex k-realizable* is for a planar embedding.

Attention 11.1.1. By observing that if a 4-valent vertex u has its three adjacent vertices on the same quadrant when u is dealt as the origin in a rectilinear embedding, then this embedding is not convex. The existence of such a vertex should be considered for the investigation of rectilinear convexity.

In a grid point embedding of a graph $G = (V, E)$ with the valencies of vertices not greater than four, if a 4-valent inner vertex has its three adjacent vertices in the same one of the four quadrants into which the plane is divided by the horizontal and the vertical lines passing through the vertex, then it is called a *CF-point*.

Lemma 11.1.3. *A grid point convex embedding of a graph with valencies of vertices not greater than four is rectilinearly convex extensible with the location of vertices fixed if, and only if, it has no CF-point.*

Proof. Because it is easy to show that a CF-point causes at least one of its incident faces rectilinearly non-convex in any rectilinear embedding with the location of vertices fixed, the necessity is obvious.

Conversely, since the outer face of a grid point convex embedding can always be extended to a rectilinearly convex one for the location of vertices fixed no matter what happen in the inner domain, it suffices to discuss the inner vertices only.

Case 1. If an inner vertex v is 3-valent, it is easy to check that all the three faces incident with v can always be made rectilinearly convex in the corners near the vertex v.

Case 2. If an inner vertex v is 4-valent, from the convexity and that v is not a CF-point, one of only four possibilities of the distribution of its adjacent vertices has to occur as shown in Figure 11.1.4(a), (b), (c) and (d) without those in symmetry. However, the rectilinearly convex extensions of the faces incident with v at the corners near v are also shown by broken lines in Figure 11.1.4(a), (b), (c) and (d), respectively.

In consequence, all the edges in the inner part of the outer-face boundary can be represented by monotone broken lines and hence a rectilinearly convex embedding is obtained. □

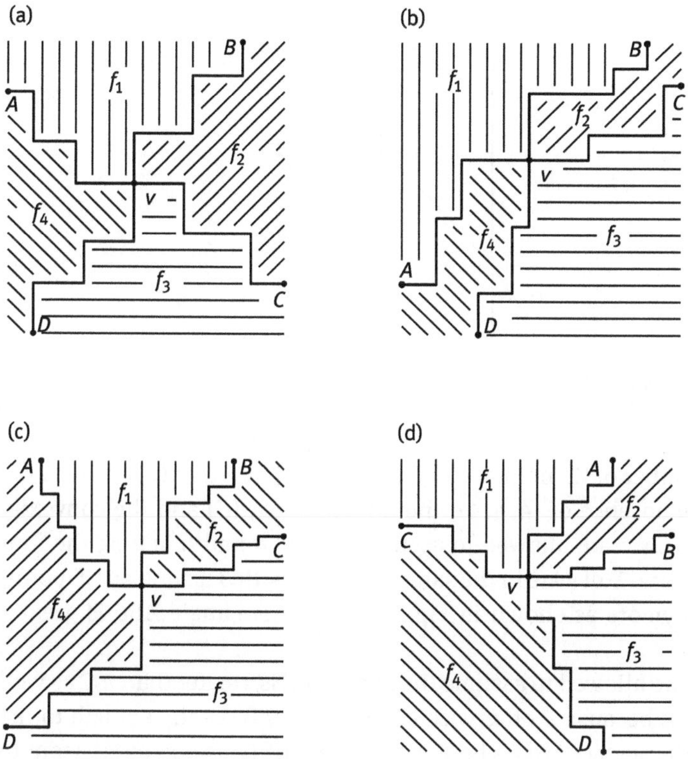

Figure 11.1.4: Four possibilities.

By no means any grid point convex embedding $\mu(G)$ of a graph can always be transformed into another without CF-point and hence rectilinearly convex extensible. If a circuit C in $\mu(G)$ has all its vertices 4-valent such that each of exactly three vertices on C has one edge in the outer and one in the inner domain, respectively, and that each of the others has two edges in the inner domain, then it is called a *4-valent 3-separating circuit* denoted by $\langle 4, 4, 4\rangle$. It is easily checked that any grid point convex embedding $\mu(G)$ with a configuration $\langle 4, 4, 4\rangle$ can never be extended into a rectilinearly convex one because there always has a CF-point on $\langle 4, 4, 4\rangle$ in $\mu(G)$.

Lemma 11.1.4. *If a planar graph G of vertex valencies not greater than four is without a convex embedding with* $\langle 4, 4, 4\rangle$, *then any convex embedding of G can be extended into such a grid point convex embedding that no vertex is a CF-point.*

Proof. First, from the discussion in Section 5.5, since G has a convex embedding, it has a grid point convex embedding $\mu(G)$. If $\mu(G)$ has no CF-point, then it is just what we need.

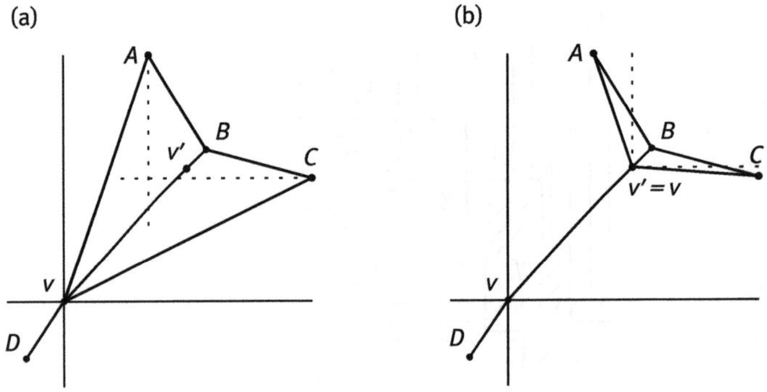

Figure 11.1.5: Two parts of an embedding.

By induction on the number of CF-points, suppose $\mu(G)$ has a CF-point. Let v be a CF-point of $\mu(G)$. From the absence of $\langle 4,4,4\rangle$, the finiteness of $\mu(G)$ and the convexity of the outer face in $\mu(G)$, we are allowed to assume that A, B and C among the four adjacent vertices A, B, C and D of v are in the same quadrant with B having no CF-point, if B is still 4-valent, anymore as shown in Fig. 11.1.5(a) where angle ABC is greater than $\pi/2$.

If B is 3-valent, we are allowed to assume the angle ABC is greater than $\pi/2$ as well as shown in Fig. 11.1.5(b) because B otherwise is replaced by B' nearly enough to the mid-point M of $\tilde{A}\tilde{C}$ on the segment BM where $\tilde{A}$ and $\tilde{C}$ are the two vertices adjacent to B on the boundary of the face incident with the angle ABC.

However, another grid point convex embedding $\mu'(G)$ can always be obtained by adding a new grid point v' near B enough in the common point set of the inner domains of the quadrangles $AvCB$ and $BA'B'C'$, where B', A' and C' are the common points of the horizontal line on C and the vertical line on A, the horizontal line on B and the vertical line on A, and the horizontal line on C and the vertical line on B, respectively, and then joining the straight segments $v'A, v'B, v'C$ and $v'D$ in $\mu(G-v)$.

Of course, v' is not a CF-point of $\mu'(G)$. Because (A, C) is not allowed to be an edge when both A and C are 4-valent from $\mu(G)$ without $\langle 4,4,4\rangle$ it can be made that none of them becomes a new CF-point in any case. By hypothesis, G has a grid point convex embedding without CF-point. The theorem is proved. □

Theorem 11.1.2. *Any 3-connected planar graph G of vertex valencies less than five without $\langle 4,4,4\rangle$ is rectilinearly convex realizable.*

Proof. From Theorem 5.5.2, G has a grid point convex embedding independent of the choice of the outer face. From Lemmas 11.1.4 and 11.1.3, the embedding is rectilinearly convex realizable. Therefore, from the arbitrariness of the choice of the embedding, G is rectilinearly convex realizable. This is the theorem. □

Although a graph with separating pair of vertices is generally not convex embeddable, it is in many cases allowed to have a rectilinearly convex embedding. More general results can be seen in Liu [29, 212].

11.2 On surfaces of genus zero

From the last section, we have known that any planar graph with the valencies of vertices not greater than four is k-embeddable for an integer k big enough. The purpose of this section is to find the minimum of k such that all kinds of graphs considered are k-embeddable. The result is that the minimum is 3. Because an edge which is incident with an articulate vertex can always be rectilinearly embedded in the plane without a bend and the resultant graph of subdividing some edges of a k-embeddable graph is also k-embeddable, we are allowed only to discuss graphs without 1-valent or 2-valent vertex here. We call a planar graph with 3-valent and 4-valent vertices a *standard graph*. For a vertex v on a circuit C in a k-embedding $\mu_k(G)$ of $G = (V, E)$, if v is at an angle of $\frac{3}{2}\pi$, π or $\frac{\pi}{2}$ in the inner domain of C, then it is called an *inner*, *straight* or *outer vertex* of C. Of course, any vertex on C is always on one of the three *states:* inner, straight or outer. Let

$$\mathrm{St}(v; C_{\mathrm{in}}) = \begin{cases} 2, & \text{if } v \text{ is an inner vertex of } C; \\ 1, & \text{if } v \text{ is a straight vertex of } C; \\ 0, & \text{if } v \text{ is an outer vertex of } C, \end{cases} \tag{11.2.1}$$

then it is called the *state function* of v on C. Further, let

$$\bar{\mathrm{St}}(v; C_{\mathrm{in}}) = 2 - \mathrm{St}(v; C_{\mathrm{in}}), \tag{11.2.2}$$

which is said to be the *complement* of $\mathrm{St}(v, C)$, then we may see

$$\mathrm{St}(v; C_{\mathrm{out}}) = \bar{\mathrm{St}}(v; C_{\mathrm{in}}), \tag{11.2.3}$$

where $\mathrm{St}(v; C_{\mathrm{out}})$ is the state function of v on C when considering the outer domain of C instead of the inner domain. Moreover, if C is the boundary of a face f, then $\mathrm{St}(v; C_{\mathrm{in}})$ and $\mathrm{St}(v; C_{\mathrm{out}})$ are denoted by $\mathrm{St}(v; f) = \mathrm{St}(v; f_{\mathrm{in}})$ and $\mathrm{St}(v; f_{\mathrm{out}})$, respectively. In fact, $\mathrm{St}(v; C_{\mathrm{in}})$ (or $\mathrm{St}(v; C_{\mathrm{out}})$) for a circuit C indicates the number of ways to connect an edge at v in the inner (or outer) domain of C.

For an edge e incident to a vertex v, if e is from the North N, the West W, the South S or the East E to v, then it is denoted by $e(v) \to N$, $e(v) \to W$, $e(v) \to S$ or $e(v) \to E$, respectively. Let us write

$$\mathrm{Dir}(e;v) = \begin{cases} N, & \text{when } e(v) \to N; \\ W, & \text{when } e(v) \to W; \\ S, & \text{when } e(v) \to S; \\ E, & \text{when } e(v) \to E, \end{cases} \tag{11.2.4}$$

which is called the *direction* of e at v.

Let f_0 be the outer face of a k-embedding $\mu_k(G)$ of G, a straight vertex v of f_0 is said to be *exposed* if the half-line with origin v perpendicular to the boundary of f_0 in f_0 does not intersect $\mu_k(G)$ at any point different from v. If an exposed vertex v of f_0 has the half-line be to the North, the West, the South or the East from v, then it is, respectively, called a *north, west, south* or *east vertex* of f_0.

Lemma 11.2.1. *If $G = (V, E)$ is k-embeddable, $k \geq 1$, then G has such a k-embedding $\mu_k(G)$ that all 3-valent vertices on the outer-face boundary are straight vertices.*

Proof. By induction on the number of 3-valent vertices that are not straight ones on the boundary of the outer face in a k-embedding $\mu_k(G)$, $k \geq 1$ of G. Of course, if the number is zero, $\mu_k(G)$ is just a embedding satisfying the desired condition. Suppose v is a 3-valent vertex which is not a straight one on the outer-face boundary in $\mu_k(G)$ of G. If one of the two edges incident with v on the boundary of the outer face has a bend s, then we may assume that vs is a segment of the horizontal or the vertical line at v and that $\mathrm{St}(s; f_0) = 2$ as defined by eq. (11.2.1) in its own right. Because v is not straight for f_0, $\mathrm{St}(v; f_0) = 0$ and we always assume $\mathrm{St}(v, f_1) = 1$ for f_1 being the finite face incident to s. By parallel moving the half-line perpendicular to vs with origin v from v to s along vs such that all its points intersecting with segments perpendicular to itself in $\mu_k(G)$ are replaced by their trails of the moving as shown in Figure 11.2.1. The resultant k-embedding $\mu'_k(G)$ has v be a straight one for f'_0 on whose boundary the number of straight vertices is one less than that for f_0 in $\mu_k(G)$. Otherwise, i.e. s is a

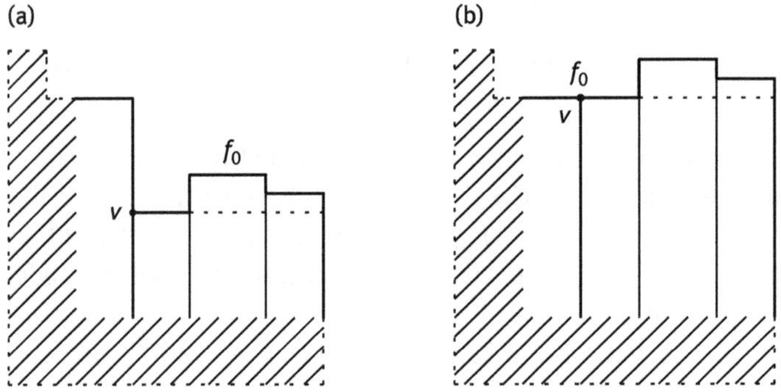

Figure 11.2.1: (a) $\mu_k(G)$ and (b) $\mu'_k(G)$.

vertex, by moving the segment sv parallel along a small segment opposite to the other edge incident to v on the boundary of f_0 such that $\mathrm{St}(t; f_0') = 0$, a new k-embedding $\mu_k'(G)$, $k \geq 1$, in which the edge (v, s) having a bend t can also be found to have the same property as above. Therefore, by the hypothesis, the lemma follows. □

Lemma 11.2.2. *For $k \geq 3$, if a graph $G = (V, E)$ has a k-embedding, then the embedding can be extended into such a k-embedding that all the 3-valent vertices on the boundary of the outer face are exposed.*

Proof. By induction on the order of G, when G is small, it is easy to check. In general, suppose v is a vertex on the outer-face boundary of a k-embedding $\mu_k(G)$ of G. If v is 3-valent, let v_1, v_2 and v_3 be the vertices adjacent to v. We may assume all the valencies of v_1, v_2 and v_3 are four without loss of generality because $\mu_k(G - v)$ is a k-embedding. From hypothesis, $\mu_k(G - v)$ can be extended into $\mu_k'(G - v)$ in which all the 3-valent vertices on the boundary of the outer-face f_0' are exposed. Let

$$\mathrm{Ar}(v_1, v_2, v_3; f_0') = (\mathrm{Ar}(v_1; f_0'), \mathrm{Ar}(v_2; f_0'), \mathrm{Ar}(v_3; f_0')), \tag{11.2.5}$$

where $\mathrm{Ar}(v; f)$ for v on the boundary of a face f represents the direction in which an edge can be arranged at v. Then, by the symmetry, there are four different cases for $\mathrm{Ar}(v_1, v_2, v_3; f_0')$ as

$$(N, N, N), (N, E, N), (N, S, N), (N, E, W).$$

In each case, we are allowed to restore the vertex v in $\mu_k'(G - v)$ to find, probably after doing operations as in the proof of Lemma 11.2.1, $\mu_k'(G)$ which has the required conditions as shown in Figure 11.2.2. If v is 4-valent. Let v_1, v_2, v_3 and v_4 be the adjacent vertices of v in $\mu_k(G)$, and $\rho(v_i) = 4$, $i = 1, 2, 3$ and 4 without loss of generality. By the same reason, we are allowed to assume $\mu_k'(G - v)$ is a k-embedding with the required conditions. In all the possible cases of $\mathrm{Ar}(v_1, v_2, v_3, v_4; f_0')$, (N, N, S, E) and (N, E, S, W) are two examples of the worst cases in causing more bends used on an edge. However, they are allowed by restoring v as shown in Figure 11.2.3 probably after suitable moving as in the proof of Lemma 11.2.1 to find a k-embedding, $\mu_k(G)$, $k \geq 3$, with required conditions from $\mu_k'(G - v)$. The lemma is obtained. □

Theorem 11.2.1. *Any planar graph $G = (V, E)$ with vertex valency not greater than four is 3-realizable.*

Proof. By induction on the order of G, for small graph under the condition, it is easy to check. As mentioned above, it suffices to discuss vertices in G have their valencies as 3 or 4. Let $v \in V$. By hypothesis, $G - v$ is 3-embeddable. Suppose $\mu_3(G - v)$ is a 3-embedding of $G - v$ with all the adjacent vertices of v being on the boundary of the

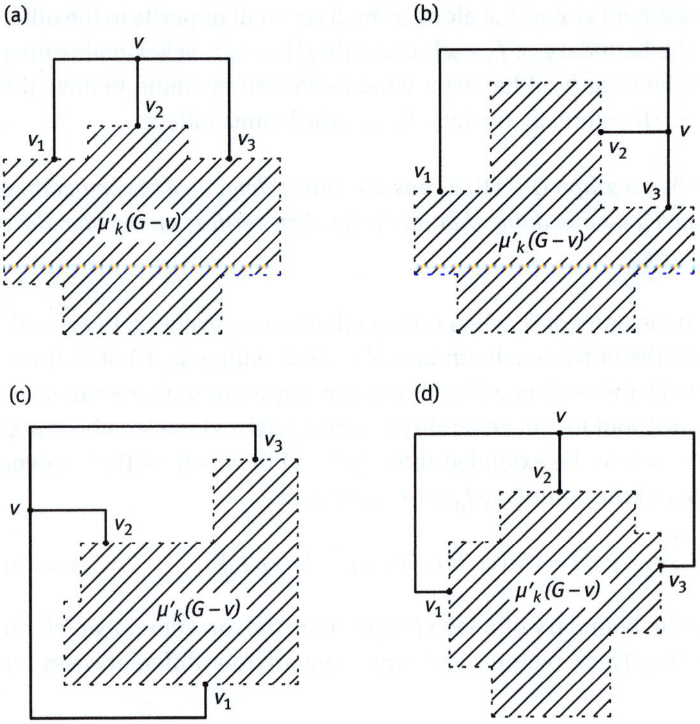

Figure 11.2.2: (a) (N,N,N); (b) (N,E,N); (c) (N,S,N); and (d) (N,E,W).

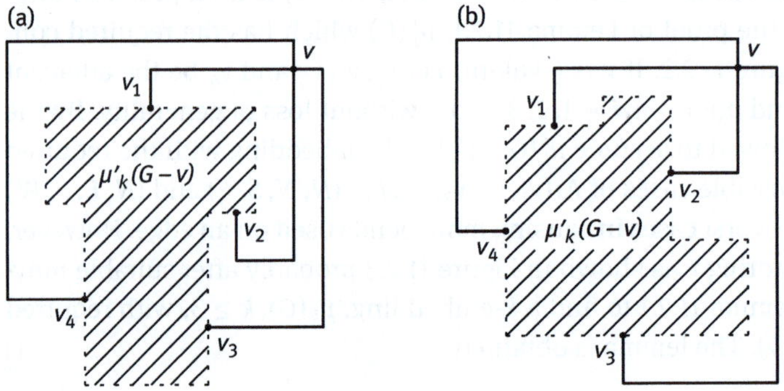

Figure 11.2.3: (a) (N,S,E,N) and (b) (N,E,S,W).

outer face $f_0\langle v\rangle$. From Lemmas 11.2.1–11.2.2, we may assume $\mu_3(G-v)$ has all the 3-valent vertices on the boundary of $f_0\langle v\rangle$ be exposed. Thus, it is allowed to put v, for $\rho(v)=3$, in the inner domain of $f_0\langle v\rangle$ such that (v,v_1), (v,v_2) and (v,v_3) are embedded with at most three bends each as shown in Figure 11.2.2. Similarly, for $\rho(v)=4$, by the method used in the proof of Lemma 11.2.2 as shown in Figure 11.2.3, a 3-embedding of G from $\mu_3(G - v)$ can also be found. When not all $v_i, i = 1, 2, 3$, are on the boundary of $f_0\langle v\rangle$, we can easily find a 3-embedding of G from $\mu_3(G - v)$. The theorem is proved from the arbitrariness of the choice of the infinite face. □

The octahedron Π_8 as shown in Figure 11.2.4(a) has a 3-embedding as shown in Figure 11.2.4(b).

Lemma 11.2.3. *Any planar embedding of Π_8 cannot be extended into a 2-embedding.*

Proof. From Theorem 5.3.3 and the symmetry, for any face being chosen to be the outer face, the resultant embedding is not different from what is shown in Figure 11.2.4(a) in rectilinear extension. By enumeration, there are 12 different cases of $\mathrm{Ar}(a, b, c, d; f_0(\Pi_8 - A))$ as follows:

$$\begin{array}{lll} (N, N, N, S), & (N, N, E, S), & (N, N, E, W), \\ (N, N, E, E), & (N, E, N, S), & (N, N, S, W), \\ (N, N, S, S), & (N, S, N, E), & (N, S, N, S), \\ (N, E, S, W), & (N, E, W, S), & (N, S, E, W) \end{array}$$

for k-embeddings, $k \le 2$, of the circuit $C = abcd$. However, it is easy to check that none of them is allowed to find a 2-embedding of Π_8 by putting B and A in the inside and outside of C, respectively. Therefore, the lemma holds. □

The lemma tells us that the result of Theorem 11.2.1 is best possible. We may also extend Theorem 11.2.1 into the following corollary by the similar method used in the proof of Theorem 11.2.1.

Corollary 11.2.1. *A planar graph $G = (V, E)$ with vertex valency at most k, $1 \le k \le 4$, is $(k - 1)$-realizable.*

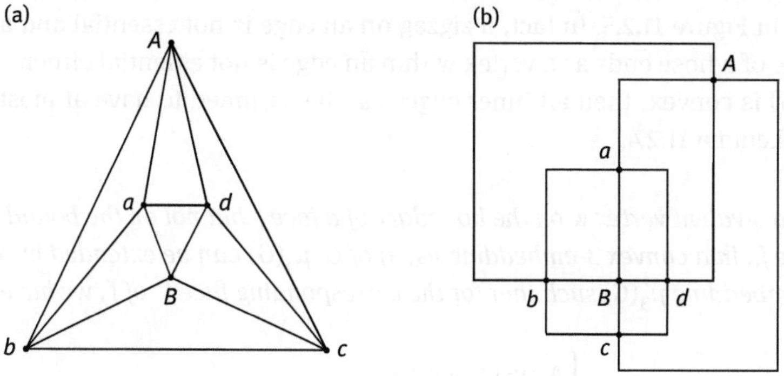

Figure 11.2.4: (a) Π_8 and (b) $\mu_3(\Pi_8)$.

Proof. When $k = 1, 2$, it is trivial. When $k = 4$ it is just Theorem 11.2.1. The only case remained is that of $k = 3$. By induction on the order of G, in view of the procedure of the proof of Theorem 11.2.1 from considering Fig. 11.2.2, the conclusion can be obtained. □

Let $Z_1 = Z(v, u)$ be the area conformed by a segment vu on a horizontal or a vertical line with the two half-lines perpendicular to vu from v and u, respectively, and let

$$Z_0 = Z(v, w)$$

for w being an inner point of vu near enough to v. A 3-embedding $\mu_3'(G)$ is said to be the *compressing* of $\mu_3(G)$ from Z_1 into Z_0 if it is obtained by the linear transformation from Z_1 to Z_0 such that any common point p on the half-line at u and on a segment of $\mu_3(G)$ perpendicular to the half-line becomes the segment which is the trail of p obtained by moving the half-line at u to that at w.

Lemma 11.2.4. *If uv is a straight segment on the boundary of an inner face f of a convex 3-embedding $\mu_3(G)$ of G, then the resultant embedding of compressing $Z_1 = Z(u, v)$ to $Z_0 = Z(u, w)$ for w being an inner point of uv in $\mu_3(G)$ is still a convex 3-embedding denoted by $\mu_3'(G)$ of G.*

Proof. From the definitions of the rectilinear convexity and the operation of compressing, $\mu_3'(G)$ can be easily checked to be a convex 3-embedding whenever $\mu_3(G)$ is a convex 3-embedding of G by the fact that no new bend on an edge is produced from compressing. □

Attention 11.2.1. In fact, the linear transformation in compressing is parallel to moving a half-line, or its particular segment, in the case of rectilinearity. This guarantees to form a rectilinear embedding to another rectilinear embedding whenever no intersection appears.

In a 3-embedding $\mu_3(G)$, an edge with more than one bend contains either a *zigzag* or a *handle* as shown in Figure 11.2.5. In fact, a zigzag on an edge is not essential and a handle having none of whose ends as a vertex within an edge is not essential either.

Further, if $\mu_3(G)$ is convex, then all inner edges can be assumed to have at most one bend by using Lemma 11.2.4.

Lemma 11.2.5. *For a 3-valent vertex w on the boundary of a face f but not on the boundary of the outer face f_0 in a convex 3-embedding $\mu_3(G)$ of G, $\mu_3(G)$ can be extended into another convex 3-embedding $\mu_3'(G)$ such that for the corresponding face f' of f, we have*

$$\mathrm{Ar}(v; f') = \begin{cases} \mathrm{Ar}(v; f), & v \neq w; \\ \mathrm{Ar}(v; f) \pm \pi/2, & v = w. \end{cases} \tag{11.2.6}$$

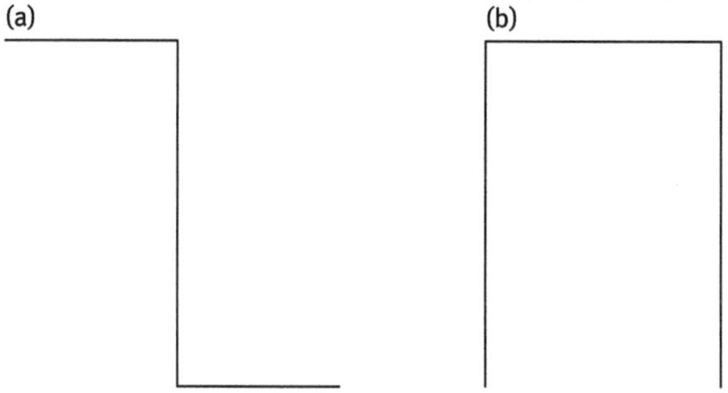

Figure 11.2.5: (a) Zigzag and (b) handle.

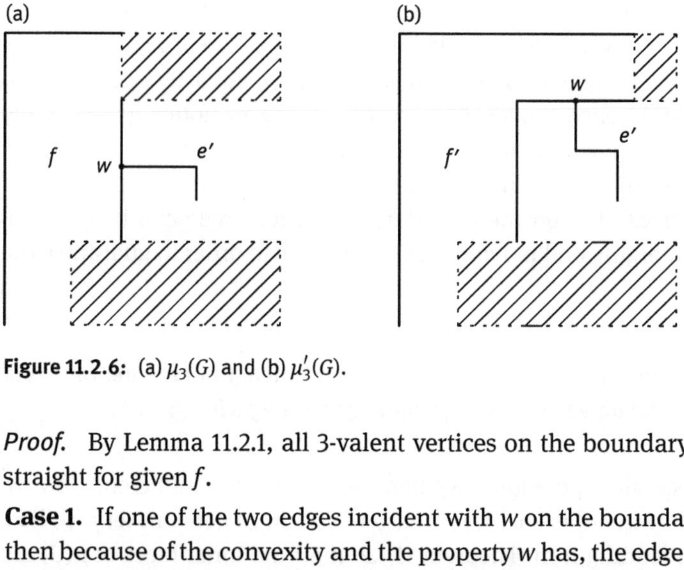

Figure 11.2.6: (a) $\mu_3(G)$ and (b) $\mu_3'(G)$.

Proof. By Lemma 11.2.1, all 3-valent vertices on the boundary of f are allowed to be straight for given f.

Case 1. If one of the two edges incident with w on the boundary of f is without bend, then because of the convexity and the property w has, the edge e' not on the boundary of f at w does not have more than one bend necessary in $\mu_3(G)$. By introducing a new bend each for e' if e' is without a bend and the edge without a bend on the boundary of f at w and then by properly moving, a convex 3-embedding $\mu_3'(G)$ is found from $\mu_3(G)$ such that eq. (11.2.6) is satisfied. If e' is with a bend as shown in Figure 11.2.6, it can always be transformed by Lemma 11.2.4 to have no bend at the same time.

Case 2. Otherwise, because both of the edges at w on the boundary of f have a bend, it is also allowed to transform $\mu_3(G)$ into $\mu_3'(G)$ which is still a convex 3-embedding as shown in Figure 11.2.7 as an example (by the same reason described above) such that eq. (11.2.6) is satisfied or it is transformed into that in case 1 so the bend of e' is in the other direction. □

Lemma 11.2.6. *Let $\mu(G)$ be a planar embedding with all inner faces adjacent to the outer face f_0 of $G = (V, E)$ non-separable with the vertex valency less than five. Then $\mu(G)$ is convex 3-extensible.*

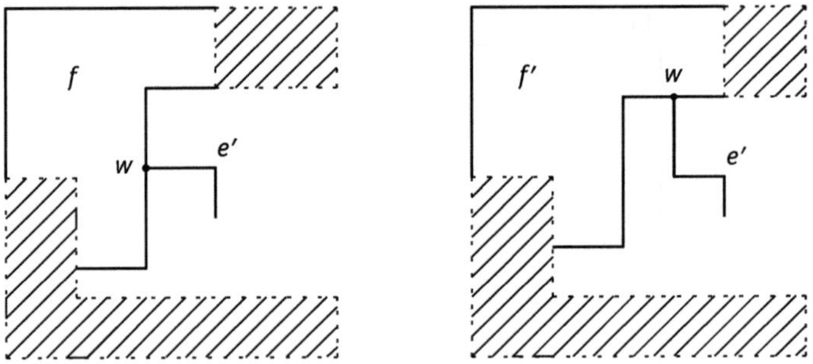

Figure 11.2.7: Two convex 3-embeddings.

Proof. Because the resultant graph of deleting all the edges on the boundary of f_0 from G is a forest and any forest is 2-embeddable such that all the articulate vertices are on the boundary of a rectangle, from the non-separability it is easily seen that $\mu(G)$ is convex 2-extensible and hence convex 3-extensible whenever noticing that at any 4-valent vertex incident to f_0, two imagined articulate edges corresponding to those on the boundary of f_0 are introduced on the forest, so that the broken segments which do not correspond to edges of f_0 on the boundary of the rectangle can be avoided, and then that properly merging and moving have to be done in order to guarantee the convexity of f_0. □

Theorem 11.2.2. *Any 3-connected planar graph G of vertex valency not greater than four is convex 3-realizable if, and only if, G has no planar embedding with $\langle 4, 4, 4\rangle$.*

Proof. Because the necessity is obvious, we here only prove the sufficiency. By induction on the size of G, when it is small, it is easy to check. In general, let $\mu(G)$ be a rectilinearly convex embedding from Theorem 11.1.2. If all the inner faces of $\mu(G)$ are adjacent to the outer face f_0, then from Lemma 11.2.6, its 3-extension $\mu_3(G)$ is a convex 3-embedding of G. Otherwise, we may suppose $e = (u, v)$ is an edge such that both u and v are not on the boundary of f_0. By hypothesis, $\mu(G - e)$ can be extended into a convex 3-embedding $\mu_3'(G-e)$ in which f' corresponds to the two faces incident with u and v in $\mu(G)$. In the same way used in the proof of Lemma 11.2.1, we may assume that all 3-vertices on the boundary of f' are straight for f'.

Case 1. If $\mathrm{Ar}(u, f') \neq \mathrm{Ar}(v; f')$. We may assume that $\mathrm{Ar}(u; f') = \mathrm{Ar}(v; f') \pm \pi/2$ because it is trivial otherwise for restoring e without bend in $\mu_3'(G - e)$ to find $\mu_3'(G)$ of G from the convexity of f' and Lemma 11.2.4. Since the two respective half-lines from u and v along $\mathrm{Ar}(u; f')$ and $\mathrm{Ar}(v; f')$ have a common segment each in the inner domain of f' by the convexity. Then from Lemma 11.2.4, the two common segments of the two half-lines with the inner domain of f' can be assumed to have a common point and hence a convex 3-embedding $\mu_3'(G)$ is found by restoring e with one bend in $\mu_3'(G - e)$.

Case 2. If $\mathrm{Ar}(u; f') = \mathrm{Ar}(v; f')$, then by Lemma 11.2.5, it can be transformed into case 1.

Because of the arbitrariness of the choice of $\mu(G)$, the theorem is obtained. □

Because any graph with a subgraph non-2-embeddable is not 2-embeddable itself, we are allowed only to discuss minimal non-2-embeddable graphs which are called 2-*obstacles*. By minimal here, we shall mean without proper subgraph non-2-embeddable anymore.

Lemma 11.2.7. *The octahedron Π_8 is a 2-obstacle.*

Proof. From Lemma 11.2.3, it is known that Π_8 is not 2-embeddable. Moreover, it is easily checked that all proper subgraphs of Π_8 are 2-embeddable. In fact, it only needs to observe the subgraph $\Pi_8 - e$ for an edge e because any proper subgraph of Π_8 is isomorphic to one of the subgraphs of $\Pi_8 - e$ from the symmetry. Figure 11.2.8 shows $\Pi_8 - e$ and its 2-embedding $\mu_2(\Pi_8 - e)$. □

For a 2-embedding $\mu_2(G)$ of a graph $G = (V, E)$, let $\mathrm{Oc}(v; f)$ be the subset of $Or = \{N, W, S, E\}$, in which each direction is occupied by an edge incident with vertex v which is on the boundary of the face f. Of course, only vertices of valency less than four have the subsets not equal to Or. Further, $\mathrm{Ar}(v; f)$ defined above is the subset of $\bar{\mathrm{O}}\mathrm{c}(v; f) = Or - \mathrm{Oc}(v; f)$, in which each direction is allowed to connect an edge in the inner domain of the face f for v being on the boundary of f. It is easily seen that if $G = (V, E) = G_1 \cup G_2$ with $G_1 \cap G_2 = v \in V$, then a 2-embedding $\mu_2(G)$ can be composed by identifying v in the two 2-embeddings $\mu_2(G_1)$ and $\mu_2(G_2)$ if, and only if, v has an incident face f_1 in $\mu_2(G_1)$ and an incident face f_2 in $\mu_2(G_2)$ such that at least one of them, assumed to be f_1 without loss of generality, is the outer face and that

$$\mathrm{Oc}(v; f_1) \subseteq \mathrm{Ar}(v; f_2). \tag{11.2.7}$$

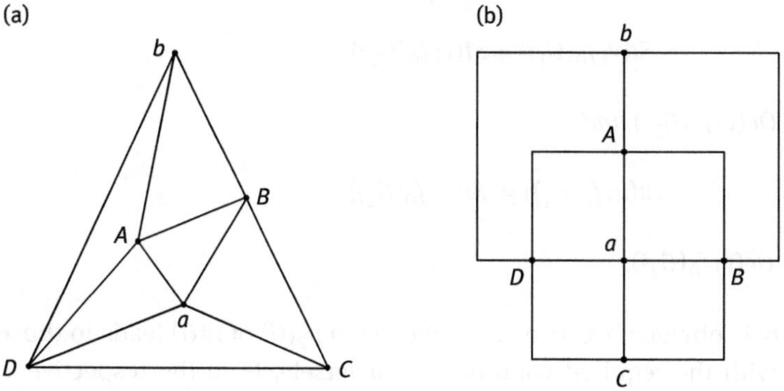

Figure 11.2.8: (a) $\Pi_8 - e$ and (b) $\mu_2(\Pi_8 - e)$.

A vertex under condition (11.3.1) is said to be *composable* for the two 2-embeddings $\mu_2(G_1)$ and $\mu_2(G_2)$. Similarly, if $G = G_1 \cup G_2$ with $G_1 \cap G_2 = u+v$, i.e. $\{u, v\} \subseteq V$ is a splitting pair of G, then a 2-embedding $\mu_2(G)$ can be composed by identifying v in $\mu_2(G_1)$ and $\mu_2(G_2)$, and u in $\mu_2(G_1)$ and $\mu_2(G_2)$ if, and only if, both u and v are composable for the faces f_i ind $\{u, v\}$ in $\mu_2(G_i)$, $i = 1, 2$, such that the rotations at u and v are in agreement with those in $\mu_2(G)$.

Lemma 11.2.8. *For $G = (V, E) = G_1 \cup G_2$ with $G_1 \cap G_2 = v \in V$, a planar embedding $\mu(G)$ with the outer-face boundary of $\mu(G)$ being the disjoint union of those of $\mu(G_1)$ and $\mu(G_2)$ is 2-extensible if, and only if, so are both $\mu(G_1)$ and $\mu(G_2)$.*

Proof. Let $f_0 = f_0(G), f_0(G_1)$ and $f_0(G_2)$ be the outer faces of $\mu(G)$, $\mu(G_1)$ and $\mu(G_2)$, respectively. The necessity is obvious because the restricted ones of a 2-extension $\mu_2(G)$ of $\mu(G)$ on $\mu(G_1)$ and $\mu(G_2)$ are also 2-extensions of $\mu(G_1)$ and $\mu(G_2)$. Conversely, suppose $\mu_2(G_1)$ and $\mu_2(G_2)$ are 2-extensions of $\mu(G_1)$ and $\mu(G_2)$, respectively. If one of the valencies $\rho(v; G_1)$ and $\rho(v; G_2)$ is 1, and letting $\rho(v; G_1) = 1$ without loss of generality, then whenever $\rho(v; G_2) = 1, 2$ or 3, we are always allowed to have $\mu_2(G_2)$ with the properties $\mathrm{St}(v; f_0(G_2)) \geq 1$ and $\mathrm{Oc}(v; f_0(G_1)) \subseteq \mathrm{Ar}(v; f_0(G_2))$. That implies v is composable for $\mu_2(G_1)$ and $\mu_2(G_2)$. Otherwise, $\rho(v; G_1) = 2$ and $\rho(v; G_2) = 2$. Because 2-valent vertex v on the boundary of a face f in a 2-embedding can always be made to have $\mathrm{St}(v; f) = 2$, $\mu_2(G_1)$ and $\mu_2(G_2)$ are allowed to have the properties $\mathrm{St}(v; f_0(G_i)) = 2$, $i = 1, 2$, and $\mathrm{Oc}(v; f_0(G_1)) = \mathrm{Ar}(v; f_0(G_2))$. Thus, v is also composable for $\mu_2(G_1)$ and $\mu_2(G_2)$. The sufficiency is found. □

Lemma 11.2.9. *Let $G = (V, E) = G_1 \cup G_2$ with*

$$G_1 \cap G_2 = u + v, \quad u, v \in V,$$
$$\rho(u; G_1) = \rho(u; G_2) = \rho(v; G_1) = \rho(v; G_2) = 2.$$

Then, a planar embedding $\mu(G)$ of G such that both u and μ are on the boundary of the outer face f_0 is 2-extensible if, and only if, so are both $\mu(G_1)$ and $\mu(G_2)$ such that

$$St(v; f_0(G_1)) = St(v; f_0(G_2))$$

with $Ar(v; f_0(G_1)) = Oc(v; f_0(G_2))$ and

$$St(u; f_0(G_1)) = St(u; f_0(G_2))$$

with $Ar(u; f_0(G_1)) = Oc(v; f_0(G_2))$.

Proof. The necessity is obvious because any 2-extension $\mu_2(G)$ of $\mu(G)$ leads to those of $\mu(G_1)$ and $\mu(G_2)$ with the required conditions. Conversely, from the respective 2-extensions $\mu_2(G_1)$ and $\mu_2(G_2)$ of $\mu(G_1)$ and $\mu(G_2)$ with the given condition, by the method used in the proof of Lemma 11.2.2, a 2-extension $\mu_2(G)$ can be deduced. □

Although Lemmas 11.2.8 and 11.2.9 can be generalized for those 2-embeddable we are not allowed to prove them right now in order to avoid some complications and have to discuss graphs 3-connected and standard as well because 1- or 2-valent vertices are not essential here.

Lemma 11.2.10. *If G has a 2-embedding such that all 3-valent vertices on the boundary of the outer face f_0 are exposed, then all the graphs which can be seen as those obtained by introducing an r-valent, $r \leq 3$, vertex in the outer face f_0 have 2-embeddings such that all 3-valent vertices on the outer-face boundaries are exposed as well.*

Proof. By the same procedure used in the proof of Lemma 11.2.2, because an r-valent vertex can always be embedded in the outer face f_0 of $\mu_2(G)$ such that all its incident edges are with at most two bends, of course the resultant embedding can be made to satisfy the condition that all 3-valent vertices on the outer face boundary are exposed, the lemma is obtained. □

Lemma 11.2.11. *For 3-connected standard graphs, all the valencies of vertices in a 2-obstacle are four.*

Proof. Suppose H is a 2-obstacle. By contradiction, let v be a 3-valent vertex from the standardness. If $H - v$ has no 4-valent vertex, then let $H_0 = H - v$. Otherwise, let $H_1 = H - v$ and let v_1 be a vertex adjacent to a 4-valent vertex, whose valency is less than four, then $H_2 = H_1 - v_1$ has less 4-valent vertices than H_1. If H_2 has no 4-valent vertex, then let $H_0 = H_2$, otherwise H_1 is replaced by $H_1 - v_1$, do the same procedure until H_0 is found in view of the finiteness. By Corollary 11.2.1, H_0 is 2-realizable. Let $\mu_2(H_0)$ be a 2-embedding such that only the outer face is incident with those adjacent to a vertex deleted. Thus, by the reversed procedure, restoring the vertices deleted such that all incident edges are represented by at most 2-bend lines each in the way described in the proof of Lemma 11.2.2 from Lemma 11.2.10, a 2-embedding of H is found. This is a contradiction to that H is a 2-obstacle. □

Lemma 11.2.12. *For a 4-valent graph $G = (V, E)$, any triangle cannot be embedded as the boundary of the outer face in a 2-embedding.*

Proof. By contradiction, suppose a triangle $v_1v_2v_3$ forms the outer face boundary of a 2-embedding $\mu_2(G)$. Because all of v_1, v_2 and v_3 are 4-valent, we have $\mathrm{St}(v_i; f_0) = 0$, $i = 1, 2, 3$, f_0 is the outer face. It is not possible to find a 2-embedding of the triangle such that the condition is satisfied. □

Lemma 11.2.13. *If all the faces in a planar embedding $\mu(G)$ of a 4-valent graph $G = (V, E)$ are triangles, then G is isomorphic to the octahedron Π_8.*

Proof. From the 4-valentness of vertices and the 3-valentness of faces, $\mu(G)$ is a (4,3)-polyhedron on the sphere. By Theorem 5.1.1, it is the octahedron. □

Lemma 11.2.14. *Any planar embedding $\mu(G)$ with the outer face f_0 of valency greater than three of a 4-valent 3-connected graph $G = (V, E)$ is 2-extensible.*

Proof. If $\mu(G)$ does not have a configuration $\langle 4, 4, 4\rangle$, then from Theorem 11.2.2, it is convex 3-extensible. Moreover, the outer face can be embedded into the plane such that each edge has at most two bends and that

$$\forall v \in V(f_0),\ \mathrm{St}(v, f_0) = 0, \tag{11.2.8}$$

where $V(f_0)$ is the set of vertices on the boundary of f_0. By noticing that all inner edges of a convex 3-embedding are with at most one bend each, the 3-embedding is in fact a 2-embedding.

In general, by induction on the number of configurations $\langle 4, 4, 4\rangle$ in $\mu(G)$, suppose $C = (v_1, \dots, v_2, \dots, v_3)$ forms such a $\langle 4, 4, 4\rangle$ in $\mu(G)$ that there is no $\langle 4, 4, 4\rangle$ in the inner domain of C. Because C can be embedded into the plane such that

$$\forall v_i (i = 1, 2, 3),\ \mathrm{St}(v_i; C_{\mathrm{in}}) = \mathrm{St}(v_i; C_{\mathrm{out}}) = 1 \tag{11.2.9}$$

and there is one edge of two bends with others of one bend on C, from Theorem 11.2.2, C with the inner part of $\mu(G)$ can be extended into a convex 2-embedding $\mu_2(G; C_{\mathrm{in}})$. Moreover, by hypothesis, C with its outer part of $\mu(G)$ has a 2-extension $\mu_2(G; C_{\mathrm{out}})$. Whenever noticing that the 2-embeddings of C is substantially unique under the condition described by eq. (11.2.9), a 2-extension $\mu_2(G)$ of $\mu(G)$ can be composed from $\mu_2(G; C_{\mathrm{in}})$ and $\mu_2(G; C_{\mathrm{out}})$. This is the lemma. □

Theorem 11.2.3. *A 3-connected standard graph $G = (V, E)$ is 2-embeddable if, and only if, $G \not\equiv \Pi_8$.*

Proof. From Lemma 11.2.7, the necessity is obvious. Conversely, from Lemmas 11.2.11–11.2.14, the complete set of 2-obstacles for 3-connected standard graphs consists of only Π_8 itself. Therefore, the sufficiency is true. □

Lemma 11.2.15. *For any non-separable graph $G = (V, E)$ of vertex valency less than five, if the boundary C of a face f has a 2-embedding $\mu_2(C)$ such that*

$$St(v; C_{\mathrm{in}}) \begin{cases} = 2, & \mathit{for}\ \rho(v; G) = 4; \\ \geq 1, & \mathit{for}\ \rho(v; G) = 3, \end{cases} \tag{11.2.10}$$

then G has a 2-embedding $\mu_2(G)$ with $\mu_2(C)$ as the boundary of the outer face.

Proof. If G has no 4-vertex, then from Corollary 11.2.1, G has a 2-embedding with C as the outer face boundary.

Therefore, by Lemma 11.2.1 if necessary the 2-embedding $\mu_2(C)$ has to satisfy condition (11.2.10).

Otherwise, by the procedure suggested in the proof of Lemma 11.2.11 we may also allow to find such a 2-embedding of G. □

Lemma 11.2.16. *Let $G = (V, E) = G_1 \cup G_2$ provided $G_1 \cap G_2 = u, u \in V, \rho(v) \leq 4$. Then, G is 2-embeddable if, and only if, so are both G_1 and G_2.*

Proof. The necessity is obvious because any 2-embedding $\mu_2(G)$ leads to the respective 2-embeddings $\mu_2(G_1)$ and $\mu_2(G_2)$ of G_1 and G_2. Conversely, from the 2-embeddability of G_1 and G_2, G is planar of course. Moreover, we are allowed to assume both G_1 and G_2 are non-separable without loss of generality since we may otherwise discuss blocks one by one. From Lemmas 11.2.8 and 11.2.15, the sufficiency can be derived. □

Lemma 11.2.17. *Let $G = (V, E) = G_1 \cup G_2$ provided $G_1 \cap G_2 = u + v, u, v \in V, \rho(u), \rho(v) \leq 4$. Then, G is 2-embeddable if, and only if, so are both G_1 and G_2.*

Proof. The necessity is obvious by the reason as described in the proof of Lemma 11.2.16. Conversely, from the 2-embeddability, G is planar. Because G always has a planar embedding $\mu(G)$ such that the outer face f_0 is incident with u and v. Let $f_0(G_1)$ and $f_0(G_2)$ be the outer faces of $\mu(G_1)$ and $\mu(G_2)$, respectively. It can be checked in any case that the boundaries C_i of $f_0(G_i), i = 1, 2$, always have 2-embeddings with condition (11.2.10) such that the composable condition for the splitting pair $\{u, v\}$ is also satisfied. Therefore, by Lemmas 11.2.8 and 11.2.15, the lemma is obtained. □

Corollary 11.2.2. *Any connected graph $G = (V, E)$ of vertex valency less than five is 2-embeddable if, and only if, it is planar and non-isomorphic to Π_8.*

Proof. From Lemmas 11.2.16 and 11.2.17 and Theorem 11.2.3, the conclusion is derived. □

Corollary 11.2.3. *A 3-connected planar graph $G = (V, E)$ of vertex valency less than five is 2-realizable if, and only if, any planar embedding of G has no triangle face with (4, 4, 4) of the valencies of its incident vertices.*

Proof. Because a triangle face with (4,4,4) can never be the outer face boundary of any 2-embedding, the necessity is true. Conversely, because of none of faces with (4,4,4) in any planar embedding, from Lemma 11.2.15 and Corollary 11.2.2, any face can be taken as the outer face for a 2-extension. Therefore, the sufficiency is also true. □

In what follows, we discuss the characterization of the convex 2-embeddability of a graph.

Lemma 11.2.18. *A planar embedding $\mu(G)$ of a 3-connected graph $G = (V, E)$ of vertex valency less than five is convex 2-extensible if, and only if, $\mu(G)$ has no configuration $\langle 4, 4, 4\rangle$ and the outer face is not a triangle $(4, 4, 4)$.*

Proof. The necessity is obvious as usual. Conversely, from no configuration $\langle 4, 4, 4\rangle$, by Theorem 11.2.2, $\mu(G)$ is convex 3-extensible. Because the outer-face boundary is not with (4,4,4), $\mu(G)$ is in fact convex 2-extensible. The sufficiency is obtained. □

Lemma 11.2.19. *A 3-connected planar graph $G = (V, E) \not\equiv \Pi_8$ of vertex valency less than five is convex 2-embeddable if, and only if, there is a planar embedding $\mu(G)$ of G without configuration $\langle 4, 4, 4\rangle$.*

Proof. Since a configuration $\langle 4, 4, 4\rangle$ causes without convex 3-extension and hence without convex 2-extension, the necessity is obtained. Conversely, because $G \not\equiv \Pi_8$, G is 2-embeddable. From Theorem 11.2.3, let $\mu_2(G)$ be a 2-embedding. However, the outer-face boundary is not with (4,4,4). From Lemma 11.2.18, the convex 3-extension of $\mu_2(G)$ is in fact a convex 2-embedding of G. This is the sufficiency. □

Theorem 11.2.4. *A 3-connected planar graph $G = (V, E) \not\equiv \Pi_8$ of vertex valency less than five is convex 2-realizable if, and only if, there is neither face $(4, 4, 4)$ nor configuration $\langle 4, 4, 4\rangle$ in any planar embedding $\mu(G)$.*

Proof. The necessity is obvious as before. Conversely, because of the uniqueness of planar embeddings discussed in Section 5.3, from Lemma 11.2.18, any planar embedding $\mu(G)$ of G is convex 2-realizable. Therefore, we have the sufficiency. □

We mainly investigate some special cases here to find a complete set of forbidden configurations for the 1-embeddability.

Lemma 11.2.20. *Let $\mu(G)$ be a planar embedding of a cubic graph G. If the outer face is not a triangle, then it is convex 1-extensible.*

Proof. From Lemma 11.2.18, $\mu(G)$ is convex 2-extensible. Suppose $\mu_2(G)$ is a convex 2-extension of $\mu(G)$. From the convexity, all the inner edges can be taken with at most one bend. Moreover, the outer face always has a convex 1-embedding such that for any vertex v on it, $\mathrm{St}(v, f_0) = 1$. Therefore, $\mu_2(G)$ is in fact a convex 1-embedding of $\mu(G)$. □

Lemma 11.2.21. *A planar embedding $\mu(G)$ of a cubic graph $G = (V, E)$ is not 1-embeddable if, and only if, all its faces are triangle.*

Proof. Because it is easily checked that a triangle f never has a 1-embedding such that $\mathrm{St}(v;f) = 1$ for all v on f, any triangle cannot be the outer face of a 1-embedding of G. The sufficiency is true. Conversely, from Lemma 11.2.20, the necessity is obtained. □

Theorem 11.2.5. *A planar embedding $\mu(G)$ of a cubic graph $G = (V,E)$ is convex 1-embeddable if, and only if, $\mu(G) \not\cong \Pi_4$, the tetrahedron.*

Proof. Because $\mu(G)$ has all triangle faces if, and only if, $\mu(G) \cong \Pi_4$ from Theorem 5.1.1, therefore, by Lemma 11.2.21, the theorem follows. □

A standard graph whose dual graph is also standard is said to be *bistandard*. A graph G is said to be *self-dual* if the planar dual of G is isomorphic to G itself. From the duality, of course, any self-dual standard graph is bistandard. It is easily shown that the following four kinds of graphs, namely $C_{4,t}, t \geq 2$; $W_{4,t} t \geq 1$; $R_t, t \geq 1$; and $T_t, t \geq 2$, as shown in Fig. 11.2.9, are all bistandard. Of course, t is the number of parallel 4-circuits. Further, it is also easily seen that any $W_{4,t}, t \geq 1$, among them is self-dual.

Lemma 11.2.22. *Any self-dual standard graph $G = (V,E)$ has exactly four 3-valent vertices and exactly four 3-valent faces.*

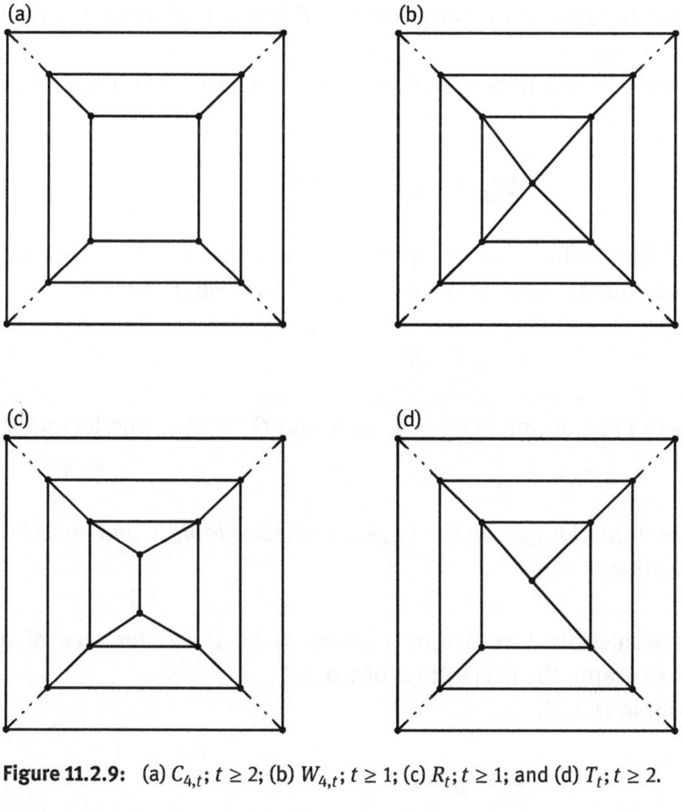

Figure 11.2.9: (a) $C_{4,t}$; $t \geq 2$; (b) $W_{4,t}$; $t \geq 1$; (c) R_t; $t \geq 1$; and (d) T_t; $t \geq 2$.

Proof. From the self-duality, we have

$$v_3 = \phi_3;\ v_4 = \phi_4$$

and hence $v = v_3 + v_4 = \phi_3 + \phi_4$, where $v_3(\phi_3)$ and $v_4(\phi_4)$ are the number of vertices (faces) of valencies 3 and 4, respectively. By the Euler formula for planar polyhedra, we have

$$2(2v_3 + 2v_4) - (3v_3 + 4v_4) = 4.$$

Therefore, $v_3 = 4$ and dually $\phi_3 = 4$. This is the lemma. □

A face of a standard graph is called (*standardly*) *admissible* if it is incident with at least four 3-valent vertices; otherwise the face is called (*standardly*) *inadmissible*.

Lemma 11.2.23. *If a self-dual standard graph $G = (V, E)$ has a 4-valent face which is admissible, then there exists an integer $t \geq 1$ such that $G \cong W_{4,t}$.*

Proof. Let f_a be the admissible face. By the admissibility, the four vertices on the boundary of f_a are all 3-valent. From Lemma 11.2.22, all vertices not on the boundary of f_a in V are 4-valent. Because of the self-duality τ, G has a 4-valent vertex $\tau(f_a)$ which is incident with four triangles.

It is allowed to have the four adjacent vertices of $\tau(f_a)$ in agreement with those incident to f_a. This is a

$$W_{4,t},\ t = 1,$$

which, of course, is self-dual. Otherwise, all the faces rather than the four incident triangles of $\tau(f_a)$ have to be quadrangles from Lemma 11.2.22 as well. It leads to

$$W_{4,t},\ t = 2.$$

And so forth, there has to exist an integer $t \geq 1$ such that $G \cong W_{4,t}$. The lemma is proved. □

Lemma 11.2.24. *A self-dual standard graph $G = (V, E)$ is 1-embeddable if, and only if, G has a 4-valent face admissible.*

Proof. Because any inadmissible face is not allowed to be the outer face of a 1-embedding of a standard graph, the necessity is obvious.

Conversely, from Lemma 11.2.23,

$$G \cong W_{4,t},\ t \geq 1.$$

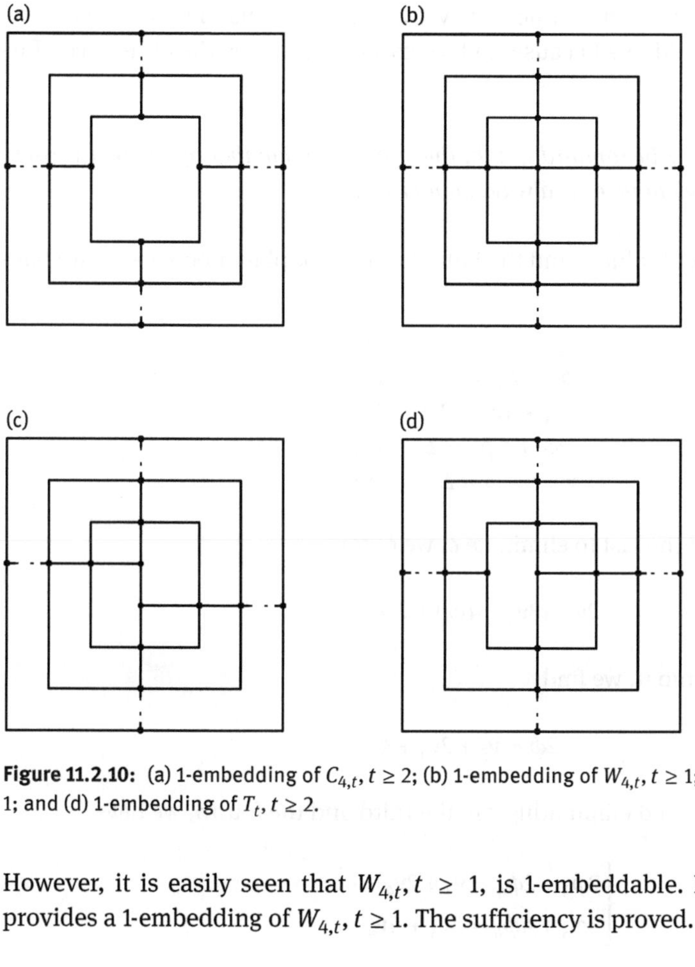

Figure 11.2.10: (a) 1-embedding of $C_{4,t}$, $t \geq 2$; (b) 1-embedding of $W_{4,t}$, $t \geq 1$; (c) 1-embedding of R_t, $t \geq 1$; and (d) 1-embedding of T_t, $t \geq 2$.

However, it is easily seen that $W_{4,t}, t \geq 1$, is 1-embeddable. In fact, Fig. 11.2.10(b) provides a 1-embedding of $W_{4,t}$, $t \geq 1$. The sufficiency is proved. □

Theorem 11.2.6. *A self-dual standard graph $G = (V, E)$ is 1-embeddable if, and only if, there exists an integer $t \geq 1$ such that*

$$G \cong W_{4,t}.$$

Proof. A direct result of Lemmas 11.2.23–11.2.24. □

Corollary 11.2.4. *A planar embedding $\mu(G)$ of a self-dual standard graph G is convex 1-extensible if, and only if, the outer face of $\mu(G)$ is admissible.*

An (*standard*) *obscurity*, denoted by Ω, is a minimal standard graph with the property that all its faces of valency not less than four are inadmissible. Naturally, a *bistandard obscurity*, denoted by B, is such an obscurity which is bistandard. Or in other words, it is a minimal bistandard graph in which every 4-valent face is inadmissible. Of course, no obscurity is 1-embeddable because no face can be chosen as the outer face of its 1-embedding.

Lemma 11.2.25. *In every bistandard graph, the sum of the number of 3-valent vertices and the number of 3-valent faces is always equal to 11.*

Proof. From the bistandardness and the Euler formula for planar polyhedra, we have the equations:

$$\begin{aligned} \nu_3 + \nu_4 &= \nu; && (1)\\ \phi_3 + \phi_4 &= \phi; && (2)\\ 3\nu_3 + 4\nu_4 &= 2\epsilon; && (3)\\ 3\phi_3 + 4\phi_4 &= 2\epsilon; && (4)\\ \nu - \epsilon + \phi &= 2. && (5) \end{aligned}$$

By using the third and the last to eliminate ϵ, we obtain

$$2\nu - (3\nu_3 + 4\nu_4) + 2\phi = 4.$$

Substituting the first into it, we find

$$2\phi = \nu_3 + 2\nu_4 + 4.$$

Then, from the second and eliminating ϵ in the third and the fourth, we have

$$\begin{cases} 2\phi_3 + 2\phi_4 = \nu_3 + 2\nu_4 + 4; \\ 3\phi_3 + 4\phi_4 = 3\nu_3 + 4\nu_4. \end{cases}$$

Because ϕ_4 and ν_4 can be eliminated from them at a time, we finally obtained that $\nu_3 + \phi_3 = 11$. This is the lemma. □

Lemma 11.2.26. *A bistandard graph $G = (V, E)$ is 1-embeddable if, and only if, G has no vertex-induced subgraph isomorphic to some B.*

Proof. Because any B is not 1-embeddable, the necessity is obvious. Conversely, since G has no vertex-induced graph isomorphic to some B, G has a quadrangle face f_a admissible. Moreover, G has no configuration $\langle 4, 4, 4\rangle$. Suppose otherwise $C_3 = (\nu_1, \nu_2, \nu_3)$ from the bistandardness is a configuration $\langle 4, 4, 4\rangle$. From the bistandardness again, all the three faces incident with an edge on C_3 in the inner domain of C_3 are either quadrangles or triangles. The last case is not allowed because of Π_4 which is, of course, an obscurity. Similarly, the three incident faces in the outer domain of C_3 are only allowed to be all quadrangles, and an obscurity also occurs. This is a contradiction to

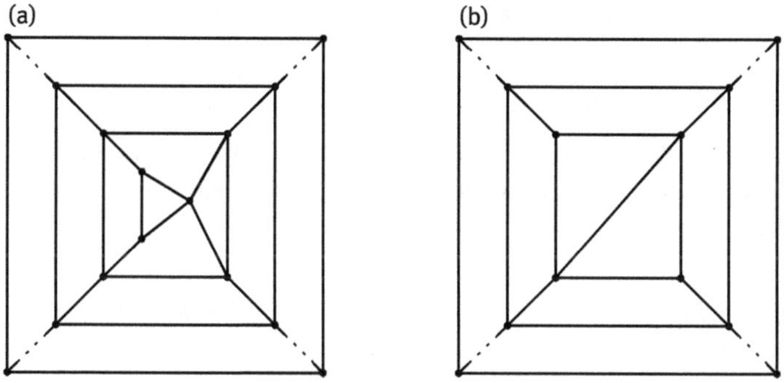

Figure 11.2.11: (a) $S_t, t \geq 1$ and (b) $U_t, t \geq 2$.

the condition of G. Hence, in view of that only bistandard graph has a separating pair of vertices and it is also an obscurity, by Lemma 11.2.18, G is convex 2-extensible with f_a as the outer face. Because of the convexity, all inner edges in the convex 2-extension can be made with at most one bend. From the admissibility of f_a, its boundary can be embedded with at most one bend of each edge. Therefore, G is 1-embeddable. The sufficiency is obtained. □

Corollary 11.2.5. *A planar embedding $\mu(G)$ of a bistandard graph G without vertex-induced subgraph isomorphic to some B is convex 1-extensible if, and only if, the outer face of $\mu(G)$ is admissible.*

Lemma 11.2.27. *A bistandard graph G is without vertex-induced subgraph isomorphic to some B if, and only if, there exists an integer $t \geq 1$ such that $G \cong W_{4,t}, C_{4,t}(t \neq 1), R_t, T_t(t \neq 1)$ as in Figure 11.2.9, and $S_t, U_t(t \neq 1)$ as shown in Figure 11.2.11.*

Proof. The sufficiency is easily checked. Conversely, because G has a 4-valent face f_a admissible, let v_1, v_2, v_3 and v_4 be the vertices on the boundary, denoted by C_1, of f_a. We may assume f_a is the outer face without loss of generality. From the bistandardness, if the four incident faces in the inner domain of C_1 are not all quadrangles, then three possibilities have to be discussed.

Case 1. Only one of the four faces is triangle. Of course, the other three are all quadrangles. It is only possible that $G \cong S_1$.

Case 2. Two of the four are triangles without common vertex. Of course, the other two are all quadrangles. This leads to $G \cong R_1$.

Case 3. Two of the four are adjacent triangles. Because they have a common edge, from the bistandardness, the other two have to be triangles too. This leads to $G \cong W_{4,1}$.

If all the four faces incident with C_1 are quadrangles, then there is a 4-circuit C_2 parallel to C_1. In general, suppose $C_t, t \geq 2$, is the 4-circuit parallel to C_1 in the inner domain of C_1. Then, three more possibilities have to be observed except for the three which lead to S_t, R_t and $W_{4,t}, t \geq 2$, as discussed above.

Case 4. Neither vertex nor edge is in the inner domain of C_t. Of course, that means $G \cong C_{4,t}$.

Case 5. A single edge is in the inner domain of C_t. This leads to $G \cong U_t$.

Case 6. Only two of the incident faces are adjacent triangles. Because of the bistandardness, one more other face which is quadrangle is allowed. This leads to $G \cong T_t$.

The proof of the lemma is complete. □

Theorem 11.2.7. *A bistandard graph $G = (V, E)$ is 1-embeddable if, and only if, there exists an integer $t \geq 1$ such that*

$$G \cong W_{4,t}, R_t, S_t, C_{4,t}(t \neq 1), T_t(t \neq 1) \text{ or } U_t(t \neq 1).$$

Proof. A direct result of Lemmas 11.2.26 and 11.2.27. □

In fact, all the 1-embeddings of $C_{4,t}, t \geq 2$; $W_{4,t}, t \geq 1$; $R_t, t \geq 1$; $T_t, t \geq 2$, as shown in Figure.11.2.10(a–d) and $S_t, t \geq 1$, $U_t, t \geq 2$ as shown in Figure 11.2.12(a–b) are convex.

Corollary 11.2.6. *A bistandard graph G is convex 1-embeddable if, and only if, G is 1-embeddable.*

Proof. From Corollary 11.2.4 and Theorem 11.2.7, it is derived directly. □

Furthermore, for general 1-embeddable graphs, we have the following theorem.

Theorem 11.2.8. *A 3-connected graph $G = (V, E)$ is 1-embeddable if, and only if, G is convex 1-embeddable.*

Proof. It is only needed to prove the necessity because the sufficiency is natural. According to the 3-connectedness of G, from Theorems 5.5.1 and 11.1.2, the necessity is

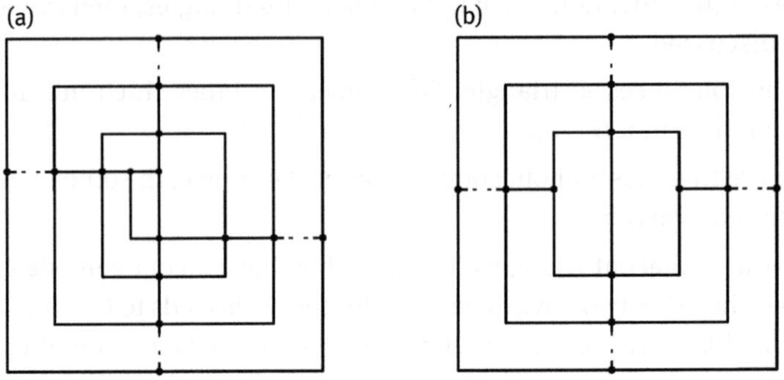

Figure 11.2.12: (a) 1-embedding of $S_t, t \geq 1$ and (b) 1-embedding of $U_t, t \geq 2$.

true because G has neither triangle (4,4,4) of the outer face nor configuration $\langle 4,4,4\rangle$ in view of the 1-embeddability and the convexity which is related to the admissibility of the outer face boundary (it can be seen in a similar manner to those mentioned above for 2-extensions). The theorem is obtained. □

Because more general obscurity has to be further studied and the idea is much related to that for the net embeddability, or 0-embeddability, the characterization of 1-embeddability for general standard graphs is left to discuss in the next chapter.

11.3 Surface models

Because an orientable surface of genus 0 is equivalent to the plane which is well studied, only surfaces of genus not zero are concentrated on. In this section, three types of surface models are observed for the observation of orthogonality.

Surface model 1. A square with a certain number of disjoint small squares within it. First problem encountered is how to locate the small squares in the square. This is dependent on the purpose. Suppose the aim for us is to reduce the bends in number. The model is chosen as all the small squares are on the diagonal of the square.

Let l, s and d be, respectively, the lengths of the square, small square sides and the distance (horizontal and vertical) between two nearest squares and k, the number of squares. Then, a surface in this model satisfies

$$l = (k-1)s + kd = (k-1)(s+d) + d. \tag{11.3.1}$$

Theorem 11.3.1. *An orientable surface in model 1 is of genus $p \geq 1$ if, and only if, each square has the form as $(aba^{-1}b^{-1})$ and $k = p$.*

Proof. Because of Operation 3 in Section 3.3 and Theorem 3.3.2, the orientable surface of genus p in standard form (3.2.1) is

$$\prod_{i=0}^{k-1}(a_i b_i a_i^{-1} b_i^{-1}),$$

and $k = p$. This is the necessity.

Conversely, by employing Theorems 3.3.1 and 3.3.2 in the other direction, the sufficiency is done. □

This theorem allows us to represent an orientable surface of genus $p = k$ by an $l \times l$ square S_0 and $k-1$ small $s \times s$ squares $S_1, \ldots, S_{k-1}$ on the diagonal of S_0 such that the distance between two nearest squares is d as shown in Figure 11.3.1(a).

Theorem 11.3.2. *A non-orientable surface in model 1 is of genus $q \geq 1$ if, and only if, each square has the form as (aa) and $k = q$.*

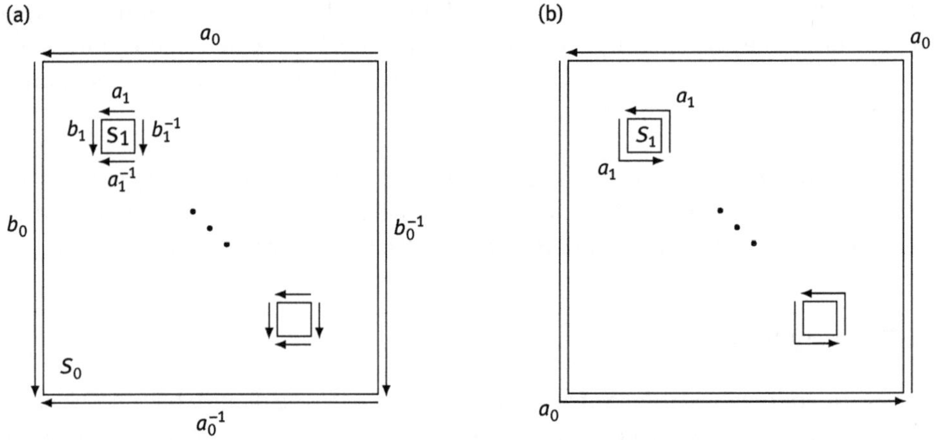

Figure 11.3.1: (a) Orientable surface and (b) non-orientable surface.

Proof. Because of Theorems 3.3.1 and 3.3.2, the non-orientable surface of genus q in standard form (3.2.1) is

$$\prod_{i=0}^{k-1}(a_i a_i),$$

and $k = q$. This is the necessity.

Conversely, by employing Theorems 3.3.1 and 3.3.2 in the other direction, the sufficiency is done. □

This theorem allows us to represent a non-orientable surface of genus $q = k$ by an $l \times l$ square S_0 and $k - 1$ small $s \times s$ squares $S_1, \ldots, S_{k-1}$ on the diagonal of S_0 such that the distance between two nearest squares is d as shown in Figure 13.3.1(b).

Surface model 2. A square, or nearest, with a β-ruler on the boundary of length 2β.

In fact, surfaces in this model are the associate polyhegons in the joint tree model of embeddings of a graph with Betti number β mentioned in Sections 9.1–9.2.

Theorem 11.3.3. *A surface is orientable with genus $p \geq 1$ if, and only if, $\delta = 0$ with the irreducible length $4p$ of the polyhegon, where δ is the binary number of β digits given in Section 9.2.*

Proof. From Theorem 3.1.4 and Lemma 3.2.1, by considering that the orientable genus is half its pregenus, the theorem is found. □

According to this theorem, we are allowed to represent an orientable surface with length 2β of genus p by a $\lfloor \beta/2 \rfloor \times \lceil \beta/2 \rceil$ rectangle of irreducible length $4p$. Of course, it is a square when $\beta = 0(\mathrm{mod}\ 2)$ shown in Figure 11.3.2(a).

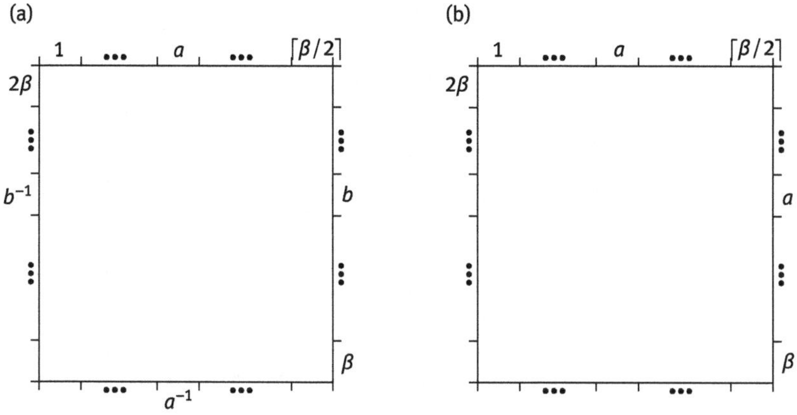

Figure 11.3.2: (a) Orientable surface and (b) non-orientable surface.

Theorem 11.3.4. *A surface is non-orientable with genus $q \geq 1$ if, and only if, $\delta \neq 0$ and the irreducible length $2q$ of the polyhegon where δ is the binary number of β digits given in Section 9.2.*

Proof. By the similar reason to the proof of Theorem 11.3.3, the theorem can also be found. □

According to this theorem, we are allowed to represent an orientable surface with length 2β of genus p by a $\lfloor \beta/2 \rfloor \times \lceil \beta/2 \rceil$ rectangle of irreducible length $2q$. Of course, it is a square when $\beta = 0 (\mathrm{mod}\ 2)$ shown in Figure 11.3.2(b).

Surface model 3. A square, or nearest, with a standard ruler on the boundary.

Theorem 11.3.5. *A surface in this model is orientable with genus $p \geq 1$ if, and only if, the $p \times p$ square is with the boundary partitioned into $4p$ equal segments in form as*

$$\left(\prod_{i=1}^{p} a_i b_i a_i^{-1} b_i^{-1} \right) .$$

Proof. This is, in fact, the standard form of an orientable surface of genus p shown in Figure 11.3.3(a). □

Theorem 11.3.6. *A surface in this model is non-orientable with genus $q \geq 1$ if, and only if, the $\lfloor q/2 \rfloor \times \lceil q/2 \rceil$ rectangle is with the boundary partitioned into $2q$ equal segments in form as*

$$\left(\prod_{i=1}^{q} a_i a_i \right) ,$$

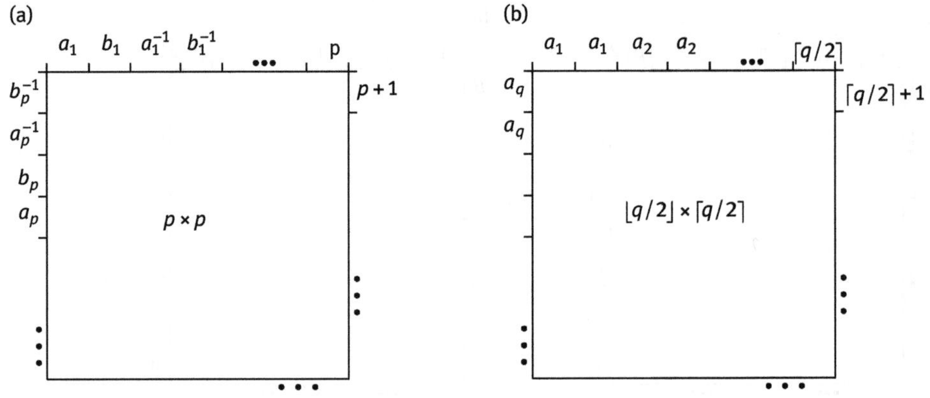

Figure 11.3.3: (a) Orientable surface and (b) non-orientable surface.

Proof. This is, in fact, the standard form of a non-orientable surface of genus q as shown in Figure 11.3.3(b). □

11.4 On surfaces of genus not zero

For the three types of models discussed in the last section, the rectilinear embeddings of a graph have to be separably investigated. However, only model 2 is particularly considered on the basis of joint trees as an example for other two in difference of complication.

The crucial idea is to transform the surface rectilinear embedding problem into a planar rectilinear embedding problem with some face boundary (or boundaries) measured by rulers.

Theorem 11.4.1. *All embeddings of a cubic graph (or 3-graph) is 2-extensible for surface model 2.*

Proof. It is seen that a joint tree of a cubic graph (not necessary to be planar) is a planar embedding of a cubic graph with the outer face boundary of a length even.

From Theorems 11.2.5 and 11.2.8, the joint tree is 1-extensible. By Lemma 11.2.1, such a 1-extension can be with all vertices straight on the outer-face boundary. By considering the arbitrariness of the boundary available, the 1-extension can be taken with the rectangle in surface model 2.

However, vertices on the boundary are by no meas all at the centre of each unit of the ruler. By observing that 1-bend edge incident to a vertex on the boundary is not essential, only straight edge has to add additional two bends(a zigzag), if necessary, for the end of the edge on the boundary at the centre.

Therefore, the theorem is proved. □

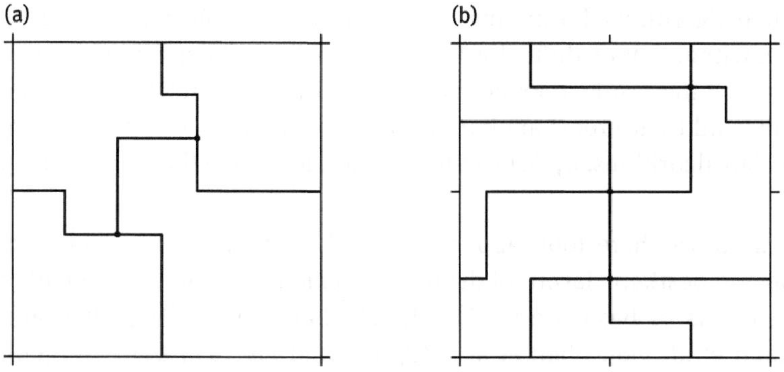

Figure 11.4.1: (a) A 2-embedding of H_1 and (b) a 2-embedding of H_2.

The conclusion of this theorem is best possible because of the instance shown in Figure 11.4.1(a), a 2-embedding of cubic graph H_1, i.e. the prism, on a surface.

Theorem 11.4.2. *All embeddings of a 4-regular graph (or 4-graph) is 2-extensible for surface model 2.*

Proof. Because joint trees of a 4-regular graph are all with vertices of valency 3 on the outer face boundary. Because of 3-regularity on the outer face boundary and no configuration ⟨4, 4, 4⟩, from Lemma 11.2.18, a 2-extension can be seen.

By considering the arbitrariness of the boundary available, the 2-extension can be taken with the rectangle in surface model 2.

However, vertices on the boundary are by no meas all at the centre of each unit of the ruler. By observing that no straight edge incident to a vertex on the boundary is not essential, only straight edge has to append additional two bends (a zigzag), if necessary, for the end of the edge on the boundary at the centre.

Therefore, the extension is still a 2-embedding and hence the theorem. □

The conclusion of this theorem is best possible because of the instance shown in Figure 11.4.1(b), a 2-embedding of the 4-regular graph H_2, a 4-graph with a spanning tree of two edges, on a surface.

11.5 Notes

11.5.1 All the results obtained in this chapter for the recognitions of the k-extensibility when the outer face of a planar embedding given the k-embeddability, and the k-realizability, $k \geq 1$, for graphs of vertex valency less than five can be used to design algorithms, in time linear of computing complexity. Moreover, they can also be used to design linear time algorithms, for finding k-embeddings of graphs known to be k-embeddable, $k \geq 1$, without much difficulty. Most of them did first appear in Liu et al. [239–247]. However, all the results on rectilinear convexity are completely new, and some known results and their proofs are improved and simplified. Of course, we are also allowed to design algorithms, by OD-tree techniques, following the clue provided in Liu [217].

11.5.2 Problems discussed here look applicable in VLSI circuit design, architectural floor plan layout, aesthetic layout of diagram schematics, algorithm animation and other areas such as in Basden and Nichols [17], Batini et al. [18], Bhatt and Leighton [30], Braun et al. [36], Chen et al. [42], Cui and Liu [63], Dai et al. [64], Du and Zhang [81], Hu and Kuh [145], Owens [276], Rosenstiehl and Tarjan [299], Sarrafzadeh and Lee [304], Storer [322], Tamassia [325], Valiant [352], Vanlier and Otten [353], Willamson [397], Wing [399], Wong [401], Yao et al. [416], Yoshida and Nakagawa [417], etc. However, only theoretical treatment is concerned here. In fact, it almost suffices to observe cubic graphs. One might think of k-embeddability of graphs of vertex valency at most $2l$, on the grid lattice, in the l-dimensional space $l \geq 3$. Although no attractive result appears, it can be imagined that some similar results might be found.

11.5.3 One might propose the similar problems for embedding graphs, in a tiling of the plane, such as the 3-gon or 6-gon tiling.

11.5.4 If the placement of vertices in the planar grid lattice is given, finding a 1-embedding, such that each edge has exactly one bend, can be transformed into a problem of determining if a quadratic Boolean equation has a solution in Liu [211] and Liu et al. [243]. In fact, the restriction of exactly one bend is not necessary. Of course, it would lead an approach to other problems, seen in Hammer and Liu [115], Hammer et al. [116].

11.5.5 Recently, some algorithms have appeared. An algorithm for finding a 1-embedding of a cubic graph in linear time can be found in Liu et al. [239]. Linear time algorithms for finding a k-embedding, $k = 2, 3$, of a standard graph are discussed in Cui and Liu [63], Liu et al. [244].

11.5.6 Further optimization problems will be discussed in Chapter 13.

11.5.7 The surface model 2 was first sketched in Liu [219] (Appendix I).

11.5.8 The part on the sphere in this chapter is basically from Liu [217] (Chapter 8). The treatment on surfaces of genus not zero extends the technique used on the sphere as a theoretical foundation.

11.5.9 If not inherited for a property considered, the characterization of this property cannot be done by forbidden minors. The k-embeddability, $k \geq 1$, in this chapter is known with inheritness. However, the forbidden configurations for characterizing embeddability of the type are all shown powerful without usage of minors.

12 Net Embeddings

12.1 Definitions

As usual, we observe 3-connected planar standard graphs first. For a planar embedding $\mu(G)$ of such a graph $G = (V, E)$, a face f is said to be *k-admissible*, $k \geq 0$, if the boundary of f has a k-embedding $\mu_k(f)$ such that

$$\forall v \in V(f), \mathrm{St}(v; f) = \begin{cases} 2, \rho(v, G) = 4; \\ 1, \text{ or } 2, \rho(v; G) = 3; \end{cases} \tag{12.1.1}$$

k-inadmissible, if otherwise. For a circuit C which is allowed to be a face boundary in $\mu(G)$, let $\rho_{\mathrm{in}}(v; \mu C)(\rho_{\mathrm{out}}(v; \mu C))$ represent the number of edges incident with v in the inner (outer) domain $C_{\mathrm{in}}(C_{\mathrm{out}})$ of C. Then, we may generalize the k-admissibility as C is k-admissible, if C has a k-embedding $\mu_k(C)$ such that $\forall v \in V(C)$,

$$\rho_{\mathrm{in}}(v; \mu C) \leq \mathrm{St}(v; C) \leq 2 - \rho_{\mathrm{out}}(v; \mu C); \tag{12.1.2}$$

k-inadmissible, if otherwise. From the definition, one might soon see that a face f is k-admissible if, and only if, the boundary of f is k-admissible when f is treated as the outer face.

Lemma 12.1.1. *For a planar embedding $\mu(G)$ of a 3-connected standard graph $G = (V, E)$, a face f is 1-admissible if, and only if,*

$$v_4(f; \mu G) \leq v(f) - 4, \tag{12.1.3}$$

where $v(f)$ and $v_4(f; \mu(G))$ are the respective numbers of vertices and 4-valent vertices on the boundary of f in $\mu(G)$.

Proof. By the observations that any rectilinear polygon has even number of straight segments and the sum of all the inner angles of a $2l$-gon is $(l + 1)\pi$, any rectilinear $2l$-gon has $l - 2$ inner and $l + 2$ outer vertices and hence contains four straight segments each of which is incident with two outer vertices. Because of the standardness, only 3-valent vertex is allowed to be straight on the outer-face boundary in a 1-embedding, there is at least one 3-valent vertex in each of the straight segments incident with two outer vertices on the boundary of a face f which is 1-admissible. However, the number of such straight segments is at least four. This implies eq. (12.1.3). The necessity is obtained. Conversely, from eq. (12.1.3), we may always find a 1-embedding of the boundary of f such that all 4-valent vertices are inner and others, of course, at least four 3-valent vertices, are straight. This is the sufficiency. □

DOI 10.1515/9783110479492-012

For a circuit C in a planar embedding $\mu(G)$ of a standard graph $G = (V, E)$, a 4-valent vertex v on C is said to be *real inner, real straight* or *real outer* if

$$\rho_{\text{in}}(v; \mu C) = 2,$$
$$\rho_{\text{in}}(v; \mu C) = \rho_{\text{out}}(v; \mu C) = 1,$$

or

$$\rho_{\text{out}}(v; \mu C) = 2,$$

because it is always on the state inner, straight or outer in any k-extension of $\mu(G)$, $k \geq 0$, respectively. Let $v_{\text{in}}(C; \mu G)$, $v_{\text{out}}(C; \mu G)$ and $v_{\text{st}}(C; \mu G)$ be the respective numbers of vertices that are real inner, real outer and real straight on C in $\mu(G)$.

Lemma 12.1.2. *A circuit C which is not the boundary of a face in a planar embedding $\mu(G)$ of a 3-connected standard graph $G = (V, E)$ is 1-inadmissible if, and only if,*

$$\begin{cases} v_{\text{out}}(C; \mu G) = 0; \\ v_{\text{st}}(C; \mu G) = 3; \\ v_{\text{in}}(C; \mu G) = v(C) - 3. \end{cases} \tag{12.1.4}$$

Proof. Suppose C is a 1-inadmissible circuit in $\mu(G)$. Because of the 3-connectedness, there are at least three vertices in $V_{\text{op}}(C) = \{v | \forall v \in V(C), \rho_{\text{out}}(v; \mu C) > 0\}$. We have the following facts.

Fact 1. $|V_{\text{op}}(C)| = 3$. If $|V_{\text{op}}(C)| > 3$ otherwise, then at least four vertices are allowed on the straight or outer state in a 1-embedding of C. From Lemma 12.1.1, C is 1-admissible. This is a contradiction to the given condition on C.

Fact 2. $\nexists\, v \in V_{\text{op}}(C), \rho_{\text{in}}(v; \mu C) = 0$. If v_1 satisfies $\rho_{\text{in}}(v_1; \mu C) = 0$ otherwise, let $v_1 v_2 v_3$ are the vertices in $V_{\text{op}}(C)$ appearing to the order in agreement with that on C. It suffices to discuss the three possibilities: **A.** $v_1 I v_2 v_3$; **B.** $v_1 I v_2 I v_3$; and **C.** $v_1 I v_2 I v_3 I$, where I represents real inner vertices involved.

However, all of them lead to C 1-admissible as shown in Figure 12.1.1.

Fact 3. All vertices on C are 4-valent in G. If there is a vertex which is not 4-valent on C otherwise, then from Lemma 12.1.1, C has to be 1-admissible.

From the facts 1–3, eq. (12.1.4) is satisfied by C. This is the necessity.

Conversely, if C is treated as the outer-face boundary of $\mu(G_{\text{in}}) = \mu(C) \cup \mu(C_{\text{in}} \cap G)$, then from Lemma 12.1.1, the sufficiency is derived because eq. (12.1.4) leading to eq. (12.1.3) is unsatisfied. □

Theorem 12.1.1. *A planar embedding $\mu(G)$ of a 3-connected standard graph $G = (V, E)$ is 1-extensible if, and only if, $\mu(G)$ has no circuit of non-boundary of a face 1-inadmissible with the outer face 1-admissible.*

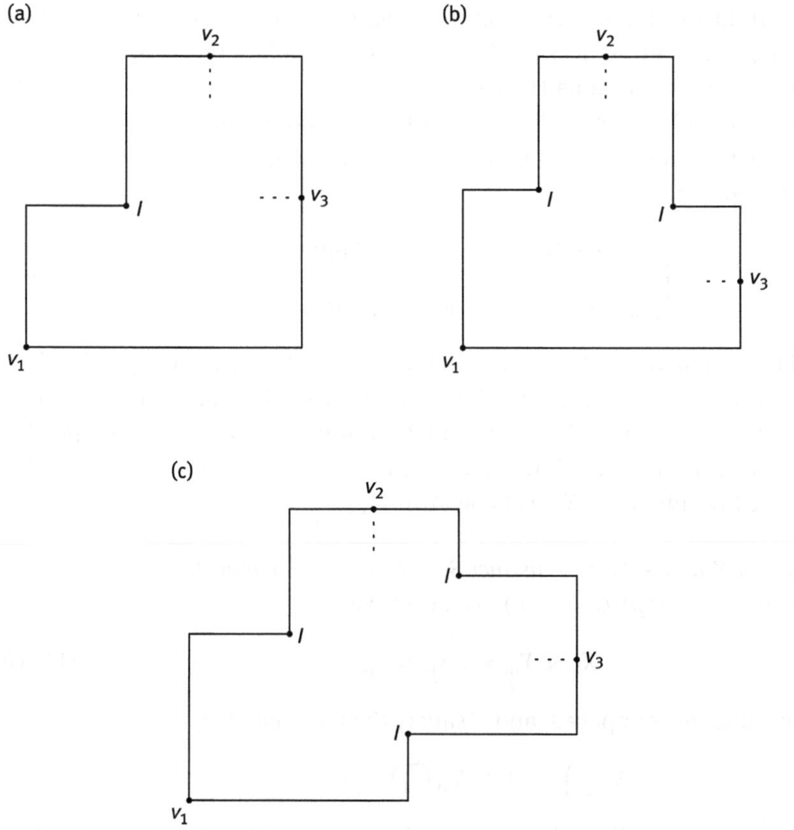

Figure 12.1.1: (a) **A**; (b) **B**; and (c) **C**.

Proof. Because eq. (12.1.4) shows that the circuit C actually associates with a configuration $\langle 4,4,4\rangle$ as described in the last chapter, the necessity is obvious.

Conversely, from the 1-admissibility of the outer face f_0 of $\mu(G)$, in view of Lemma 12.1.1, f_0 satisfies eq. (12.1.3) and hence does not associate with a triangle $(4,4,4)$. Further, by the condition of no configuration $\langle 4,4,4\rangle$, from Theorem 11.2.4, $\mu(G)$ is convex 2-extensible. Because f_0 is 1-embeddable, then the extension is in fact a convex 1-embedding and hence 1-extensible. □

In what follows, we investigate the complete set of forbidden configurations for the 1-embeddability. First, as we have known, any obscurity is a forbidden configuration. From Lemma 12.1.1, an obscurity, denoted by Ω as well, is in fact such a minimal graph (standard of course) that each of its faces has at most three 3-valent vertices. Because any 3-valent face has at most three 3-valent vertices, maximal standard planar graphs without separating triangle are all obscurities for the 1-embeddability. In fact, there are only three maximal standard graphs: Π_4, Π_8 and the one denoted by Π_{4+v} which is obtained by introducing a new vertex v in the inner domain of a face with three new edges connecting the three vertices on the boundary of the face on Π_4 by the similar reason used in the proof of Lemma 11.2.27.

If a circuit C in a planar embedding $\mu(G)$ of a standard graph $G = (V, E)$ has all its vertices of valency four such that there is three vertices v_1, v_2 and v_3 on C with the following property: $\forall v \in V(C)$,

$$\begin{cases} \rho_{\text{out}}(v; \mu C) = 2, v \neq v_i, i = 1, 2 \text{ and } 3; \\ \rho_{\text{out}}(v; \mu C) = 1, v = v_i, i = 1, 2 \text{ or } 3, \end{cases} \tag{12.1.5}$$

then C is said to be an *outer* $\langle 4,4,4\rangle$*-circuit* denoted by $\langle 4,4,4\rangle_{out}$. Correspondingly, a configuration $\langle 4,4,4\rangle$ that has appeared before is also called an *inner* $\langle 4,4,4\rangle$*-circuit* denoted by $\langle 4,4,4\rangle_{\text{in}}$ as shown in Figure 12.1.2. Both of them are called *single side* $\langle 4,4,4\rangle$*-circuits* or simply denoted by S-$\langle 4,4,4\rangle$. In fact, if X is a $\langle 4,4,4\rangle_{\text{out}}$, then for any face in X_{in} as the outer face, X has to be $\langle 4,4,4\rangle_{\text{in}}$.

Lemma 12.1.3. *Let X and Y be two distinct S-$\langle 4,4,4\rangle$ in a 3-connected planar embedding $\mu(G)$ of a standard graph $G = (V, E)$, then we have*

$$X_{\text{in}} \not\subseteq Y_{\text{in}} \Longrightarrow X_{\text{in}} \subseteq Y_{\text{out}}. \tag{12.1.6}$$

Proof. By contradiction, suppose X and Y satisfy the relations that

$$X_{\text{in}} \bigcap Y_{\text{in}} \neq \emptyset; \; X_{\text{in}} \bigcap Y_{\text{out}} \neq \emptyset.$$

From the symmetry, we are allowed to assume X is $\langle 4,4,4\rangle_{\text{in}}$ without loss of generality. Because each vertex on X has two incident edges in X_{in} except only for v_i, $i = 1, 2$ and 3, in view of Y having edges in both X_{in} and X_{out}, from Jordan curve theorem X and Y have to be with exactly two of v_i, $i = 1, 2$ and 3, in common. This is a contradiction to the 3-connectedness of G. □

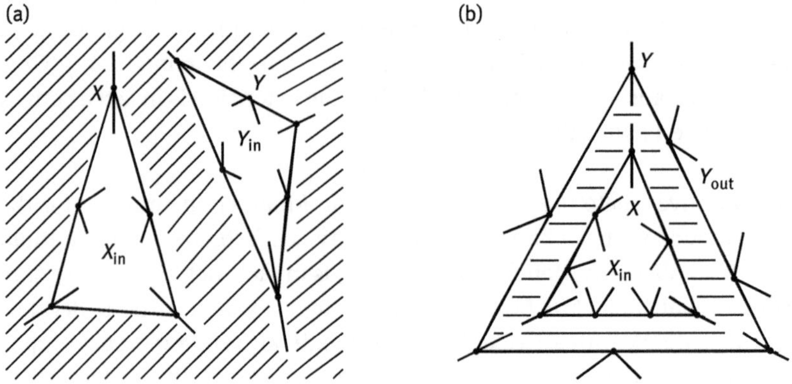

Figure 12.1.2: (a) $\Pi(X_{\text{in}}, Y_{\text{in}})$ and (b) $\Pi(X_{\text{in}} \subseteq Y_{\text{in}})$.

If X is with $\langle 4,4,4\rangle_{\text{in}}$ and Y with $\langle 4,4,4\rangle_{\text{out}}$ such that $X_{\text{in}} \subseteq Y_{\text{in}}$, then they are called a $\langle 4,4,4\rangle$-*doughnut*, as shown in Figure 12.1.2(b), denoted by $\Pi(X_{\text{in}} \subseteq Y_{\text{in}})$. If both of X and Y are with $\langle 4,4,4\rangle_{\text{in}}$ such that $X_{\text{in}} \cap Y_{\text{in}} = \emptyset$, then they are called $\langle 4,4,4\rangle$-*spectacles*, as shown in Figure 12.1.2(a), denoted by $\Pi(X_{\text{in}}, Y_{\text{in}})$. Further, if there is no $\langle 4,4,4\rangle$-doughnut in $Y_{\text{in}} - X_{\text{in}}$ for $\Pi(X_{\text{in}} \subseteq Y_{\text{in}})$ or no $\langle 4,4,4\rangle$-spectacles in $X_{\text{in}} \cup Y_{\text{in}}$ for $\Pi(X_{\text{in}}, Y_{\text{in}})$, then both $\Pi(X_{\text{in}} \subseteq Y_{\text{in}})$ and $\Pi(X_{\text{in}}, Y_{\text{in}})$ are simply denoted by Π.

Lemma 12.1.4. *In a planar embedding $\mu(G)$ of a 3-connected standard graph $G = (V, E)$, Π is a forbidden configuration for the 1-embeddability of G.*

Proof. If $\Pi = \Pi(X_{\text{in}}, Y_{\text{in}})$, because X is with $\langle 4,4,4\rangle_{\text{in}}$, from Lemma 12.1.1, X is 1-inadmissible for $X_{\text{in}}(G) = (X \cup X_{\text{in}}) \cap \mu(G)$ and hence $\mu(G)$ is not 1-extensible for any face in X_{out} as the outer face. On the other hand, because Y is with $\langle 4,4,4\rangle_{\text{out}}$, from Lemma 12.1.1, Y is 1-inadmissible for $Y_{\text{out}}(G) = (Y \cup Y_{\text{out}}) \cap \mu(G)$ and hence $\mu(G)$ is not 1-extensible for any face in Y_{in} as the outer face. However, from $X_{\text{in}} \subseteq Y_{\text{in}}$, $\mu(G)$ is not 1-extensible for any face in X_{in} as the outer face. In view of the uniqueness of $\mu(G)$ by Theorem 5.3.3, G is not 1-embeddable.

If $\Pi = \Pi(X_{\text{in}}, Y_{\text{in}})$, by the similar reason, G is never 1-extensible for any face as the outer face. Therefore, G is not 1-embeddable as well.

From the two cases discussed earlier, it implies that Π is a forbidden configuration for the 1-embeddability of G.

The lemma is obtained. □

Theorem 12.1.2. *A 3-connected standard planar graph $G = (V, E)$ is 1-embeddable if, and only if, G has neither Ω as vertex-induced subgraph nor Π.*

Proof. From Lemmas 12.1.1 and 12.1.4, the necessity is obvious. Conversely, if all the faces of the embedding $\mu(G)$ are 1-inadmissible, then G has Ω as a vertex-induced subgraph. This is a contradiction to the given condition. Suppose f is a face of $\mu(G)$, which is 1-admissible. If $\mu(G)$ has no circuit X with $\langle 4,4,4\rangle_{\text{in}}$ for f, then by Theorem 12.1.1, $\mu(G)$ is 1-extensible when f is chosen to be the outer face. Because $\mu(G)$ has no $\Pi(X_{\text{in}}, Y_{\text{in}})$, from Lemma 12.1.3, we are allowed to assume X is such a configuration $\langle 4,4,4\rangle_{\text{in}}$ that there is no $\langle 4,4,4\rangle_{\text{in}}$ in X_{out}. Thus, all the faces, including f, in X_{out} are not allowed to be chosen as the outer face of a 1-embedding. We now have to observe the faces in X_{in}. If they are all 1-inadmissible, then $X_{\text{in}}(G)$ has to contain an Ω as a vertex-induced subgraph, a contradiction to the given condition. Suppose f_1 in X_{in} is a face 1-admissible. Because G has no $\Pi(X_{\text{in}} \subseteq Y_{\text{in}})$, by Lemma 12.1.3 and the similar discussion on f, we may assume X_1 is the circuit with $\langle 4,4,4\rangle_{\text{in}}$ for f_1 in X_{in} such that there is no other circuit with $\langle 4,4,4\rangle_{\text{in}}$ in $X_{1\text{in}}$. Then all the faces, including f_1 of course, in $X_{1\text{in}}$ are not allowed to be an outer face of a 1-embedding of G. If all the faces in $X_{\text{in}} \cap X_{1\text{out}}$ are 1-inadmissible, then it has to contain an Ω, a contradiction to the given condition of G as well. Assume f_2 is a face 1-admissible in $X_{\text{in}} \cap X_{1\text{out}}$. Of course, if $\mu(G)$ has no circuit with $\langle 4,4,4\rangle_{\text{in}}$ for f_2, then from Theorem 12.1.1, $\mu(G)$ is 1-extensible for f_2 as the outer face. Otherwise, do the procedure further, from the finiteness of the number of faces, so a face has to be found such that $\mu(G)$ is 1-extensible for the face as the outer face according to that G has neither Ω nor Π. In consequence, G is 1-embeddable. The sufficiency is obtained. □

Corollary 12.1.1. *Any 4-valent planar graph $G = (V, E)$, or standard with less than four 3-valent vertices, is not 1-embeddable.*

Proof. Because all faces are 1-inadmissible, G contains Ω as a vertex-induced subgraph. From Theorem 12.1.2, it follows. □

However, a standard planar graph with four 3-valent vertices is allowed to be 1-embeddable. In fact, $W_{4,t}, t \geq 1$, is such a kind of graphs whose 1-embeddings are shown in Figure 11.2.10(b).

In order to discuss 1-embeddability of general planar graphs, of course with vertices of valencies less than five, we may generalize Theorem 12.1.1 for the 1-extensibility first. Moreover, we have to investigate all the planar embeddings in a similar way to the last chapter according to the theory described in Section 5.5 for the 1-embeddability.

12.2 Face admissibility

For 0-embeddability, or say *net embeddability* of graphs, because any standard planar graph is never 0-embeddable as will be seen, we have to observe graphs in which 2-valent vertices are allowed. Of course, a 0-embeddable graph can be obtained by subdividing some edges in a k-embeddable graph for $k \geq 1$. A graph that has a 0-embedding is called a *net graph.*

First, we investigate the 0-embeddings of a circuit. Of course, it may soon be seen that any triangle, i.e. a circuit of length three, can never have a 0-embedding because the shortest circuits in the *grid graph* that represents the grid lattice in the plane have four vertices. Therefore, graphs considered here have to be without triangle and of course with all the vertices of valencies less than five preassumed as in the last chapter.

Let $\mu_0(C)$ be a 0-embedding of a circuit C. Then, the vertices on C can be partitioned into three parts: one, denoted by S, consists of all those incident to an angle of π, the other two, denoted by I and O, are all those at a corner with the angle of $\frac{3}{2}\pi$ and $\frac{\pi}{2}$ in the inner domain of $\mu_0(C)$, respectively. If vertices of state S are not considered, then $\mu_0(C)$ determines a cyclic sequence which consists of I and O in the clockwise, for instance, on $\mu_0(C)$. Such a sequence is called the *corner sequence of* $\mu_0(C)$. For a given cyclic sequence *Seq* of symbols O and I, let $n_O(\text{Seq})$ and $n_I(\text{Seq})$ be the respective numbers of O and I which appear in Seq. If *Seq* is a corner sequence of a circuit in the grid graph then Seq is said to be a *corner sequence* itself.

Lemma 12.2.1. *A cyclic sequence Seq is a corner sequence if, and only if, it has even length and satisfies*

$$n_O(Seq) - n_I(Seq) = 4. \tag{12.2.1}$$

Proof. To prove the necessity. Let m be the length of Seq. Because Seq is a corner sequence, there is an m-gon $m \geq 4$, in the grid graph such that Seq is its corner sequence. Suppose the m-gon has k horizontal segments. Since each vertex is on exactly one of the segments each of which joins exactly two vertices, we have $m = 2k, k \geq 2$. Because the sum of inner angles is $2(k-1)\pi$ with $(k-2)$ angles of $\frac{3}{2}\pi$ and $k+2$ angles of $\frac{\pi}{2}$, by noticing that $n_O(\text{Seq})$ and $n_I(\text{Seq})$ are respective numbers of angles of $\frac{\pi}{2}$ and $\frac{3}{2}\pi$, eq. (12.2.1) is satisfied.

To prove the sufficiency. By induction on the length of Seq, when the length is four, the only possibility is that

$$\text{Seq} = OOOO.$$

It is surely a corner sequence of the square. Suppose any sequence of length $2l, l \geq 2$, which satisfies eq. (12.2.1), is a corner sequence. We observe Seq of length $2(l+1)$ with eq. (12.2.1). First, from eq. (12.2.1), there is an "OI" as a segment in Seq. If the segment is left off Seq, then the resultant sequence Seq$'$ has its length of $2l$ and satisfies eq. (12.2.1). By the hypothesis, Seq$'$ is a corner sequence. We may for convenience assume that Seq$'$ has the form

$$\text{Seq}' = X_1X_2\cdots X_{2l}.$$

Then, Seq can be seen as

$$\text{Seq} = \text{Seq}'X_{2l+1}X_{2l+2},$$

where $X_{2l+1} = O$ and $X_{2l+2} = I$. Let Seq$'$ be the corner sequence of a 0-embedding $\mu_O^r(C_{2l})$ of circuit

$$C_{2l} = v_1v_2\cdots v_{2l}$$

such that v_i is of the state $X_i, i = 1, 2, \ldots, 2l$. According to the four exhaustive possibilities of the choices of X_1 and X_{2l}:

A1. $X_1 = O$ and $X_{2l} = O$;

A2. $X_1 = I$ and $X_{2l} = O$;

A3. $X_1 = O$ and $X_{2l} = I$;

A4. $X_1 = I$ and $X_{2l} = I$,

we can always find

$$C_{2l+2} = v_1 v_2 \cdots v_{2l} v_{2l+1} v_{2l+2}$$

such that Seq is the corner sequence of a 0-embedding $\mu_0(C_{2l+2})$ as shown in Figures 12.2.1–12.2.4. □

A cyclic sequence Seq which consists of symbols O and I is said to be *I-isolated* if any I in Seq does not have a neighbour I, or in other words, II is not a segment of Seq. If a cyclic sequence of O and I is the corner sequence of a convex polygon in the grid graph, then it is called *convex* as well.

Attention 12.2.1. A corner sequence can also be seen as a binary sequence whenever O and I are, respectively, replaced by 0 and 1. Then, one might find a number of interesting properties particularly similar to what has been discussed in this section.

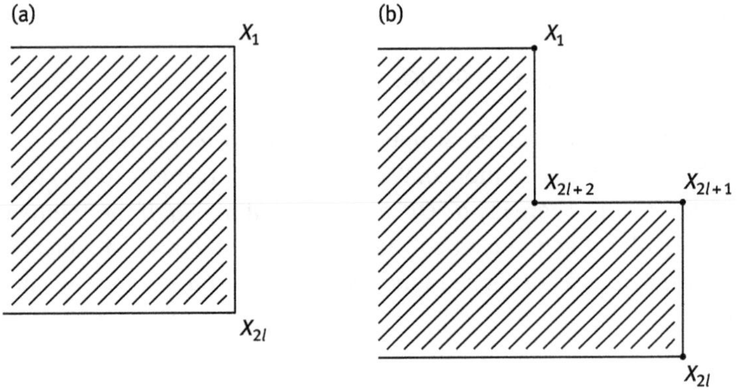

Figure 12.2.1: (a) Seq′ and (b) Seq.

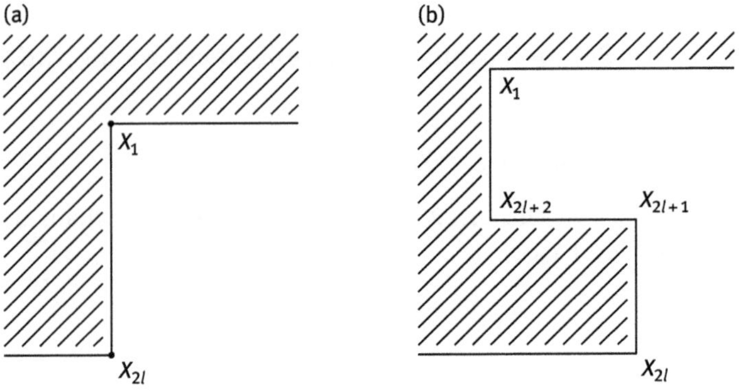

Figure 12.2.2: (a) Seq′ and (b) Seq.

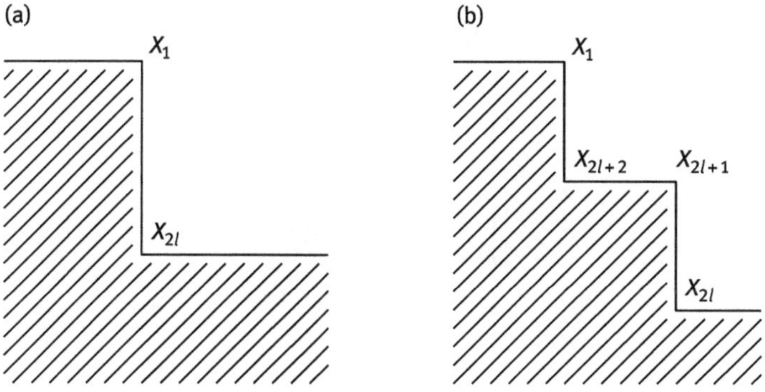

Figure 12.2.3: (a) Seq′ and (b) Seq.

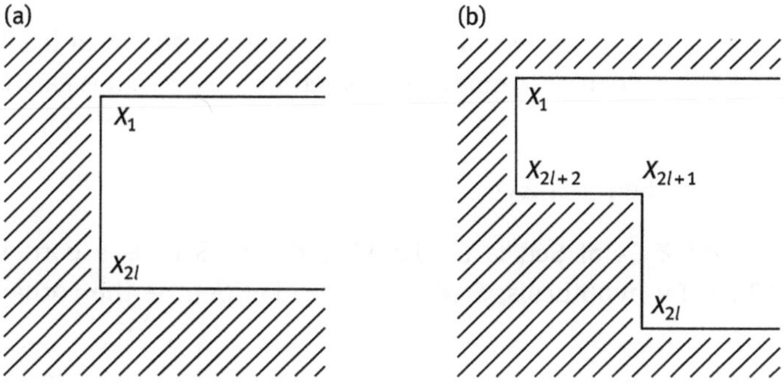

Figure 12.2.4: (a) Seq′ and (b) Seq.

Lemma 12.2.2. *A cyclic sequence Seq of O and I is convex if, and only if, it is a corner sequence which is I-isolated.*

Proof. To prove the necessity. The first statement is obvious from the definition of the convexity for Seq. By contradiction, to prove the second statement, suppose there is a segment *II* in Seq. Let Seq be the corner sequence of $C_{2l} = v_1 v_2 \cdots v_{2l}$ such that both v_1 and v_2 have the state I without loss of generality. Then, there are only two possibilities as shown in Figure 12.2.5 for the location of v_1 and v_2. They both lead to a contradiction to the convexity.

To prove the sufficiency. By induction on the length of

$$\mathrm{Seq} = X_1 X_2 \cdots X_{2l},\ l \geq 2.$$

If $l = 2$, then Seq = *OOOO* which is I-isolated. Of course, Seq is convex because it is the corner sequence of a rectangle which is convex in the grid graph. In general, suppose the sufficiency is true for sequences of length not greater than l, $l \geq 2$. We are allowed to consider

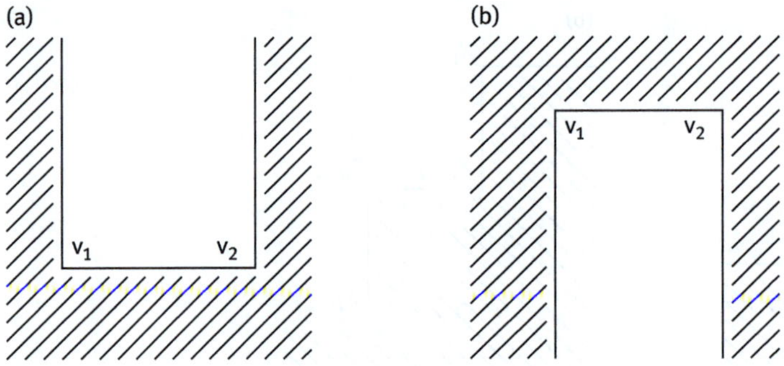

Figure 12.2.5: Two corner sequences.

$$\mathrm{Seq} = X_1X_2\cdots X_{2l+1}X_{2l+2}$$

such that $X_{2l+1} = O$ and $X_{2l+2} = I$ without loss of generality because Seq has at least one I for $l \geq 2$ from Lemma 12.2.1. Of course, from the I-isolatedness, $X_1 = O$. It is easily seen that

$$\mathrm{Seq}' = X_1X_2\cdots X_{2l}$$

obtained by leaving X_{2l+1} and X_{2l+2} of Seq is I-isolated and that Seq′ is a corner sequence from Lemma 12.2.1. By hypothesis, *Seq*' is convex. Let Seq′ is the corner sequence of $2l$-gon

$$C_{2l} = v_1v_2\cdots v_{2l}$$

in the grid graph such that v_i is on the state X_i, $1 \leq i \leq 2l$. However, Seq′ has two possibilities: $X_{2l} = O$ and $X_{2l} = I$ for each of which, Seq is shown to be convex as in Figures 12.2.6 and 12.2.7. □

Based on Lemmas 12.2.1 and 12.2.2, we are now allowed to observe the admissibility of a face for the 0-embeddability. The 0-admissibility of a face is always meant that for the face as the outer face if it is 0-extensible. For a face of a planar embedding $\mu(G)$ of a graph $G = (V, E)$ with valencies of vertices less than five, let $n_{i\text{-val}}(f)$ be the number of i-valent vertices on the boundary of the face, f, $i = 2, 3$ or 4.

Theorem 12.2.1. *A face f of a planar embedding $\mu(G)$ of a graph $G = (V, E)$ is 0-admissible if, and only if, all the vertices on the boundary of f have their valencies less than five and satisfy the relation:*

$$n_{2\text{-val}}(f) \geq n_{4\text{-val}}(f) + 4. \tag{12.2.2}$$

Further, if i,

$$0 \leq i \leq n_{2\text{-val}}(f),$$

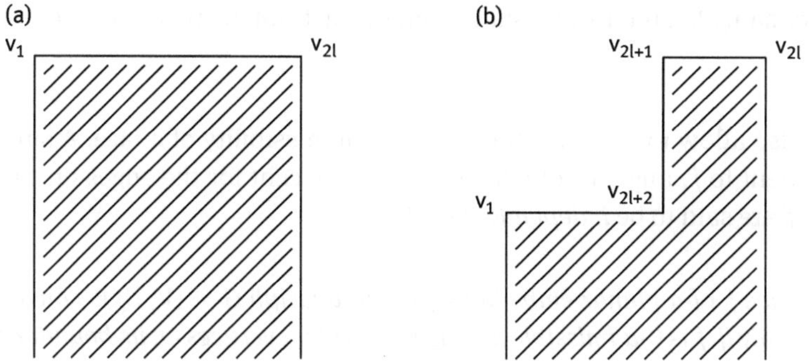

Figure 12.2.6: $X_{2l} = O$.

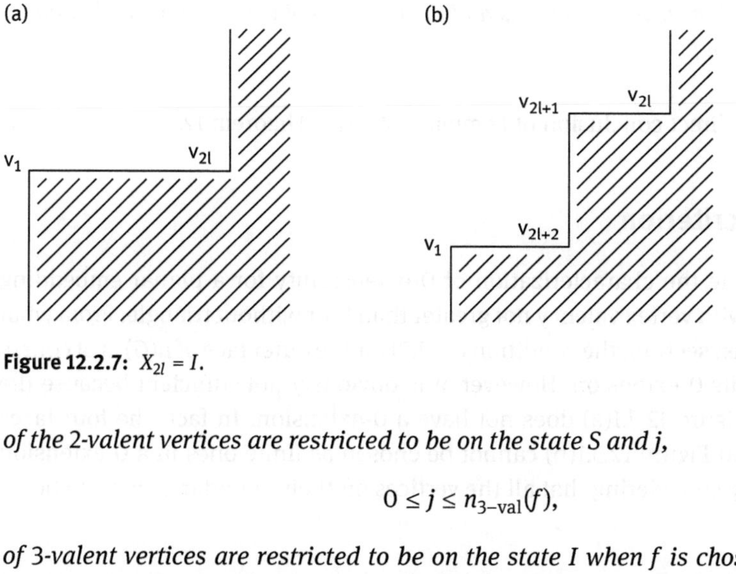

Figure 12.2.7: $X_{2l} = I$.

of the 2-valent vertices are restricted to be on the state S and j,

$$0 \le j \le n_{3\text{-val}}(f),$$

of 3-valent vertices are restricted to be on the state I when f is chosen to be the outer face, then f is 0-admissible if, and only if, all the valencies of vertices on the boundary of f are less than five and satisfy the relation

$$n_{2\text{-val}}(f) - i \ge n_{4\text{-val}}(f) + j + 4. \tag{12.2.3}$$

Proof. Because any 4-valent vertex on the boundary of the outer face in a 0-embedding has to be on the state I and only 2-valent vertices are allowed to be on the state O, from eq. (12.2.1) we have eq. (12.2.2). This is the necessity of the first statement. Conversely, by choosing

$$n_{2\text{-val}}(f) - n_{4\text{-val}}(f) - 4$$

of 2-valent vertices and all 3-valent vertices on the state S, from Lemma 12.2.1, the sufficiency of the first statement is obtained.

Further, by Lemma 12.2.1 and the first statement, the last statement of the theorem is found. □

Similarly, we may also allow to establish the convex 0-admissibility of a face. A subset of vertices, incident to f, any pair of which are not adjacent for the edges on the boundary of a face f are said to be *independent* for f.

Theorem 12.2.2. *A face f of a planar embedding $\mu(G)$ of a graph $G = (V, E)$ is convex 0-admissible if, and only if, f is 0-admissible and all the 4-valent vertices on its boundary are independent of f. Further, a face f is convex 0-admissible with the constraints: i, $0 \le i \le n_{2\text{-val}}(f)$, of 2-valent vertices are on the state S and j, $0 \le j \le n_{3\text{-val}}(f)$, of 3-valent vertices are on the state I if, and only if, f is 0-admissible with the constraints and the set which consists of all the 4-valent vertices and the j vertices of valency 3 on the boundary of f is independent of f.*

Proof. This is the direct conclusion of Lemma 12.2.2 and Theorem 12.2.1. □

12.3 General criterion

First, we investigate the characterization of 0-extensibility for a planar embedding $\mu(G)$ of a graph G with vertex valency not greater than four without triangle. From what discussed in the last section, the condition (12.2.2) for the outer face of $\mu(G)$ is of course necessary to find its 0-extension. However, it is obviously not sufficient because the graph shown in Figure 12.3.1(a) does not have a 0-extension. In fact, the four faces with the shadow in Figure 12.3.1(b) cannot be chosen as finite ones in a 0-extension from eq. (12.2.1) by considering that all the vertices on their boundaries have to be on the state O.

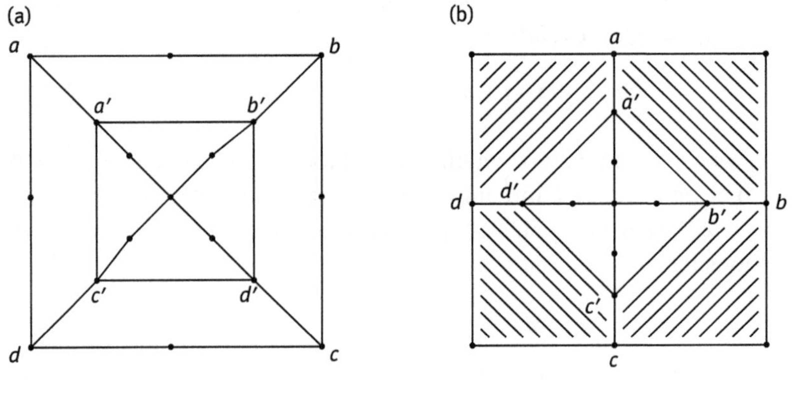

Figure 12.3.1: (a) $\mu(G)$ and (b) $\mu'(G)$.

This suggests us to observe the so-called inner 0-admissibility of a face. A face f in a planar embedding of a graph is said to be *inner 0-admissible* if it is allowable as an inner face of a 0-embedding of the graph.

Lemma 12.3.1. *A face f in a planar embedding $\mu(G)$ of a graph with vertex valency not greater than four without triangle is inner 0-admissible with the restriction that j of 3-valent vertices are on the state O for $0 \le j \le n_{3\text{-}val}(f)$ if, and only if, the following inequality holds:*

$$n_{2-\mathrm{val}}(f) \ge n_{4-\mathrm{val}}(f) + j - 4, \qquad (12.3.1)$$

where f on the two sides are not the same face.

Proof. Because any 4-valent vertex has to be on the state O of an inner face boundary, any 3-valent vertex has two choices of the state O or the state S, and 2-valent vertex is allowed to be on one of the three states, and from eq. (12.2.1) we have the necessity. Conversely, if $n_{4\text{-val}}(f) + j - 4$ of 2-valent vertices are chosen to be on the state I, j of 3-valent vertices to be on the state O and other 2- and 3-valent vertices on the state S, then from eq. (12.2.1) we may also find a rectilinear polygon which is allowable for f in a 0-embedding of G by noticing that all 4-valent vertices are of course on the state O. This is the sufficiency. □

Because a 4-valent vertex has to have its incident angles as $\pi/2$ each and a 4-valent face has to have its incident angles as $\pi/2$ as well except only for the outer face with the four angles $3\pi/2$ each if 4-valent, we are allowed to observe the distribution of the vertices of valencies less than four and the faces of valencies greater than four in a 0-embedding $\mu_0(G)$. Let $\mathrm{def}(v) = 4 - \rho(v; G)$ be called the *deficiency* of vertex v on G and $\mathrm{res}(f) = \rho(f; \mu G) - 4$ be called the *residual* of $f \in F$ in a planar embedding $\mu(G)$ of a graph $G = (V, E)$ with F being the set of all faces in $\mu(G)$.

Lemma 12.3.2. *For a planar embedding $\mu(G)$ of a graph $G = (V, E)$, we have*

$$\sum_{v \in V} \mathrm{def}(v) - \sum_{f \in F} \mathrm{res}(f) = 8. \qquad (12.3.2)$$

Proof. Because an edge has two ends, from the definition of deficiency of a vertex we have

$$\begin{aligned} 4v(G) &= \sum_{v \in V} \mathrm{def}(v) + \sum_{v \in V} \rho(v; G) \\ &= \sum_{v \in V} \mathrm{def}(v) + 2\epsilon(G). \end{aligned} \qquad (12.3.3)$$

Dually, we have

$$\begin{aligned}4\phi(G) &= \sum_{f\in F} \rho(f;\mu G) - \sum_{f\in F} \mathrm{res}(f)\\ &= 2\epsilon(G) - \sum_{f\in F} \mathrm{res}(f).\end{aligned} \tag{12.3.4}$$

In consequence, from eqs. (12.3.3) and (12.3.4) and the Euler formula for planar embeddings, we obtain

$$\sum_{v\in V} \mathrm{def}(v) - \sum_{f\in F} \mathrm{res}(f) = 4(\nu(G) - \epsilon(G) + \varphi(G)) = 8.$$

This is eq. (12.3.2). □

The relation (12.3.2) for a planar embedding $\mu(G)$ of a graph G is said to be the *equilibrium* of $\mu(G)$. The value $\sum_{v\in V} \mathrm{def}(v)$, or $8 + \sum_{f\in F} \mathrm{res}(f)$ as well, denoted by *eq* is called the *equilibrious value* of $\mu(G)$. If an equilibrium satisfies

$$\begin{cases}\forall v \in V, & \mathrm{def}(v) \geq 0;\\ \forall f \in F, & \mathrm{res}(f) \geq 0.\end{cases} \tag{12.3.5}$$

then it is called *positive*. It is easily seen that only graphs with positive equilibrium are possible to have a 0-embedding.

Next, we have to investigate the assignment of the angles at each vertex of deficiency greater than zero or in a face of residual greater than zero. Because articulate vertices and hence articulate edges are not substantial for our purpose here, all graphs are of course assumed to be without articulate vertex. Thus, any angle in a 0-embedding $\mu_0(G)$ is only allowed to be one of $\pi/2, \pi, 3\pi/2$. Let $\mathscr{A}(\mu_0 G)$ be the set of all angles in $\mu_0(G)$. And, let $\mathscr{A}(f;\mu_0 G)$ and $\mathscr{A}(v;\mu_0 G)$ be the respective sets of the angles incident with v and in f. For convenience, we introduce integral function called an *assignment* θ on $\mathscr{A}(\mu_0 G)$ as $\forall A \in \mathscr{A}(\mu_0 G)$,

$$\theta(A) = \begin{cases}0, & \text{if } A = \pi/2;\\ 1, & \text{if } A = \pi;\\ 2, & \text{if } A = \frac{3\pi}{2},\end{cases} \tag{12.3.6}$$

which is said to be the *assigned value* of A in $\mu_0(G)$.

Lemma 12.3.3. *For a* 0*-embedding* $\mu_0(G)$ *of* $G = (V, E)$ *with the face set* F*, the following relations hold*:

$$\begin{cases} \forall v \in V, & \sum_{A \in \mathscr{A}(v;\mu_0 G)} \theta(A) = def(v); \\ \forall f \in F, & \sum_{A \in \mathscr{A}(f;\mu_0 G)} \theta(A) \\ & = \begin{cases} res(f), & if\ f \neq f_0; \\ res(f) + 8, & if\ f = f_0, \end{cases} \end{cases} \tag{12.3.7}$$

where f_0 is the outer face of $\mu_0(G)$.

Proof. First, let us write

$$\tilde{\theta} = 1 - \theta, \tag{12.3.8}$$

then we have $\forall v \in V$,

$$\sum_{A \in \mathscr{A}(v;\mu_0 G)} \tilde{\theta}(A) = \begin{cases} 4, & \text{if } \rho(v; \mu_0 G) = 4; \\ 2, & \text{if } \rho(v; \mu_0 G) = 3; \\ 0, & \text{if } \rho(v; \mu_0 G) = 2. \end{cases}$$

On the other hand, from eq. (12.3.8),

$$\sum_{A \in \mathscr{A}(v;\mu_0 G)} \tilde{\theta}(A) = \rho(v; \mu_0 G) - \sum_{A \in \mathscr{A}(v;\mu_0 G)} \theta(A).$$

This implies the first relation of eq. (12.3.7).

Further, from Lemma 12.2.1, we have $\forall f \in F - \{f_0\}$,

$$\begin{aligned} 4 &= \sum_{A \in \mathscr{A}(f;\mu_0 G)} \tilde{\theta}(A) \\ &= \rho(f; \mu_0 G) - \sum_{A \in \mathscr{A}(f;\mu_0 G)} \theta(A) \end{aligned}$$

and

$$\begin{aligned} -4 &= \sum_{A \in \mathscr{A}(f_0;\mu_0 G)} \tilde{\theta}(A) \\ &= \rho(f_0; \mu_0 G) - \sum_{A \in \mathscr{A}(f_0;\mu_0 G)} \theta(A). \end{aligned}$$

This leads to the second relation of eq. (12.3.7). □

For convenience, let $A_{v,f}$ be the angle incident with $v \in V$ in $f \in F$ for a planar embedding $\mu(G)$ of a graph $G = (V, E)$ with F as the set of faces in $\mu(G)$.

Lemma 12.3.4. *For a planar embedding $\mu(G)$ of a graph $G = (V, E)$ with positive equilibrium without articulate vertex, if there is an integral function $\theta : \mathscr{A}(\mu G) \to \{0, 1, 2\}$ such that eq. (12.3.7) is satisfied, then θ as the assignment determines a 0-extension $\mu_0(G)$ of $\mu(G)$.*

Proof. By induction on the number of faces in $\mu(G)$, when $\mu(G)$ is a circuit $C = v_1 v_2 \cdots v_\nu$ in the cyclic order, $v_i \in V, i = 1, 2, \ldots, \nu$. From the first relation of eq. (12.3.7),

$$\mathrm{def}(v_i) = 2 = \nu(A_{v_i, f_0}) + \theta(A_{v_i, f_1})$$

for $i = 1, 2, \ldots, \nu$, where f_0 and f_1 are the outer and the inner faces of $\mu(G)$, respectively. Thus, $(\theta(A_{v_i, f_0}), \theta(A_{v_i, f_1}))$ has the three possibilities: (1,1), (2,0) or (0,2), which corresponds to

$$(\pi, \pi),\ (3\pi/2, \pi/2) \text{ or } (\pi/2, 3\pi/2)$$

of $(A_{v_i, f_0}, A_{v_i, f_1})$, respectively. From the second relation of eq. (12.3.7) and Lemma 12.2.1, θ as the assignment determines a 0-extension $\mu_0(G)$ of $\mu(G)$.

In general, we may assume v_0 is a 2-valent vertex on the boundary of the outer face f_0 because the outer face f_0 contains at least four angles of $3\pi/2$ from the second relation of eq. (12.3.7) and Lemma 12.2.1. Let G' be the resultant graph of deleting all the inner vertices of $P(v_1, v_2)$ which contains v_0 and some other 2-valent inner vertices with

$$\rho(v_i, G) \geq 3,\ i = 1, 2,$$

and let f_0' be the outer face of $\mu(G')$. It is easy to check that the integral function θ' defined by

$$\theta'(A) = \begin{cases} \theta(A), \text{ if } A \neq A_{v_i, f_0'}; \\ \theta(A_{v_i}, f_0) + \theta(A_{v_i}, f_1) + 1, \text{ otherwise} \end{cases}$$

for $i = 1, 2$, satisfies eq. (12.3.7) on $\mu(G')$ where f_1 is the inner face adjacent to f_0 in $\mu(G)$. By the hypothesis, θ' as the assignment determines a 0-extension $\mu_0(G')$ of $\mu(G')$. Moreover, from the second relation of eq. (12.3.7) and Lemma 12.2.1, $\mu_0(G')$ can be extended into a 0-extension $\mu_0(G)$ of $\mu(G)$ obtained by restoring the path $P(v_1, v_2)$ on $\mu(G')$. □

Theorem 12.3.1. *A planar embedding $\mu(G)$ of a graph G with positive equilibrium without articulate vertex has a 0-extension $\mu_0(G)$ if, and only if, there exists an integral function*

$$\theta : \mathscr{A}(\mu G) \to \{0, 1, 2\}$$

such that θ satisfies relation (12.3.7).

Proof. A direct result of Lemmas 12.3.3 and 12.3.4. □

For a planar embedding $\mu(G)$ of $G = (V, E)$ with positive equilibrium, we introduce a bipartite graph, denoted by

$$\mathrm{Qui} = \mathrm{Qui}(\mu G) = (X_Q, Y_Q; E_Q)$$

associated with $\mu(G)$, determined in the following way:

$$\begin{cases} X_Q &= \bigcup_{\substack{v\in V\\ \mathrm{def}(v)>0}} \Big\{v(i) | \forall i, 1 \le i \le \mathrm{def}(v)\Big\}; \\ Y_Q &= \bigcup_{\substack{f\in F-\{f_0\}\\ \mathrm{res}(f)>0}} \Big\{f(i) | \forall i, 1 \le i \le \mathrm{res}(f)\Big\} \\ & \bigcup \{f_0(i) | \forall i, 1 \le i \le \mathrm{res}(f_0) + 8\} \end{cases} \tag{12.3.9}$$

and

$$\begin{aligned} E_Q = \bigcup_{\substack{v\in V\\ \mathrm{def}(v)>0}} \bigcup_{\substack{f \in F - \{f_0\}\\ \mathrm{res}(f) > 0\\ f \text{ ind } v}} & \Big\{(v(i), f(j)) | \forall i, j, \\ & 1 \le i \le \mathrm{def}(v), 1 \le j \le \mathrm{res}(f)\Big\} \\ & \bigcup \bigcup_{\substack{v\in V(f_0)\\ \mathrm{def}(v)>0}} \Big\{(v(i), f_0(j)) | \forall i, j, \\ & 1 \le i \le \mathrm{def}(v), 1 \le j \le \\ & \mathrm{res}(f_0) + 8\Big\}. \end{aligned} \tag{12.3.10}$$

We call the bipartite graph Qui the *equilibrious graph* of $\mu(G)$. From equilibrium (12.3.2), we always have

$$\mathrm{eq} = |X_Q| = |Y_Q|. \tag{12.3.11}$$

Lemma 12.3.5. *For a planar embedding $\mu(G)$ of a graph $G = (V, E)$, there exists an integral function $\theta : \mathscr{A}(\mu G) \to \{0, 1, 2\}$ such that eq.* (12.3.7) *is satisfied for $\mu(G)$ if, and only if, the equilibrious graph Qui(μG) has a perfect matching.*

Proof. The necessity can be derived from equilibrium (12.3.2) and the definition of the equilibrious graph. Conversely, suppose

$$M = \{(x_i, y_i) | \forall i, 1 \le i \le \mathrm{eq}, x_i \in X_Q, y_i \in Y_Q\} \tag{12.3.12}$$

is a perfect matching on Qui. let

$$\begin{aligned} M(A_{v,f}) = \{(x_i, y_i) | \forall (x_i, y_i) \in M(1 \le i \le eq), \\ (x_i \text{ ind } v) \wedge (y_i \text{ ind } f)\} \end{aligned} \tag{12.3.13}$$

for $A_{v,f} \in \mathscr{A}(\mu G), v \in V, f \in F$. It is easily checked that

$$\forall A \in \mathscr{A}(\mu G),\ \theta(A) = |M(A)| \tag{12.3.14}$$

is an integral function which satisfies condition (12.3.7) from the definition of Qui. □

Theorem 12.3.2. *A planar embedding $\mu(G)$ of a graph $G = (V, E)$ of positive equilibrium without articulate vertex has a 0-extension $\mu_0(G)$ if, and only if, its equilibrious graph* Qui *has a perfect matching.*

Proof. A direct result of Theorem 12.3.1 and Lemma 12.3.5. □

12.4 Special criteria

First, we specify the general criterion described by Theorem 12.3.2 in the way of finding forbidden configurations for the 0-embeddability.

For a planar embedding $\mu(G)$ of a graph $G = (V, E)$ of positive equilibrium with F as the set of faces in $\mu(G)$, let C^* be a cocircuit. From Jordan curve theorem discussed in Section 4.2, $\mu(G \backslash C^*)$ has exactly two connected components denoted by $\mu^*_{\text{in}} = \mu_{\text{in}}(G; C^*)$ and $\mu^*_{\text{out}} = \mu_{\text{out}}(G; C^*)$. Let V^*_{in} and V^*_{out} be the respective vertex sets of μ^*_{in} and μ^*_{out}. If $\tilde{V} = V^*_{\text{in}}$ (or V^*_{out}) satisfies

$$\sum_{v \in \tilde{V}} \mathrm{def}(v) > \sum_{\substack{f \in F \\ f \text{ ind } v}} \mathrm{res}_0(f) \tag{12.4.1}$$

and no cocircuit C^*_1 has $\mu_{\text{in}}(G; C^*_1) \subset \mu^*_{\text{in}}$ (or $\mu_{\text{out}}(G; C^*_1) \subset \mu^*_{\text{out}}$) and satisfies eq. (12.4.1), then C^* is said to be *vertex unequilibrious* and denoted by Ω_{ver}. In eq. (12.4.1),

$$\mathrm{res}_0(f) = \begin{cases} \mathrm{res}(f) + 8, & \text{if } f = f_0; \\ \mathrm{res}(f), & \text{otherwise}, \end{cases} \tag{12.4.2}$$

where f_0 is the outer face of $\mu(G)$ as usual.

On the other hand, let C be a circuit in $\mu(G)$. From Jordan curve theorem, let

$$\mu_{\text{in}} = \mu_{\text{in}}(G; C) = C \cup (\mu(G) \cap C_{\text{in}})$$

and

$$\mu_{\text{out}} = \mu_{\text{out}}(G; C) = C \cup (\mu(G) \cap C_{\text{out}}).$$

The face sets of μ_{in} and μ_{out} in F are, respectively, denoted by F_{in} and F_{out}. Of course, $F = F_{\text{in}} + F_{\text{out}}$. If $\tilde{F} = F_{\text{in}}$ (or F_{out}) satisfies

$$\sum_{f \in \tilde{F}} \text{res}_0(f) > \sum_{v \in V(\mu_{\text{in}})} \text{def}(v) \tag{12.4.3}$$

and no circuit C' has $\mu_{\text{in}}(G; C') \subset \mu_{\text{in}}$ (or $\mu_{\text{out}}(G; C') \subset \mu_{\text{out}}$) with eq. (12.4.3), then C is said to be *face unequilibrious* and denoted by Ω_{fac}.

Lemma 12.4.1. *For a planar embedding $\mu(G)$ of a graph G with positive equilibrium, $\mu(G)$ has an Ω_{ver} if, and only if, $\mu(G)$ has an Ω_{fac}.*

Proof. Suppose C^* is the cocircuit determined by Ω_{ver} and assume $\mu^*_{\text{in}} = \mu_{\text{in}}(G; C^*)$ satisfies eq. (12.4.1) without loss of generality from the symmetry between μ^*_{in} and μ^*_{out}. By equilibrium (12.3.2), let $C = \sum_{f \subset \mu^*_{\text{out}}} \partial f$, then $\mu_{\text{in}} = \mu(G; C_{\text{in}}) = \mu^*_{\text{out}}$, if $f_0 \not\subset \mu^*_{\text{out}}$; or $\mu_{\text{out}} = \mu(G; C_{\text{out}}) = \mu^*_{\text{out}}$, if $f_0 \subset \mu_{\text{out}}$, satisfies eq. (12.4.3). Moreover, if C is not an Ω_{fac}, then μ_{in} (or μ_{out}) has a circuit C' such that $C'_{\text{in}} \subset C_{\text{in}}$ (or $C'_{\text{out}} \subset C_{\text{out}}$) and $\mu'_{\text{in}} = \mu(G; C'_{\text{in}})$ (or $\mu'_{\text{out}} = \mu(G; C'_{\text{out}})$) satisfies eq. (12.4.3). In consequence, an Ω_{fac} can always be found. The necessity is obtained. Conversely, in the similar way, the sufficiency can also be obtained. □

Lemma 12.4.2. *Both Ω_{ver} and Ω_{fac} are forbidden configurations for the 0-extensibility.*

Proof. From Lemma 12.4.1, it is only necessary to verify for one of Ω_{ver} and Ω_{fac}. If a planar embedding $\mu(G)$ has Ω_{ver}, then from Theorem 1.3.8, the equilibrious graph $\text{Qui}(\mu G)$ does not have a perfect matching. In view of Theorem 12.3.2, $\mu(G)$ is not 0-extensible. Hence, from the minimality, Ω_{ver} is a forbidden configuration. □

Theorem 12.4.1. *A planar graph $G = (V, E)$ of vertex valency less than five without triangle is 0-embeddable if, and only if, there is no Ω_{ver} (or Ω_{fac}) in a planar embedding $\mu(G)$ of G.*

Proof. The necessity is obtained by Lemma 12.4.2. Conversely, because a planar embedding $\mu(G)$ has no Ω_{ver}, we have that for any $B \subset V$,

$$\sum_{v \in B} \text{def}(v) \le \sum_{\substack{f \text{ ind } v \\ v \in B}} \text{res}_0(f). \tag{12.4.4}$$

If there exists $B \subset V$ such that eq. (12.4.4) is not satisfied otherwise, then the vertex-induced subgraph of $X(B) \cup Y(B)$ in the equilibrious graph Qui(μG) has a connected component, denoted by

$$H = (X_0, Y_0; E_0),\ X_0 \subseteq X(B),\ Y_0 \subseteq Y(B)$$

such that

$$\begin{aligned}\sum_{v \text{ ind } V(X_0)} \mathrm{def}(v) &= |X_0| > |Y_0| \\ &= \sum_{f \text{ ind } F(Y_0)} \mathrm{res}_0(f),\end{aligned} \tag{12.4.5}$$

where $X(B)$ and $Y(B)$ are the sets of vertices in X_Q and Y_Q corresponding to the vertices in B and faces incident with B, and conversely, $V(X_0)$ and $F(Y_0)$ are the sets of vertices and faces of $\mu(G)$ corresponding to X_0 and Y_0, respectively. Thus, there exists a cocircuit C^* such that μ^*_{in} or μ^*_{out} is a planar embedding of the subgraph induced by B in G and satisfies eq. (12.4.1). This leads to a contradiction that $\mu(G)$ has no Ω_{ver}. By virtue of eq. (12.4.4), from Theorem 1.3.8, Qui(μG) has a perfect matching. In consequence, the sufficiency follows from Theorem 12.3.2. □

According to those obtained earlier, we may find a large number of planar graphs with positive equilibrium, all of which are not 0-embeddable. Only a few types of them are listed here.

Corollary 12.4.1. *For a planar embedding $\mu(G)$ of a positive equilibrium graph G, if no face is admissible, i.e. any face f satisfies*

$$n_{2\text{-val}}(f) < n_{4\text{-val}}(f) + 4, \tag{12.4.6}$$

then $\mu(G)$ is not 0-extensible. Moreover, if G is homeomorphic to a 3-connected graph, then eq. (12.4.6) implies G is not 0-embeddable.

Proof. A direct result of Theorem 12.2.1. □

Corollary 12.4.2. *Any triangle-free graph G with at most three of 2-valent vertices on each circuit is not 0-embeddable.*

Proof. A direct conclusion of Corollary 12.4.1. □

Corollary 12.4.3. *Any planar embedding $\mu(G)$ of a triangle-free graph G with one of the configurations:*

$$H_i,\ i \ge 1,\ \text{and}\ Q_i,\ i \ge 2,$$

as shown in Fig. 12.4.1, is not 0-embeddable.

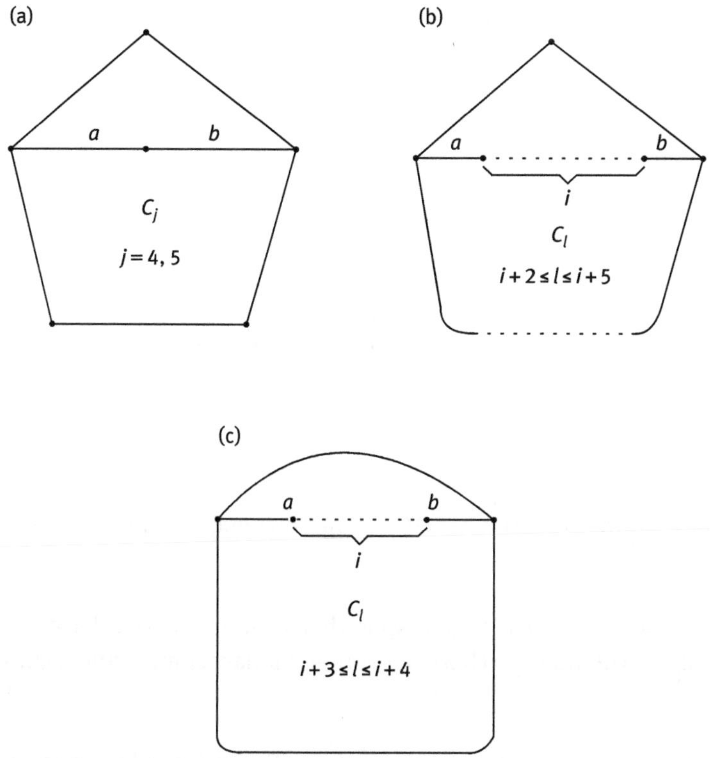

Figure 12.4.1: a) H_1; (b) $H_i, i \geq 2$; and (c) $Q_i, i \geq 2$.

Proof. Because $H_i, i \geq 1$, or $Q_i, i \geq 2$, has Ω_{ver} by the observation that the edge pair $\{a, b\}$ as marked in Figure 12.4.1 forms a cocircuit C^*, from Theorem 12.4.1, the corollary is obtained. □

Corollary 12.4.4. *Any planar embedding $\mu(G)$ of a triangle-free graph G with one of the configurations:*

$$\Phi_i,\ i \geq 1,\ \textit{and}\ \Phi_i',\ i \geq 2,$$

as shown in Figure 12.4.2 is not 0-embeddable.

Proof. Because any of $\Phi_i, i \geq 1$, as shown in Figure 12.4.2(a), and $\Phi_i', i \geq 2$, as shown in Figure 12.4.2(b) have three of 2-valent vertices, from Corollary 12.4.2, the corollary follows. □

Corollary 12.4.5. *If each planar embedding $\mu(G)$ of a positive equilibrium graph G has a face f with the relation*

$$n_{4\text{-val}}(f) > n_{2\text{-val}}(f) + 4, \qquad (12.4.7)$$

then G is not 0-embeddable.

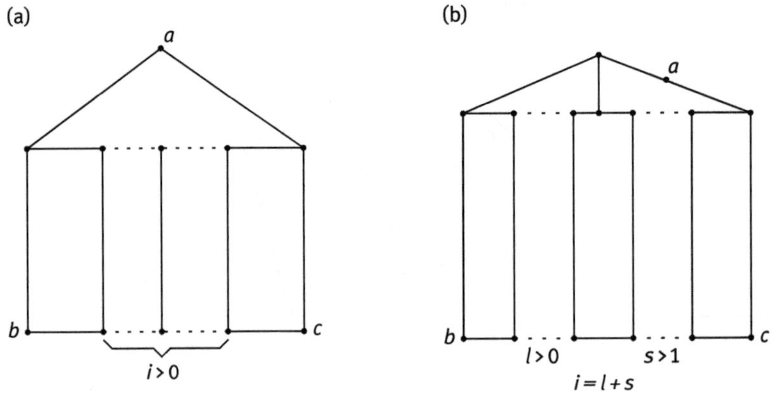

Figure 12.4.2: (a) $\Phi_i, i \geq 1$ and (b) $\Phi'_i, i \geq 2$.

Proof. Because the face boundary forms an Ω_{fac}, from Theorem 12.4.1, the proof is obtained. □

Further, we investigate two kinds of planar graphs that have much simpler characterizations for the 0-embeddability. Those are outer planar graphs and Halin graphs.

Lemma 12.4.3. *A triangle-free outer planar graph has two independent edges, each of which is incident with two 2-valent vertices.*

Proof. First, it is easily seen that any outer planar graph has at least two 2-valent vertices that are not adjacent and are on the boundaries (non-common, if adjacent) of two distinct inner faces. Moreover, if a 2-valent vertex is not adjacent to a 2-valent vertex, then from the outer planarity it has to be on a triangle. Therefore, one of the two adjacent vertices of each 2-valent vertex is 2-valent. Because two 2-valent vertices incident to different inner faces are never adjacent, the lemma is found. □

Theorem 12.4.2. *An outer planar graph $G = (V, E)$ without 4-valent vertex is 0-embeddable if, and only if, it is triangle free.*

Proof. Because the necessity is obvious, we only prove the sufficiency. By induction on the edge number of G, when it is with a few edges or without an inner edge of the outer-face boundary, it is easy to check.

In general, because G is without triangle, we are allowed to suppose (u, v) is an edge such that both u and v are 2-valent from Lemma 12.4.3. Let L be the longest vertex-induced path that contains the edge (u, v) with all its inner vertices 2-valent. Assume u_0 and v_0 are the vertices adjacent to the two ends of L. Because G has inner edges of the outer-face boundary, we are also allowed to suppose (u_0, v_0) is an inner edge of G.

Let $G_1 = G - L$. Then, G_1 is an outer planar graph without 4-valent vertex as well. By the hypothesis, G_1 is 0-embeddable. From the outerplanarity, both u_0 and v_0 are on the outer-face boundary. Because of the longestness of L and no 4-valent vertex in G, both u_0 and v_0 are 2-valent in G_1. However, u_0 and v_0 can always be made that none of u_0 and v_0 is on the state I of the boundary of the outer face in a 0-embedding $\mu_0(G_1)$. Hence, a 0-embedding $\mu_0(G)$ can be extended by restoring u and v from $\mu_0(G_1)$. □

As mentioned in Section 2.1 of Liu [34], a Halin graph is constructed by a tree and the outer boundary circuit. We here assume the tree has no 2-valent vertex. In fact, this is not essential for our purpose here because we only discuss Halin graphs without 4-valent vertex for the simplicity.

For a planar embedding $\mu(G)$ of a graph $G = (V, E)$, let us write

$$\mathrm{dif}(S) = \sum_{\substack{f \text{ ind } S \\ f \in F}} \mathrm{res}_0(f) - \sum_{v \in S} \mathrm{def}(v), \tag{12.4.8}$$

where $S \subseteq V$.

Lemma 12.4.4. *A triangle-free Halin graph with inner vertex valency greater than two has at least two 2-valent vertices on the outer boundary circuit.*

Proof. Let L be the longest path on the tree in a Halin graph $G = (V, E)$ with the given condition. Suppose u and v are the two ends of L and hence on the outer boundary circuit from the longestness. If u_1 and v_1 are the respective vertices adjacent to u and v on L, then from the longestness of L and the valencies of u_1 and v_1 greater than two u_1 and v_1 have to be incident to a face each with at least two edges in common with the outer boundary circuit in accordance with no triangle in G. Therefore, there are at least two 2-valent vertices on the outer boundary circuit. □

Theorem 12.4.3. *A triangle-free Halin graph $G = (V, E)$ with all inner vertices 3-valent is 0-embeddable if, and only if, the outer boundary circuit has at least four 2-valent vertices.*

Proof. As the necessary is obvious from Corollary 12.4.2, we only prove the sufficiency. From Theorem 12.4.1, it suffices to verify that G has no Ω_{ver}. By contradiction, suppose C^* is a cocircuit vertex-unequilibrious, then one of the following two cases may happen: $\mu_{\mathrm{in}}^* = \mu_{\mathrm{in}}(G; C^*)$ has a vertex on the outer boundary circuit $B = \partial f_0$ and otherwise. For convenience, let F_{in}^* be the set of all the faces incident with a vertex of μ_{in}^* in G.

Case 1. $V_{\mathrm{in}}^* \cap V(B) = \emptyset$, where $V_{\mathrm{in}}^* = V(\mu_{\mathrm{in}}^*)$. Because μ_{in}^* is a tree, we have

$$|F_{\mathrm{in}}^*| = |E(C^*)|. \tag{12.4.9}$$

Moreover, because all vertices in V_{in}^* are 3-valent in G, we have

$$|F_{\text{in}}^*| = |V_{\text{in}}^*| + 2. \tag{12.4.10}$$

Since the valency of a face incident only (otherwise) with an articulate vertex of μ_{in}^* is at least two (one) greater than the number of its boundary edges incident with μ^* or C^*, and if the number $\nu_1(\mu_{\text{in}}^*)$ of articulate vertices in μ_{in}^* is less than four and each of them is with a quadrangle, then at least $4 - \nu_1(\mu_{\text{in}}^*)$ other faces have their valencies each at least two greater than the number of the incident edges with μ_{in}^* or C^*, we find

$$\begin{aligned}\sum_{f \in F_{\text{in}}^*} \rho(f; G) &\geq 2(\epsilon(\mu_{\text{in}}^*) + |V_{\text{in}}^*| + 2) + |V_{\text{in}}^*| \\ &\quad +2 + \max\{4, \nu_1(\mu_{\text{in}}^*)\} \\ &\geq |V_{\text{in}}^*| + 4(|F_{\text{in}}^*| - 1) + 4 \\ &= |V_{\text{in}}^*| + 4|F_{\text{in}}^*|\end{aligned} \tag{12.4.11}$$

by considering eqs. (12.4.9) and (12.4.10) and $\epsilon(\mu_{\text{in}}^*) = |V_{\text{in}}^*| - 1$. On the other hand, from the 3-valentness of inner vertices of G, we know

$$\sum_{v \in V_{\text{in}}^*} \text{def}(v) = |V_{\text{in}}^*|. \tag{12.4.12}$$

Thus, from eqs. (12.4.11) and (12.4.12), $\text{dif}(V_{\text{in}}^*) \geq 0$. This is a contradiction to the vertex-unequilibriousness of C^*.

Case 2. $V_{\text{in}}^* \cap V(B) \neq \emptyset$. Let $V_0 = V_{\text{in}}^* \cap V(B)$ and $V_1 = V_{\text{in}}^* \backslash V_0$. From what is discussed in case 1, we have

$$\text{dif}(V_{\text{in}}^*) = \text{dif}(V_1) + \text{res}_0(f_0) - \sum_{v \in V(B)} \text{def}(v). \tag{12.4.13}$$

According to eqs. (12.4.11) and (12.4.12), we know that

$$\text{dif}(V_1) \geq \max\{0, \nu_2(f_0; \mu_{\text{in}}^*) - 4\}, \tag{12.4.14}$$

where $\nu_2(f_0; \mu_{\text{in}}^*)$ is the number of 2-valent vertices of G in $V(B)$, which correspond to articulate vertices in $\mu_{\text{in}}^*[V_1]$. Moreover, let

$$\triangle = \text{res}_0(f_0) - \sum_{v \in V(B)} \text{def}(v),$$

then

$$\triangle \geq |V(f_0)| + 4 - (|V_0| + \nu_2(f_0; \mu_{\text{in}}^*)). \tag{12.4.15}$$

By virtue of eqs. (12.4.13)–(12.4.15), we obtain

$$\mathrm{dif}(V_{\mathrm{in}}^*) \geq |V(f_0)| - |V_0| = 0.$$

This is a contradiction to the vertex-unequilibriousness of C^*. □

12.5 Notes

12.5.1 Many algorithms, most of which are linear time in computing complexity, can be designed for recognizing if a graph is homeomorphic to a 3-connected one, or its planar embedding is 0-extensible, 0-embeddable, even 0-realizable, and further presenting the realization by computers. In spite of the results in Liu [209], the problems on convexity are completely solved too. However, the recognition of 0-embeddability for a general graph is absolutely not easy, in view of the computing complexity.

12.5.2 The problems discussed here can be extended to higher dimension, one might also like to see what happens on a surface although the underlying structure is no longer seen as an integral lattice.

12.5.3 On the net embeddability, one may find some fundamental discussion in Liu [209]. However, that paper only provides some partial results. This chapter completes the task, proposed in Liu [209].

12.5.4 In fact, the net extensibility is equivalent to the following Diophantine equations: for a planar embedding $\mu(G)$ of $G = (V, E)$, with F as the set of faces in $\mu(G)$, find $x_A \in \{-1, 0, 1\}, A \in \mathscr{A}(\mu G)$ such that

$$\begin{cases} \forall v \in V, \quad \sum\limits_{A \in \mathscr{A}(v;\mu G)} x_A = \begin{cases} 4, \ \rho(v; G) = 4; \\ 2, \ \rho(v; G) = 3; \\ 0, \ \text{otherwise}, \end{cases} \\ \forall f \in F, \quad \sum\limits_{A \in \mathscr{A}(f;\mu G)} x_A = \begin{cases} -4, f = f_0; \\ 4, \ \text{otherwise}. \end{cases} \end{cases} \tag{12.5.1}$$

Although the possibilities for the choices of $x_A, A \in \mathscr{A}(\mu G)$, are exponential type of the order or the size of G, Sections 12.3 and 12.4 show that eq. (12.5.1) is polynomially solvable.

12.5.5 On the realization of a net embedding on the plane, horizontal and vertical diagrams have been employed to determine the coordinates of vertices under certain conditions. Details about them can be seen in Liu [211, 212].

12.5.6 On the basis of Sections 11.3 and 11.4, net embeddings of a graph available on a general surface can also be investigated for one of the three orthogonal models of surfaces described in Section 11.3.

12.5.7 In a similar way to Sections 11.3 and 11.4, net embedding on a specific surface of genus given, not 0, can also be discussed on one of the three surface models.

12.5.8 If not inherited for a property considered, the characterization of this property cannot be done by forbidden minors. The net embeddability in this chapter is known with inheritness. However, the forbidden configurations for characterizing net embeddability is also shown powerful without the usage of minor(s) as well, as indicated in item 11.5.9.

13 Extremality on Surfaces

13.1 Maximal genus

The theoretical idea of this section is reductions initiated from Liu [196]. A *reducible operation* (or in short, a *reduction*) was defined as follows. Let G be a connected graph of valencies not less than 3 and x and y, two adjacent edges. Such a transformation, denoted by $\pi(v';x,y)$, at vertex v' for x and y from G into G' of valencies not less than 3 is called a reducible operation whenever $G' = \pi(v';x,y)G$ is connected as well where v' is the common vertex v of x and y when its valency is greater than 3; or the vertex adjacent to the common vertex otherwise and

$$G' = \begin{cases} G - \{x, y\}, & \text{if } v' \text{ is } v; \\ G - \{v\}, & \text{otherwise.} \end{cases} \tag{13.1.1}$$

A pair of adjacent edges satisfying eq. (13.1.1) in a graph is said to be *available*.

In fact, for a spanning tree T on a graph $G = (V, E)$, let a and b be an adjacent pair of cotree edges, or a adj b, the operation from G to G' such that $G' \sim_{\text{top}} G - \{a, b\}$ when their common vertex v is of valency greater than 3; or $G' = G - \{v\}$ otherwise is just the reduction.

Because of the connectedness of G', each time of reduction in deleting a vertex has to be with an edge deleted in a spanning tree of G. This enables us to consider reduction $\pi(v';x,y)$ as deleting a pair of cotree edges, denoted simply by $\pi_{x,y}$ for a spanning tree.

Let $n_G(\pi)$ be the maximum number of reductions (or in short, the *reduction number*) that can be done on G and written as

$$\Pi_G = \{(x_i, y_i) | 1 \le i \le n_G(\pi), \pi(v_i'; x_i, y_i) \text{ available}\}. \tag{13.1.2}$$

A graph is called a *cascade* if all circuits are independent, i.e. pairwise vertex-disjoint (or V-disjoint). Because of no circuit on a tree, trees are all cascades as degenerate.

On the basis of the theory described in Liu [196], for a Eulerian circuit $C(v, e, e^{-1})$ with edges and vertices (or $C(e, e^{-1})$ as an edge set) obtained by the E-C procedure on double graph GG, a *maximum genus Eulerian circuit* is defined as a regular Eulerian circuit with γ_{Max} instead of β.

Lemma 13.1.1. *Given a graph $G = (V, E)$ connected and not a cascade. Let $G' = \pi(v' : x, y)G$, then for any $(x, y) \in \Pi_G$,*

$$\max_{S \in \mathcal{S}} g_S(G) = \max_{S \in \mathcal{S}'} g_S(G') + 1, \tag{13.1.3}$$

DOI 10.1515/9783110479492-013

where $\mathscr{S}$ and $\mathscr{S}'$ are the sets of all orientable surfaces G and G' can, respectively, be embedded on, and g_S is the genus of S.

Proof. Because of at most one genus reduced by a reduction, we have

$$\gamma_{\text{Max}}(G') \leq \gamma_{\text{Max}}(G) - 1.$$

On the other hand, from the maximality of Π_G, a maximum genus Eulerian circuit $C(v, e, e^{-1})$ can be constructed and $C(e, e^{-1})$ is in form as $(Axx^{-1}yy^{-1}BCD)$ where C is the segment other than $x^{-1}y$ between the two reflective vertices. By the operation $\Delta_{x,y}$, for the sake of brevity, instead of $\Delta^{p}_{\xi,\eta,\xi',\eta'}$ in Liu [196] on $C(e, e^{-1})$, we have

$$\begin{aligned}\Delta_{x,y}C(e, e^{-1}) &= (AxCy^{-1}Bx^{-1}yD)\\ &\sim_{\text{top}} (ABCDxyx^{-1}y^{-1}).\end{aligned}$$

This shows that genus is reduced by at least one from a reduction and hence

$$\gamma_{\text{Max}}(G') \geq \gamma_{\text{Max}}(G) - 1.$$

Thus, the lemma is done. □

A graph obtained from a number of reductions on G which cannot yet do the reduction is called a *irreducible* graph of G.

Lemma 13.1.2. *All irreducible graphs of a given graph G are cascades.*

Proof. First, it is seen that for a reduction $\pi(v'; x, y)$ on G, by virtue of the connectedness of $G' = \pi(v'; x, y)G$, x and y are both not a cut edge. That implies each of x and y is on a circuit and the two circuits have a common vertex.

Then by considering that a graph can always be reduced by a reduction whenever it has two circuits with a vertex in common, each irreducible graph of G is a cascade.

This is the lemma. □

Lemma 13.1.3. *The maximum genus of a cascade is* 0.

Proof. Because any Eulerian circuit obtained from a cascade by the E-C procedure in Liu [196] cannot do the operation $\Delta_{x,y}$ for any pair of edges x and y, the cascade is then with its maximum genus 0. □

Theorem 13.1.1. *For any connected graph G, its orientable maximum genus*

$$\gamma_{\text{Max}}(G) = n_G(\pi), \tag{13.1.4}$$

the reduction number of G for $\pi(v'; x, y)$.

Proof. From Lemma 13.1.1,

$$\gamma_{\text{Max}}(G) = n_G(\pi) + \gamma_{\text{Max}}(G_0),$$

where G_0 is an irreducible graph of G.

On account of Lemma 13.1.2, G_0 is a cascade. Then from Lemma 13.1.3,

$$\gamma_{\text{Max}}(G_0) = 0.$$

Therefore, the theorem is obtained. □

This theorem has an advantage in simplifying the procedure for determining the maximum genus particularly of a cubic graph because a reduction in this case is corresponding only to delete a vertex.

In Xuong [408], the Betti deficiency $\xi(G)$ is defined to be the minimum number of odd size components among all cotrees to determine the orientable maximum genus of a graph.

A spanning tree T of G which has its cotree with ξ odd size components is called *reasonable*. A graph whose edge set can be partitioned into a spanning tree and a matching is called a*snuff*.

Lemma 13.1.4. *All irreducible graphs of G for a reasonable tree are snuffs.*

Proof. Because each even size component are pairwise reduced and each odd size component are pairwise reduced except exactly for one edge, the irreducible graph has its cotree edges pairwise vertex-disjoint, and hence a matching. Thus, the irreducible graph is a snuff. The proof is done. □

Attention 13.1.1. The irreducible graph is not independent of the choice of reductions but is independent of their order. In Figure 13.1.1, (a) shows a graph H with a tree represented by bold lines, (b), the irreducible graph H_1 for the reductions

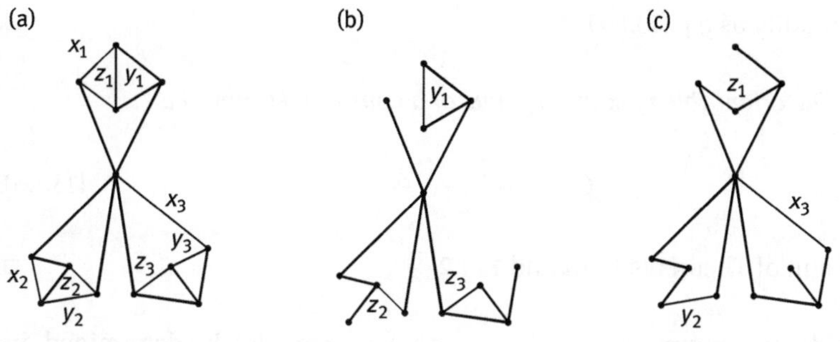

Figure 13.1.1: $(a)H; (b)H_1; and (c)H_2$.

$\Pi_1 = \{\pi_{x_1,z_1}, \pi_{x_2,y_2}, \pi_{x_3,y_3}\}$ and (c), the irreducible graph H_2 for the reductions $\Pi_2 = \{\pi_{x_1,y_1}, \pi_{x_2,z_2}, \pi_{y_3,z_3}\}$. In H_1, there is no way to reduce the number of odd size components. However, H_2 shows that the number of odd size components can be reduced by two due to the existence of two fundamental circuits with a vertex in common. Although H_1 and H_2 are both with three odd size components, only H_2 can be transformed to have one odd size component by employing the following method. For two fundamental circuits C_x and C_y with a vertex in common where x and y cotree edges, suppose x' and y' at a common vertex are on, respectively, C_x and C_y without loss of generality. By putting x' as a cotree edge and x as a tree edge if $x \neq x'$ while for y' and y as well, the resultant tree T' has its cotree with at least two odd size components less than T does. The reason is that new cut edge except for articulate edge occurs in H_1.

Lemma 13.1.5. *Let T be a spanning tree of G such that the number of odd size components in its cotree is the Betti deficiency $\xi(G)$, then the reduction number $n'_G(\pi)$ for $\pi(x,y)$, where both x and y are in the cotree, is $n_G(\pi)$.*

Proof. First, because a reduction $\pi_{x,y}$ can always correspond to a reduction $\pi(v';x,y)$, we have $n'_G(\pi) \leq n_G(\pi)$.

Then, to show $n'_G(\pi) \nless n_G(\pi)$. By contradiction, assume $n'_G(\pi) < n_G(\pi)$. Let G'_0 be the graph obtained by doing $\pi_{x,y}$ for $n'_G(\pi)$ times on G. Because $n'_G(\pi) < n_G(\pi)$, we can construct $n_G(\pi)$ reductions $\pi_{x,y}$ for T' to find G'_1 which has at least two odd size components than G'_0 does. This is a contradiction to the minimality of T.

Therefore, the lemma is true. □

Theorem 13.1.2. *For a connected graph G, its Betti deficiency*

$$\xi(G) = \beta(G) - 2n_G(\pi). \tag{13.1.5}$$

Proof. From Lemma 13.1.5, the Betti number, i.e. the number of cotree edges,

$$\beta(G) = 2n_G(\pi) + \xi(G).$$

This is the same quality as eq. (13.1.4). □

Corollary 13.1.1. *The orientable maximum genus of a connected graph G is*

$$\gamma_{\text{Max}}(G) = \frac{\beta(G) - \xi(G)}{2}. \tag{13.1.6}$$

Proof. A direct result of Theorems 13.1.1 and 13.1.2. □

The non-orientable maximum genus of a graph was completely determined in Liu [195].

13.2 Minimal genus

Two ways are considered for determining the minimum genus of a general graph and a specific graph. This section concerns only with orientable case because non-orientable case can be done similarly whereas with additional choices of cotree edges.

Because of the difficulty, only a procedure of accessing the minimum genus of a general graph is first provided in decreasing the upper bounds of the minimum step by step until the minimum genus based on the theory described in Section 9.3.

Let $G = (V, E)$ be a graph and T, a spanning tree of G. For a rotation system σ on G, $\hat{T}_\sigma$ is the expanded tree of T. Let x_i, $1 \leq i \leq \beta$, be all the cotree edges where β is the Betti number of G.

Suppose the associate surface S, a polyhegon on $X = \{x_i, x_i^{-1} | 1 \leq i \leq \beta\}$, is divided into k layers: $0, 1, \ldots, k-1$.

Let Π_T^{+j} be the set of all exchangers on the layers at least j, $0 \leq j \leq k-1$, for the given associate surface S on T of G. The value

$$\gamma_T^{+j} = \min_{\pi \in \Pi_T^{+j}} g(\pi S)$$

is called a *minimal j-genus* of G.

Lemma 13.2.1. *The sets of Π^{+j} for $1 \leq j \leq k-1$ have the following relation as*

$$\Pi_T^{+0} \supset \Pi_T^{+1} \supset \cdots \supset \Pi_T^{+(k-1)}. \tag{13.2.1}$$

Proof. It is apparently seen from Section 9.3. □

Lemma 13.2.2. *All the minimal j-genera of G satisfy the following inequality as*

$$\gamma_T^{+0} \leq \gamma_T^{+1} \cdots \leq \gamma_T^{+(k-1)}. \tag{13.2.2}$$

Proof. A direct result of Lemma 13.2.1. □

Lemma 13.2.3. *For any associate surface S of G, γ_T^{+0} is a constant, i.e. independent of T.*

Proof. In virtue of Theorem 9.1.5, Π^{+0} is independent of the choice of a spanning tree on G and hence the lemma. □

This lemma enables us to adopt γ^{+0} instead of γ_T^{+0}.

Theorem 13.2.1. *The minimum genus of G is $\gamma_{\min}(G) = \gamma^{+0}$.*

Proof. A direct result of Lemmas 13.2 and 13.3. □

Another way is for determining the minimum genus of a graph G_n when G_n can be constructed from G_{n-1} by a local operation such as addition of a few of vertices, or edges for $n \geq 2$ and the minimum genus of G_1 is easily determined. Only take Q_n, the n-cube; $K_{m,n}$, the complete bipartite graph of order $m+n$; and K_n, the complete graph of order n as examples. Of course, the genera of Q_1, $K_{1,1}$ and K_1 are all known to be 0.

For the n-cube Q_n, $n \geq 3$, shows that the minimum genus $\gamma_n = g_{\min}(Q_n)$ with γ_{n-1} satisfies the relation

$$g_{\min}(Q_n) = 2^{n-4}(n-3) + g_{\min}(Q_{n-1}) \tag{13.2.3}$$

from an associate surface of Q_{n-1} with genus γ_{n-1} to get an associate surface of Q_n with genus γ_n.

Theorem 13.2.2. *The minimum genus of the n-cube Q_n determined by eq.* (13.2.3) *is*

$$\gamma_{\min}(Q_n) = 2^{n-3}(n-4) + 1$$

for $n \geq 3$.

Proof. By induction on n, for $n = 3$, it is easy to see that $\gamma_{\min}(Q_3) = 0 = 2^0(-1) + 1$ because Q_3 is the cube which is planar. Then by the hypothesis,

$$\gamma_{\min}(Q_{n-1}) = 2^{n-4}(n-5) + 1$$

for $n > 3$. From eq. (13.2.3), we have

$$\gamma_{\min}(Q_n) = 2^{n-4}(n-3) + 2^{n-4}(n-5) + 1 = 2^{n-3}(n-4) + 1.$$

This is the conclusion of the theorem. □

For the complete bipartite graph $K_{m,n}$, $m \geq n \geq 4$, shows that the minimum genus $\gamma_{m,n} = g_{\min}(K_{m,n})$ with $\gamma_{m,n-1}$ satisfies the relation

$$g_{\min}(K_{m,n}) = \left\langle \frac{m-2}{4} \right\rangle + g_{\min}(K_{m,n-1}) - 1, \tag{13.2.4}$$

where

$$\left\langle \frac{m-2}{4} \right\rangle = \begin{cases} \left\lceil \dfrac{m-2}{4} \right\rceil, & m = 0(2 \nmid n), 1(2 \nmid n; 2|n, 2|\lfloor n/2 \rfloor), \\ & \quad 3(2 \nmid n, 2 \nmid \lfloor n/2 \rfloor; 2|n, 2|\lfloor n/2 \rfloor); \\ \dfrac{m-2}{4}, & m = 2(\mathrm{mod}\ 4); \\ \left\lfloor \dfrac{m-2}{4} \right\rfloor, & m = 0(2|n), 1(2|n, 2 \nmid \lfloor n/2 \rfloor), \\ & \quad 3(2 \nmid n, 2|\lfloor n/2 \rfloor; 2|n, 2 \nmid \lfloor n/2 \rfloor) \end{cases}$$

from an associate surface of $K_{m,n-1}$ with genus $\gamma_{m,n-1}$ to get an associate surface of $K_{m,n}$ with genus $\gamma_{m,n}$.

Theorem 13.2.3. *The minimum genus of the complete bipartite graph $K_{m,n}$ determined by eq. (13.2.4) is*

$$\gamma_{\min}(K_{m,n}) = \left\lceil \frac{(m-2)(n-2)}{4} \right\rceil .$$

for $m \geq n \geq 2$.

Proof. By induction on n for any $m \geq n$. When $n = 2$, it is easy to see that for any $m \geq n = 2$, $\gamma_{\min}(K_{m,2}) = 0$ because of $K_{m,2}$, a graph of m parallel paths with same ends, which is planar. Then from eq. (13.2.4), by the hypothesis

$$\gamma_{\min}(K_{m,n}) = \left\langle \frac{m-2}{4} \right\rangle + \left\lceil \frac{(m-2)(n-3)}{4} \right\rceil - 1$$

for $m \geq n > 2$. By checking each case, the theorem can be done. □

For the complete graph K_n, $n \geq 5$, shows that the minimum genus $\gamma_n = g_{\min}(K_n)$ with γ_{n-1} satisfies the relation

$$g_{\min}(K_n) = \left\langle \frac{n-4}{6} \right\rangle + g_{\min}(K_{n-1}), \tag{13.2.5}$$

where

$$\left\langle \frac{n-4}{6} \right\rangle = \begin{cases} \left\lceil \dfrac{n-4}{6} \right\rceil, & n = 2, 1(2 \nmid \lfloor n/6 \rfloor), 3(2 | \lfloor n/6 \rfloor), \\ & 5(2 \nmid \lfloor n/6 \rfloor); \\ \dfrac{n-4}{6}, & n = 4 (\text{mod } 6); \\ \left\lfloor \dfrac{n-4}{6} \right\rfloor, & n = 0, 1(2 | \lfloor n/6 \rfloor), 3(2 \nmid \lfloor n/6 \rfloor), \\ & 5(2 | \lfloor n/6 \rfloor) \end{cases}$$

from an associate surface of K_{n-1} with genus γ_{n-1} to get an associate surface of K_n with genus γ_n.

Theorem 13.2.4. *The minimum genus of the complete graph K_n determined by eq. (13.2.4) is*

$$\gamma_{\min}(K_n) = \left\lceil \frac{(n-3)(n-4)}{12} \right\rceil$$

for $n \geq 4$.

Proof. By induction on n for any $n \geq 4$. When $n = 4$, it is easily seen that for any $n = 4$, $\gamma_{\min}(K_4) = 0$ because of K_4, the tetrahedron, which is planar. Then from eq. (13.2.5), by the hypothesis

$$\gamma_{\min}(K_n) = \left\langle \frac{n-4}{6} \right\rangle + \left\lceil \frac{(n-4)(n-5)}{12} \right\rceil$$

for $n > 4$. By checking each case, the theorem can be obtained. □

13.3 Shortest embedding

From what discussed in Section 5.5, it is known that a graph which has a convex embedding in the plane with f_0 as the infinite face is always convexly embeddable for any convex polygon chosen as the boundary of f_0. We are here allowed to assume the convex polygon is given as the boundary of the infinite face for which the graph considered is convexly embeddable.

Let $G = (V, E)$ be a graph which is convexly embeddable with a circuit B_0 as the infinite face boundary. Suppose each vertex $v \in V$ has a weight $w_v > 0$ and B_0 is a convex polygon given in the plane. For a convex embedding $\mu(G)$ of G with the boundary condition that the infinite face boundary is a given convex polygon B_0 whose vertex set $V(B_0) = \{a_1, a_2, \ldots, a_k\}$, $k \geq 3$, let

$$f(\mu G) = \sum_{u \in V} w_u \sum_{\substack{(u,v) \in E \setminus E(B_0) \\ v \in V_u}} w_v\, l_\mu(u, v), \tag{13.3.1}$$

where $l_\mu(u, v)$ is the length of the edge $(u, v) \in E$ in the embedding $\mu(G)$ and $E(B_0)$ is the edge set of B_0. The problem we are concerned with is to find the coordinates of the vertices, which are not in $V(B_0)$, in the inner domain of B_0 such that a convex embedding $\mu_0(G)$ of G can be constructed with the property that

$$f(\mu_0 G) = \min_{\mu} f(\mu G), \tag{13.3.2}$$

where μ is taken over all possible convex embeddings with the boundary condition. In eq. (13.3.2), μ_0 is called an *optimum solution*, or a *minimum convex embedding* of G.

Because G is always assumed to be simple, of course without multi-edges, all minimum convex embedding has to be straight line embedding in the plane. This enables us to consider the length of an edge in an embedding $\mu(G)$ as the Euclidean distance anyway. Let x_v and y_v be the coordinates of a vertex $v \in V$ in the plane. Then function

$$g(\mu G) = \sum_{u \in V} w_u \sum_{(u,v) \in E \setminus E(B_0)} w_v l_\mu^2(u, v) \tag{13.3.3}$$

for a planar embedding μ of G has the same optimum solutions as $f(\mu, G)$ does, where

$$l_\mu(u, v) = \sqrt{(x_u - x_v)^2 + (y_u - y_v)^2}. \tag{13.3.4}$$

One might think of the simplest case that there is only one inner vertex of a planar embedding $\mu(G)$, or in other words $G = W_{s+1}$ is the wheel of order $s+1$ first to see what happens to the optimum solution for the problem.

Suppose v is the inner vertex and $a_1, a_2, \ldots, a_s$ are the other vertices which are of course on the outer-face boundary of $\mu(W_{s+1})$. In this case, $p_i = x_{a_i}, q_i = y_{a_i}, i = 1, 2, \ldots, s$, are all known and the only unknowns are x_v and y_v. Because

$$\begin{aligned} g(\mu W_{s+1}) &= \sum_{1\le i\le s} w_v w_{a_i} + \sum_{1\le i\le s} w_{a_i} w_v l_\mu^2(a_i, v) \\ &= 2 \sum_{1\le i\le s} w_v w_{a_i} l_\mu^2(v, a_i) \end{aligned}$$

is a convex function on the plane, it has the unique minimum solution determined by the equation system:

$$\begin{cases} \dfrac{\partial g}{\partial x_v} = 4 \sum\limits_{i=1}^{s} w_v w_{a_i}(x_v - p_i) = 0, \\ \dfrac{\partial g}{\partial y_v} = 4 \sum\limits_{i=1}^{s} w_v w_{a_i}(y_v - q_i) = 0, \end{cases} \tag{13.3.5}$$

which is equivalent to the following equation system:

$$\begin{cases} \sum\limits_{i=1}^{s} w_{a_i}(x_v - p_i) = 0, \\ \sum\limits_{i=1}^{s} w_{a_i}(y_v - q_i) = 0 \end{cases} \tag{13.3.6}$$

called the *critical equations* of the problem.

It is easily seen that the critical equations given by eq. (13.3.6) is well defined for some i, $i = 1, 2, \ldots, s$, $w_{a_i} > 0$ and the solution is as follows:

$$x_v = \frac{\sum_{i=1}^{s} w_{a_i} p_i}{\sum_{i=1}^{s} w_{a_i}}; \; y_v = \frac{\sum_{i=1}^{s} w_{a_i} q_i}{\sum_{i=1}^{s} w_{a_i}}, \tag{13.3.7}$$

which is just the barycenter of a_i, $i = 1, 2, \ldots, s$. It is also easily seen that the straight-line embedding of W_{s+1}, determined by eq. (13.3.7) in the plane is convex if, and only if, the outer-face boundary is a convex polygon.

Moreover, this simpler idea can be extended with a slight difference causing from the occurrence of more than one inner vertices into the general case when the number of inner vertices is without the restriction of only one being allowed.

For the sake of brevity, let us write

$$v = (x_v, y_v), \tag{13.3.8a}$$

which also represents the coordinates in the plane for $v \in V - V(B_0)$ and likewise

$$a = (x_a, y_a) \tag{13.3.8b}$$

for $a \in V(B_0)$ which are all known.

Let us write

$$\begin{cases} W_u = \sum_{\substack{(u,v)\in E\setminus E(B_0)\\ v\in V\setminus V(B_0)}} w_v + \sum_{\substack{(u,a)\in E\setminus E(B_0)\\ a\in V(B_0)}} w_a, \\ d_u = \sum_{\substack{(u,a)\in E\setminus E(B_0)\\ a\in V(B_0)}} w_a a \end{cases} \tag{13.3.9}$$

for $u \in V - V(B_0)$.

Then, we are allowed to introduce the following equation system for any planar embedding $\mu(G)$ given:

$$W_u u - \sum_{\substack{v\in V\setminus V(B_0)\\ (u,v)\in E}} w_v v = d_v. \tag{13.3.10}$$

Lemma 13.3.1. *For vertex weight w_v, $v \in V$ positive of a graph $G = (V, E)$ and an outer-face boundary chosen, the equation system* (13.3.10) *is well defined.*

Proof. Because we always have

$$W_u - \sum_{\substack{v\in V\setminus V(B_0)\\ (u,v)\in E}} w_v = \sum_{\substack{a\in V(B_0)\\ (u,a)\in E\setminus E(B_0)}} w_a \geq 0$$

from eq. (13.3.9) for any $u \in V \setminus V(B_0)$, the coefficient matrix of eq. (13.3.10) is without a characteristic value which is zero. Therefore, the equation system (13.3.10) is well defined. □

Because the judgement of the planarity of a graph with a circuit given as the outer-face boundary can be polynomially done based on the theory which is established in Chapter 8, we are here allowed to assume a graph with a circuit as the boundary mentioned below is always planar with the boundary condition that the circuit is chosen to be the outer-face boundary of its planar embeddings. If the solution of eq. (13.3.10) is found, then a straight-line planar embedding can be uniquely constructed in the way that a straight-line segment is connected between two points if, and only if, the two corresponding vertices are adjacent in G. We call the embedding a *critical embedding* of G with B_0 as the outer-face boundary.

Lemma 13.3.2. *The critical embedding determined by the solution of eq.* (13.3.10) *reaches at the minimum of $f(\mu, G)$ expressed in eq.* (13.3.1) *among all the planar embeddings $\mu(G)$ with the same outer-face boundary B_0.*

Proof. Because $f(\mu G)$ has the same minimum as $g(\mu G)$ does, we are allowed to discuss $g(\mu G)$ only. From the convexity of $g(\mu G)$ which is easily checked by the positive definiteness of the Jacobian determinant of $g(\mu G), g(\mu G)$ has the unique minimum at the solution of eq. (13.3.10) for G with the boundary B_0. Therefore, the lemma is obtained. □

It is obviously seen that the critical embedding is not always a convex one because if the boundary B_0 is chosen to be not a convex polygon in the plane then the critical embedding is never convex. However, we have

Lemma 13.3.3. *The critical embedding of a graph G with a boundary B_0 is convex if, and only if, B_0 is a convex polygon in the plane.*

Proof. The necessity is straightforward. For the sufficiency, we have to prove the following two claims first. □

Claim 1. In the critical embedding, each inner vertex is in the inner domain of the convex hull of its neighbours.

Proof. Let v be an inner vertex of the outer-face boundary in the critical embedding and $v_1, v_2, \ldots, v_l$ be all the neighbours of v. Because all the inner vertices satisfy eq. (13.3.10), we soon find that v is just on the barycenter of $v_1, v_2, \ldots, v_l$ with their respective weights

$$w_1 = w_{v_1}, w_2 = w_{v_2}, \ldots, w_l = w_{v_l}$$

from the discussion on the wheels mentioned above. This implies that v is in the inner domain of the convex hull of $v_1, v_2, \ldots, v_l$. The claim is obtained. □

Claim 2. If a face f in a planar embedding is not a convex polygon, then there exists a vertex on the boundary of the face f, which is not in the inner domain of the convex hull of its neighbours.

Proof. Because f is not a convex polygon, there exists a vertex v on the boundary of f such that the two adjacent vertices u and w on the boundary are not adjacent in G and that the straight-line segment between u and w is without common point in the inner domain of the face f. However, such an vertex v is never allowed to be in the inner domain of the convex hull of all its neighbours. This is the claim.

By contradiction, if the critical embedding is not convex, then it has to have a face f which is not a convex polygon. From Claim 2, there is a vertex v on the boundary of f such that v is not in the inner domain of the convex hull of all its neighbours. This is a contradiction to Claim 1. The sufficiency is obtained. □

Theorem 13.3.1. *The function $f(\mu G)$ expressed by eq.* (13.3.1) *for a planar embedding $\mu(G)$ of a graph $G = (V, E)$ with the boundary B_0 of the outer face given has its minimum at what determines a convex embedding $\mu_0(G)$ of G in the plane if, and only if, B_0 forms a convex polygon, and the convex embedding is just the critical embedding determined by the solution of eq.* (13.3.10).

Proof. A direct result of Lemmas 13.3.1–13.3.3. □

Now, from the theorem, we are allowed to call eq. (13.3.10) the *critical equation* for a general graph as that for the wheels.

If all the weights at vertices are taken to be a constant, particularly 1, then eq. (13.3.10) becomes

$$\rho_G(u)u - \sum_{\substack{v \in V \setminus V(B_0) \\ (u,v) \in E}} v = \sum_{\substack{a \in V(B_0) \\ (u,a) \in E}} a \tag{13.3.11}$$

for $u \in V \setminus V(B_0)$.

Corollary 13.3.1. *For a planar embedding of a graph $G = (V, E)$ with the outer-face boundary B_0, if B_0 is a convex polygon, then the straight-line embedding of G determined by the solution of eq.* (13.3.11) *is a convex embedding whenever it is convexly embeddable with the outer-face boundary B_0.*

Proof. A direct conclusion of Theorem 13.3.1 when all $w_v = 1, v \in V$. □

This is a theorem obtained first by Tutte in Tutte [344].

If the boundary condition is restricted as some vertices of B_0 among which there are at least three corresponding points not collinear fixed in the plane, then the corresponding problem for such kind of boundary condition can also be solved in the similar way as shown above.

The problem of determining the minimum of $f(\mu G)$ above is in fact combinatorially equivalent to that of determining the minimum of

$$h(\mu G) = \sum_{e \in E \setminus E(B_0)} w(e) l_\mu(e), \tag{13.3.12}$$

where $w(e) > 0$, $e = (u, v)$, $u, v \in V$, is a weight function on $E \setminus E(B_0)$ for $G(V, E)$ with a given boundary B_0.

Suppose a straight-line embedding $\mu(G)$ of a graph $G = (V, E)$ in the plane is given. Of course, it is allowed to assume $\mu(G)$ is the minimum embedding with all the weights at vertices being 1 for the given outer-face boundary from what discussed above. Moreover, it is not necessary to assume the embedding is convex as mentioned there.

If an inner face of $\mu(G)$ is not a triangle, then it can be partitioned into triangles called *tiles* of the face by joining some straight-line segments between non-adjacent vertices on the face boundary. This procedure is called *tiling* the face. The resultant construction of tiling a face is said to be a *triangulation*, or a *tiling* as well, of the face. If all the non-triangle faces in $\mu(G)$ are tiled, then the tiled construction is called a *triangulation* or a *tiling* of $\mu(G)$ as well.

A triangulation of $\mu(G)$, which has the sum of all the lengths of peripheries of tiles minimum among all the possible tiling, is said to be *shortest*. How to find the shortest tiling of a given planar embedding $\mu(G)$ or a graph G efficiently is the main problem that we provide an answer in this section.

Mathematically, because the planar embedding $\mu(G)$ is known, from what we discuss in the last section, the only thing we have to do here for the problem is to find the shortest tiling of a polygon which is either convex or not on the plane.

Suppose Pol $= (a_1, a_2, \dots, a_n)$ is a convex polygon and a tiling $T_1(\text{Pol})$ of Pol is given. Let (a_i, a_j, a_k) and (a_j, a_k, a_l) be two tiles with the common boundary which consists of the single edge (a_j, a_k). From the convexity of Pol, the quadrangle (a_i, a_j, a_k, a_l) is always convex. Thus, we are allowed by deleting the common edge (a_j, a_k) and adding (a_i, a_l) on $T_1(\text{Pol})$ to obtain another tiling $T_2(\text{Pol})$ of Pol. The operation of transforming $T_1(\text{Pol})$ to $T_2(\text{Pol})$ is called a *diagonal replacement* on $T_1(\text{Pol})$. The tiling $T_{\text{st}}(\text{Pol})$ obtained by adding all the edges (a_i, a_1), $i = 3, 4, \dots, n-2$, on Pol is called the *standard tiling* of Pol with a_1 as the *base point*. Of course, any convex polygon has a standard tiling for any of its vertices as the base point.

Lemma 13.3.4. *Any tiling* $T(\text{Pol})$ *of a convex polygon* Pol *can be transformed into a standard tiling with any vertex chosen on Pol as the base point by a series of diagonal replacements.*

Proof. Let Pol $= (v_1, v_2, \dots, v_s)$ and v_1 be chosen as the base point without loss of generality. If a tiling $T(\text{Pol})$ has v_1 of valency $s-1$, then it is the standard one with v_1 being the base point itself. Otherwise, there is a triangle face

$$(v_1, v_k, v_l),\ 2 \le k < l-1 \le s-2$$

on $T(\text{Pol})$, because (v_k, v_l) is not an edge of Pol. There is another triangle (v_k, v_t, v_l), $k < t < l$, on $T(\text{Pol})$. From the convexity of Pol, the quadrangle (v_1, v_k, v_t, v_l) is convex. After the replacement of (v_k, v_t) by (v_1, v_l) the resultant tiling $T_1(\text{Pol})$ has the valency of v_1 be one greater than that $T(\text{Pol})$ does. From the finiteness of the valency of v_1 on $T(\text{Pol})$, a standard tiling with v_1 as the base point can be obtained by the procedure on what obtained is not a standard one yet. □

From Lemma 13.3.4, we may further see that the standard tiling with v_1 as the base point can be found by $s-1-\rho(T; v_1)$ times of the diagonal replacement, where $\rho(T; v_1)$ is the valency of v_1 on the tiling $T(\text{Pol})$ of an s-gon Pol. Because the inverse of a diagonal replacement is also a diagonal replacement, any tiling $T(\text{Pol})$ can be obtained from a standard tiling $T_{\mathrm{st}}(\mathrm{Pol})$ of an s-gon Pol by $(s-1)-\rho(T;b)$ times of the diagonal replacement where b is the base point of $T_{\mathrm{st}}(\mathrm{Pol})$.

Lemma 13.3.5. *For any two tilings* $T_1(\mathrm{Pol})$ *and* $T_2(\mathrm{Pol})$ *of a convex s-gon* Pol, $T_1(\mathrm{Pol})$ *can be found from* $T_2(\mathrm{Pol})$ *by at most*

$$2s-2-\max_{1\le i\le s}\left(\rho(T_1,v_i)+\rho(T_2,v_i)\right)$$

times of the diagonal replacement.

Proof. From Lemma 13.3.4, $T_1(\mathrm{Pol})$ can be obtained from a standard tiling $T_{\mathrm{st}}(\mathrm{Pol})$ with v as the base point by

$$s-1-\rho(T_1,v)$$

times of the diagonal replacement. And, by the same reason, $T_2(\mathrm{Pol})$ can be obtained from the standard tiling $T_{\mathrm{st}}(\mathrm{Pol})$ by doing

$$s-1-\rho(T_2,v)$$

diagonal replacements. Because of the arbitrariness of the choice of v, $T_1(\mathrm{Pol})$ can always be derived from $T_2(\mathrm{Pol})$ by the times of diagonal replacements as mentioned above. The lemma is found. □

Now, we introduce two more lemmas that allow us implicitly to employ the diagonal replacement for finding the shortest tiling of a convex polygon on the plane.

Lemma 13.3.6. *For a convex polygon* $Pol=(v_1,v_2,\dots,v_s)$, $s\ge 4$, *there exists a vertex* v_i, $1\le i\le s$, *such that the length* $l(v_{i-1},v_{i+1})$ *is not greater than the minimum between* $l(v_i,v_{i-2})$ *and* $l(v_i,v_{i+2})$.

Proof. Suppose $i\ne 1$. We may assume that $l(v_1,v_3)<l(v_s,v_2)$ without loss of generality from the symmetry. If $l(v_2,v_4)\ge l(v_1,v_3)$, then $i=2$. Otherwise, we have $l(v_2,v_4)<l(v_1,v_3)$ and do the same procedure at v_2 instead of v_1. From the finiteness of s, the assumption and the convexity of Pol in the plane, a v_i which is required has to be found. □

A vertex that has the property described in the lemma on a convex polygon is said to be *truncatible.*

Lemma 13.3.7. *A vertex* v_i *on a convex polygon* $\mathrm{Pol}=(v_1,v_2,\dots,v_s)$ *is truncatible if, and only if,*

$$l(v_{i-1},v_{i+1})\le l(v_i,v_j) \tag{13.3.13}$$

for $j=i-2,i-3,\dots,i+2$ *in the cyclic order on Pol.*

Proof. The sufficiency is obvious because the truncatibility of a vertex is only for $j = i-2$ and $i + 2$ in eq. (13.3.13).

The necessity is obtained from the fact that on a convex polygon Pol=$(v_1, \ldots, v_s)$, we always have that the polygon Pol$\langle v_i \rangle$ obtained by deleting the vertices v_{i-1}, v_i, v_{i+1} and adding the straight segment (v_{i-2}, v_{i+2}) from Pol is still convex. Thus, all the vertices except only for v_{i-1}, v_i and v_{i+1} are in the common area formed by the two rays: (v_i, v_{i-2}) and (v_i, v_{i+2}) and the side of the line (v_{i-2}, v_{i+2}) different from that v_i is in. This implies eq. (13.3.13). □

According to Lemmas 13.3.5–13.3.7, we are allowed to design a procedure for finding a tiling of a given convex polygon

$$\text{Pol} = (v_1, v_2, \ldots, v_s).$$

For brevity, we may assume that no two segments (v_{i-1}, v_{i+1}) and (v_i, v_{i+2}) for $i = 1, 2, \ldots, s$ in the cyclic order on Pol have the same length without loss of generality. We call this case *non-degenerate*.

Ta-procedure Choose a truncatible vertex v on Pol and form a new polygon Pol$[v]$ by deleting v and adding a new straight segment from Pol. If Pol$[v]$ is a triangle, then stop; otherwise, do it on Pol$[v]$ instead of Pol.

Because Pol$[v]$ obtained each time in the procedure is always a convex polygon from the convexity of *Pol*, the procedure works and provides a tiling of Pol, which is induced by the final triangle and all the edges deleted. A tiling T(Pol) which is not allowed to do a diagonal replacement anymore for making the resultant tiling shorter is called a *minimal tiling.*

Theorem 13.3.2. *The tiling obtained by Ta-procedure on a convex polygon Pol given is a minimal tiling of* Pol.

Proof. From Lemmas 13.3.4–13.3.7 and the definition of Ta-procedure, the theorem is derived directly. □

Moreover, by following what is called *Tb-procedure*, a variation of the Ta-procedure, a minimal tiling is produced as well.

Tb-procedure Choose a truncatible vertex v_i which has

$$(v_{i-1}, v_{i+1})$$

as the least in the order defined beforehand and find Pol$[v_i]$. If Pol$[v_i]$ is a triangle, then stop; otherwise, do it on Pol$[v_i]$ instead of Pol.

This is a kind of greedy algorithm without the convention of the non-degenerateness.

The tiling obtained by the procedures above is shown to be the shortest one for Pol regular in any case. However, it is not right in general even geometrically. The simplest example is the convex pentagon

$$P_5 = ABCDE$$

such that the lengths

$$l(AC) = l(AD) = 5,\ l(CE) = l(BD) = 5$$

and $l(BE) = 4$. The shortest tiling of P_5 is formed by

$$(A, C), (A, D)$$

with the boundary. The length of what obtained by the procedures above on P_5 is one greater than the shortest tiling.

In what follows, we provide a way to find the shortest tiling of a polygon combinatorially. That is all the lengths are replaced by any weights (non-negative of course) given.

Let Tri(Pol) be the set of all triangles on the given polygon Pol. We introduce a graph denoted by $G(\mathrm{Pol})$ whose vertex set is Tri(Pol) with the binary relation of two triangles having just one common edge as the adjacentness of two vertices. A vertex of $G(\mathrm{Pol})$ which corresponds to a triangle with two edges on the boundary of Pol is said to be a *terminal*. The subset of the vertex set of $G(\mathrm{Pol})$ which consists of all the terminals is the *terminal set* of $G(\mathrm{Pol})$.

For a polygon Pol of order s, a tree of size $s - 3$ in $G(\mathrm{Pol})$ is said to be *feasible* if all the valencies of vertices on the tree are not greater than three and all its articulate vertices are in the terminal set. Let Ftr(Pol) denote the set of all feasible trees in $G(\mathrm{Pol})$.

Lemma 13.3.8. *Let* Til(Pol) *be the set of all tiles of a given polygon* Pol. *Then,* Til(Pol) *and* Ftr(Pol) *have the same cardinality, i.e. we have*

$$|\,\mathrm{Til(Pol)}\,| = |\,\mathrm{Ftr(Pol)}\,|\,. \tag{13.3.14}$$

Proof. First, it is easily seen from the definition of a feasible tree that for a tiling $Ti \in \mathrm{Til(Pol)}$, there is a unique feasible tree $\mathrm{Tr} \in \mathrm{Ftr(Pol)}$ whose edges correspond to the common edges of triangles in Til(Pol). Conversely, the unique tiling can always be constructed from a feasible tree by all the corresponding straight segments of the edges of the tree on Pol in a natural way. Thus, we have a 1-to-1 correspondence between Til(Pol) and Ftr(Pol). This implies the lemma. □

According to the definition of $G(\mathrm{Pol})$, we may restrict ourselves to that all the triangles that correspond to vertices of $G(\mathrm{Pol})$ are inner ones of the polygon Pol in the plane in order to be suitable for Pol which is not convex. In the general case, Lemma 13.3.5 is remained to be true.

Theorem 13.3.3. *For a polygon* Pol (*not necessary to be convex here*), *the minimization problem of finding a shortest tiling in* Til(Pol) *is equivalent to that of finding a shortest feasible tree in* Ftr(Pol). *That is*

$$\min_{Ti \in Til(\mathrm{Pol})} W(Ti) = \min_{Tr \in Ftr(\mathrm{Pol})} W(Tr), \tag{13.3.15}$$

where $W(Ti)$ and $W(Tr)$ represent the sums of weights (lengths as a special case) of all edges on $Ti \in Til(Pol)$ and $Tr \in Ftr(Pol)$, respectively.

Proof. A direct result of Lemma 13.3.5. □

Although the problem of finding a shortest tree with the size given in a graph is certainly not easy in view of computing complexity, we are allowed in this case to design an efficient algorithm for finding a shortest feasible tree by observing particular properties of G(Pol) in general. Because more pages should be occupied, we are not allowed to explain in detail.

Now, we have seen that the problem of finding a shortest tiling of a straight-line planar embedding of a graph can be polynomially solved in the way of determining a shortest tiling face by face either geometrically or combinatorially. The aim of this section is reached.

13.4 Thickness

In the mid-twentieth century, most papers on layouts and routings of VLSI circuit design concerned with *planar decomposition* of a graph $G = (V, E)$, i.e. partition of the edge set as $E = E_1 + E_2 + \cdots + E_k$ such that all the edge-induced subgraphs $G_i = G[E_i]$, $1 \le i \le k$, are planar. The minimum of k among all possible planar decompositions is called the *thickness* of G, denoted by $t(G)$.

Because of the hardness to determine $t(G)$ for a graph G, several approaches used to be tried to access the problem shown as in Beineke and Harary [20–22], Beineke et al. [23], Harary [122], Alekseev and Gonchakov [4], etc. A common way is to fix a spanning tree as oriented. However, this section is working on a circuit oriented, particularly a Hamiltonian circuit oriented.

Let G be a graph and C, one of its circuits. For an embedding $\mu(G)$ of G in a space, let B_i, $1 \le i \le k$, be all connected components of $\mu(G) - \mu(C)$, then $[B_i]$ is called a *bridge* (mod C). This definition is from the lectures in *Academia Sinica* in Liu [225]. However, the origin of such a bridge is in combinatorial version with some sophistication in Tutte [334].

Given a bridge $B \in \mathscr{B}_C$, the set of all bridges (mod C) of G, C is divided into pieces by boundary points of B. Each piece is called a *B-section* of C. Two bridges B_1 and B_2 are said to be *parallel* if all boundary points of B_1 are in one of B_2-sections; otherwise, *alternative*.

It is easily seen that the parallelism between bridges is an equivalence on $\mathscr{B}$. Let $\mathscr{B}_i$, $1 \le i \le s$, be all classes under this equivalence.

Theorem 13.4.1. *A graph G is planar if, and only if, for each circuit C, $\mathscr{B}_C$ has at most two classes.*

Proof. Necessity. It is easily checked that every circuit C has all of its bridges in at most two classes from the Jordan closed curve axiom.

Sufficiency. Because of the given condition, a procedure can be established by following the fundamental circuits one by one to embed the bridges in the manner that one class is put in one domain of the circuit, and the other class, if any, in other domain. Finally, by the finite recursion principle, a planar embedding is obtained. □

This theorem can be used to deduce a number of corollaries among which the following two are available to our case here.

Corollary 13.4.1. *A Hamiltonian graph G is planar if, and only if, $\mathscr{B}_C$ has at most two classes where C is a given Hamiltonian circuit.*

Proof. Because of each bridge as an edge, the fact that C has all its bridges in at most two classes leads to that every circuit has its all bridges in at most two classes. From Theorem 13.4.1, the corollary follows. □

In fact, as one class of the bridges is put in one domain of the Hamiltonian circuit and the other, in the other domain, a planar embedding is found.

Corollary 13.4.2. *A Hamiltonian graph G is outer planar if, and only if, $\mathscr{B}_C$ has at most one class, where C is a given Hamiltonian circuit.*

Proof. A particular case of Corollary 13.4.1. □

For a given subgraph H of a graph $G = (V, E)$, the *H-thickness* of G is the least number k such that $E = E_1 + E_2 + \cdots + E_k$ with $G[E_i]$ planar and $E_H \subseteq E_i$ for all $1 \le i \le k$.

Theorem 13.4.2. *The C-thickness of a Hamiltonian graph with such a circuit C is*

$$t_C(G) = \left\lceil \frac{s}{2} \right\rceil, \tag{13.4.1}$$

where s is the number of classes in $\mathscr{B}_C$.

Proof. On account of Corollary 13.4.1, the theorem is deduced. □

Let $\mu(G)$ be an embedding of graph G determined by a joint tree $\hat{T}_\delta$. Two cotree edges are said to be *interlaced* if the two occurrences of one are alternative with the two occurrences of the other. If a partition of the set of all cotree edges is $E(\bar{T}) = E_1 + E_2 + \cdots + E_k$ with all E_i, $1 \le i \le k$, then no pair of edges interlaced is called a *cotree decomposition*, or precisely cotree k-decomposition.

For an embedding $\mu(G)$ of graph G with a spanning tree T given on a surface S, the *embedding thickness* of G is

$$t_\mu(G) = \min\{k | \text{for all } k\text{-compositions}\}.$$

And, the *surface thickness* is

$$t_S(G) = \min t_\mu(G)$$

among all possible surface embeddings.

Given an embedding $\mu(G)$ of graph G by a joint tree $\hat{T}_\delta$, let s_μ be the number of classes on the set of cotree edges on the basis of the equivalence that two cotree edges are equivalent if, and only if, without interlace.

Theorem 13.4.3. *For an embedding $\mu(G)$ of graph G with a spanning tree T given on a surface S, the embedding thickness is determined by*

$$t_\mu(G) = \left\lceil \frac{s_\mu}{2} \right\rceil. \tag{13.4.2}$$

Proof. Let $A(G)$ be the *associate graph* produced from the polyhegon of a joint tree in such a way that vertices are all occurrences of letters (or numbers) and edges, all pairs of occurrences of a letter on the polyhegon, then the embedding thickness is the C-thickness of the Hamiltonian graph $A(G)$. In virtue of Theorem 13.4.2, the theorem follows. □

Although both eqs. (13.4.1) and (13.4.2) provide an upper bound of $t_S(G)$, the exact value of $t_S(G)$ for a graph is still less known.

13.5 Crossing number

Although the determination of the crossing number of a graph is difficult without general result obtained up to now, in spite of estimation and qualitative discussion as shown in Tutte [346], Levow [175], Lerda and Majoranic [174], etc., only some specific types of graphs are concerned with us to determine the number, seen in Liu and Liu [191], Hao and Liu [117], Lu et al. [254], and this section concerns with another type of crossing numbers for a general graph for accessing to its crossing number.

First, let $\theta_\pi(P)$ be the number of interlace pairs on $\pi(P)$ where $\pi = (\pi_1, \pi_2, \ldots)$ is a sequence of Transform 2 π_i, $i \ge 1$, from P. Then the *interlace number*, denoted by $\theta(P)$, of a polyhegon P is defined to be the minimum length of sequences $\pi = (\pi_1, \pi_2, \ldots)$ of Transform 2 π_i, $i \geq 1$, from P one by one until none can be done by Transform 2, because the number of interlace pairs never increased by doing Transform 2 once on a polyhegon.

Theorem 13.5.1. *For an orientable polyhegon P, its genus is*

$$g(P) = \min_{\pi \in \Pi} \theta(\pi P). \tag{13.5.1}$$

Proof. Let Π be the set of sequence of operations (Transform 2) that can be done on P one by one in order and Π_0, the set of $\pi \in \Pi$ such that on $\pi(P)$ Transform 2 cannot be done anymore. On the basis of what discussed in Section 3.1, because of $\theta_{\pi_0}(P)$ independent of the choice of $\pi_0 \in \Pi_0$,

$$\max_{\pi_0 \in \Pi} \theta(\pi_0 P) = \min_{\pi_0 \in \Pi} \theta(\pi_0 P).$$

Hence, the conclusion is done. □

Based on this theorem, it enables us to write $\theta(P) = g(P)$.

Let $G = (V, E)$ be a graph and T, a spanning tree of G. For a rotation system σ on G, $\hat{T}_\sigma$ is the expanded tree of T. Let x_i, $1 \leq i \leq \beta$, be all the cotree edges where β is the Betti number of G.

Suppose the associate surface S, a polyhegon on $X = \{x_i, x_i^{-1} | 1 \leq i \leq \beta\}$, is divided into k layers: $0, 1, \ldots, k-1$.

Let Ω^{+j} be the set of all exchangers on the layers at least j, $0 \leq j \leq k-1$, for the given associate surface S of G. The value

$$\theta_T^{+j} = \min_{\omega \in \Omega^{+j}} \theta(\omega S)$$

is called a *minimal j-crossing number* of G.

Lemma 13.5.1. *All the minimal j-crossing numbers of G satisfy the following inequality as*

$$\theta_T^{+0} \leq \theta_T^{+1} \leq \ldots \leq \theta_T^{+(k-1)}. \tag{13.5.2}$$

Proof. A direct result of Lemma 13.2.1. □

For a spanning tree T, the crossing number over all T-immersions is called the *T-crossing number*, or *tree crossing number*. Because there is a 1-to-1 correspondence between the set of all T immersions and the set of all joint trees on T, it is soon seen that the crossing number of an T-immersion $\mu_T(G)$ is just the interlace number of its corresponding polyhedron written as $\theta(\mu_T(G))$ as well.

Lemma 13.5.2. *For a given spanning tree T of a graph G, the T-crossing number of G is*

$$\mathrm{cr}_T(G) = \theta_T^{+0}.$$

Proof. By considering the relationship between T-immersions and joint trees, the lemma is deduced from Lemma 13.5.1. □

The *cotree crossing number* of a graph is the minimum among the T-crossing numbers over all possible spanning tree T.

Lemma 13.5.3. θ_T^{+0} *is independent of the choice of spanning tree T.*

Proof. Let $\mathscr{U}_T$ be the set of all T immersions of a graph, because $\mathscr{U}_T$ is seen to be Π^{+0}. In virtue of Theorem 9.1.5, Π^{+0} is independent of the choice of a spanning tree on G and hence the lemma. □

In virtue of this theorem, θ_T^{+0} is allowed to replace by θ^{+0}.

Theorem 13.5.2. *The cotree crossing number of a graph is* θ^{+0}.

Proof. A direct result of Lemmas 13.5.2 and 13.5.3. □

13.6 Minimal bend

The problem we concern with in this section is called the *bend minimization*: to find a rectilinear embedding $\gamma(G)$ of a graph G such that the total number of bends on all the edges is the minimum. According to what discussed in Section 11.1, we here investigate the minimization restricted to rectilinear extensions of a planar embedding $\mu(G)$ of G. A rectilinear extension of $\mu(G)$ with the minimum of total number of bends is said to be the *minimum bend extension* $r_0(G)$ of $\mu(G)$, which is of course a minimal (but not minimum in general) bend embedding of G.

Lemma 13.6.1. *No minimum extension* $\gamma_0(G)$ *of a planar embedding* $\mu(G)$ *has an edge which is with a zigzag as shown in Figure* 11.2.5(a).

Proof. If $\gamma_0(G)$ has an edge e with a zigzag otherwise, then by the treatment mentioned in Section 8.2, we may always find a rectilinear embedding $\gamma_1(G)$ in which all the edges except for e which is with one bend less than that in $\gamma_0(G)$ has the same number of bends. And a contradiction to the minimality of $\gamma_0(G)$ occurs. □

This lemma tells us that only handles as shown in Figure 11.2.5(b) are allowed to be with an edge which has more than one bend in a minimum extension.

For a handle on an edge, each of the two bends on the handle is incident with two angles: one is $\pi/2$ and the other, $3\pi/2$. Because the two $\pi/2$ (or $3\pi/2$) angles have to be in the same face whose boundary includes the edge the handle is in, we may define the direction of the handle to be from the face of the two $\pi/2$ angles are in to that the two $3\pi/2$ angles are in. It is easily seen that if an edge without zigzag has more than one handle, then all of them have to be with the same direction. This allows us to construct an associate graph called the *generalized equilibrious graph* and denoted by $\widetilde{\text{Qui}}(\gamma)$ with weights on vertices and edges for any rectilinear embedding $\gamma(G)$ without zigzag of a graph G. Let V, E and F be the sets of vertices, edges and faces in $\gamma(G)$, respectively. Then the vertex set $\tilde{V}$ and the edge set $\tilde{E}$ of the generalized equilibrious graph $\widetilde{\text{Qui}}(\gamma)$ are defined to be as follows:

$$\begin{cases} \tilde{V} = V_+ + F, \\ \tilde{E} = \tilde{E}(V_+, F) + \tilde{E}(F), \end{cases} \tag{13.6.1}$$

where

$$\begin{cases} V_+ = \{v \mid \text{def}(v) > 0, v \in V\}; \\ \tilde{E}(V_+, F) = \{\prec v, f \succ \mid v \in V_+ \cap V(f), \\ \qquad\qquad f \in F\}; \\ \tilde{E}(F) = \{(f, g) \mid E(f) \cap E(g) \neq \emptyset, \\ \qquad\qquad f, g \in F\}. \end{cases} \tag{13.6.2}$$

On $\widetilde{\text{Qui}}(\gamma)$, we further choose subsets S and T of $\tilde{V}$ as

$$\begin{cases} S = V_+ + F_-; \\ T = F_+, \end{cases} \tag{13.6.3}$$

where

$$\begin{cases} F_- = \{f \mid \text{res}(f) < 0, f \in F\}; \\ F_+ = \{f \mid \text{res}(f) > 0, f \in F\}. \end{cases} \tag{13.6.4}$$

and assign weights on S and T as

$$\begin{cases} \forall s \in S, \quad a(s) = \begin{cases} \text{def}(s), \ s \in V_+; \\ -\text{res}(f), \ s \in F_-; \end{cases} \\ \forall t \in T, \quad d(t) = \text{res}_0(f). \end{cases} \tag{13.6.5}$$

Lemma 13.6.2. *For S and T defined by eq. (13.6.3) with the weights $a(s), s \in S$, and $d(t), t \in T$, on the generalized equilibrious graph $\widetilde{\mathrm{Qui}}(\gamma)$, we have*

$$\sum_{s \in S} a(s) = \sum_{t \in T} d(t). \tag{13.6.6}$$

Proof. A direct result of Lemma 9.3.2. □

The generalized equilibrious graphs $\widetilde{\mathrm{Qui}}(\gamma)$ look like the equilibrious graphs Qui in Section 12.3. However, $\widetilde{\mathrm{Qui}}$ has all the edges in $\tilde{E}(V_+, F)$ directed and the vertices involving faces in the embedding γ. Let $b(e)$ be the number of bends on edge $e \in E$ and $b(A)$, the sum of $b(e), e \in A$ for a set $A \subseteq E$ in the embedding γ. Then we may assign weights on an edge $e \in \tilde{E}$ of $\widetilde{\mathrm{Qui}}$ as

$$\lambda(e) = \begin{cases} \theta(A_e), & e = \prec v, f \succ \in \tilde{E}(V_+, F); \\ b(e_f, e_{f,g}), & e = (f, g) \in \tilde{E}(f), \end{cases} \tag{13.6.7}$$

where θ is the assignment on angles defined by eq. (13.6.3) while A_e, $e = \prec v, f \succ$, is the angle in f at v and $e_{f,g} = E(f) \cap E(g)$, the common boundary of f and g in γ.

By what discussed in Section 12.3, it is easy to see the following relations hold:

$$\sum_{e \in \tilde{E}(v)} \lambda(e) = a(v), \quad v \in S \tag{13.6.8}$$

for $\tilde{E}(v)$ being the set of edges incident to v in $\widetilde{\mathrm{Qui}}$, and

$$\sum_{e \in \tilde{E}(v)} \lambda(e) = d(v), \quad v \in T \tag{13.6.9}$$

in which $\lambda(e) > 0$ for $e = (f, g) \in \tilde{E}$ is defined to be the direction from $\pi/2$ to $3\pi/2$ at each bend on the common boundary of f and g; $-\lambda(e)$, in the opposite direction. Further, it can also be verified that

$$\sum_{e \in \tilde{E}(v)} \lambda(e) = 0, \; v \in \tilde{V} \setminus (S + T) = F_0, \tag{13.6.10}$$

where $F_0 = \{f \mid \mathrm{res}(f) = 0, f \in F\}$, from the relation on the sum of inner angles in a polygon.

Similarly to the description in Section 9.3, we also know that for any function $\lambda(e)$ on $\tilde{E}$ such that $\lambda(e) = 0, 1$ or 2 and eqs. (13.6.8) and (13.6.9) are satisfied, a rectilinear embedding without zigzag can always be found such that $\lambda(e)$ is just what determined by eq. (13.6.7) for the embedding.

Because any planar embedding $\mu(G)$ of a graph G without a vertex of valency greater than four is allowed to construct the generalized equilibrious graph $\widetilde{\mathrm{Qui}}(\mu)$. If unknowns $x_e, e \in \tilde{E}$, are introduced, then the following equations

$$\sum_{e \in \tilde{E}(v)} x_e = \begin{cases} a(v), & v \in S; \\ d(v), & v \in T; \\ 0, & \text{otherwise} \end{cases} \tag{13.6.11}$$

can be shown to have a non-negative solution with the constraint:

$$\forall e \in \tilde{E}(V_+, F), \quad x_e = 0, 1 \text{ or } 2 \tag{13.6.12}$$

from Lemma 13.6.2.

Lemma 13.6.3. *The minimum bend extension $\gamma_0(G)$ of a planar embedding $\mu(G)$ can be constructed from a non-negative solution of eq.* (13.6.11) *on the generalized equilibrious graph $\widetilde{\text{Qui}}(\mu)$ with the constraint* (13.6.12) *such that*

$$\sum_{e \in \tilde{E}(F)} x_e = \min! \tag{13.6.13}$$

Proof. From what mentioned above, we know that for any rectilinear extension $\gamma(G)$ without zigzag of $\mu(G)$, there is a non-negative solution of eq. (13.6.11) with the constraint (13.6.12), from which the extension can be constructed. And conversely, from any solution of eq. (13.6.11) with eq. (13.6.12) for $\mu(G)$, a rectilinear extension without zigzag can be constructed either. This implies that the minimum bend extension can be obtained from a non-negative solution of eq. (13.6.11) with eq. (13.6.12) by Lemma 13.6.1.

Moreover, from eq. (13.6.7) we see that the minimum shown in eq. (13.6.13) is that of total number of bends among all possible rectilinear extensions of $\mu(G)$. □

Although the consistency of eq. (13.6.11) with eq. (13.6.12) is known and a non-negative solution can be found according to a rectilinear extension without zigzag, which is always existential from Section 8.1 for any planar embedding $\mu(G)$ of G without vertex of valency greater than four in the way of determining $\lambda(e)$ on $\tilde{E}$ based on eq. (13.6.7), we here provide a direct way to find a non-negative solution of the equation with constraint.

First, choose a spanning tree $\tilde{T}$ on $\widetilde{\text{Qui}}(\mu)$ for the planar embedding $\mu(G)$ such that all $v \in V_+, a(v) = 1$, are articulate vertices. It can be shown that such kind of tree exists whenever G is connected. Then solve the equation which is restricted to be on $\tilde{T}$ in the way of solving equations at articulate vertices in V_+ with weight 1 first. Because $\tilde{T}$ corresponds to a basis of the cycle space on $\widetilde{\text{Qui}}(\mu)$, from Lemma 13.6.2 the solution obtained on $\tilde{T}$ is just one on the whole $\widetilde{\text{Qui}}(\mu)$.

In a solution $x_e = \tau_e, e \in \tilde{E}$, obtained by the procedure above, the *positive direction* of $\tau_e, e = (u, v)$, is defined as from u to v (or from v to u) if τ_e is evaluated from the equation at u (or v). Of course, if $\tau_{(u,v)}$ has the positive direction from u to v, then $\tau_e = \tau_{<u,v>} > 0$ while $\tau_{(v,u)} = -\tau_{<u,v>} < 0$. According to eq. (13.6.13), a *characteristic function*, denoted by $\text{ch}(e)$, of edge e on $\tilde{E}$ of $\widetilde{\text{Qui}}(\mu)$ can be defined as

$$\mathrm{ch}(e)=\begin{cases}1, & e\in\tilde{E}(F),\\ 0, & \text{otherwise.}\end{cases}\tag{13.6.14}$$

For a circuit C on $\widetilde{\mathrm{Qui}}(\mu)$, let $\tilde{E}_+(C,\tau)$ and $\tilde{E}_-(C,\tau)$ be the subsets of edges with clockwise and anticlockwise directions of the values of variables in the solution on C. Noticing that $\tilde{E}_+(C,\tau)\cap\tilde{E}_-(C,\tau)=\emptyset$ because each edge in C has at most one direction of the corresponding value of the variable. If

$$\begin{cases}\sum\limits_{\substack{e\in\tilde{E}_+(C,\tau)\\ \tau_e>0}}\mathrm{ch}(e)\Bigg(\text{or}\sum\limits_{\substack{e\in\tilde{E}_-(C,\tau)\\ \tau_e>0}}\mathrm{ch}(e)\Bigg)>\frac{1}{2}\sum\limits_{e\in\tilde{E}(C)}\mathrm{ch}(e);\\ \Big\{e\in\tilde{E}(C)-\tilde{E}_+(C,\tau)\Big(\text{or }\tilde{E}_-(C,\tau)\Big)\mid\tau_e=0\Big\}\\ \qquad\qquad\bigcap\tilde{E}(V_+,R)=\emptyset,\end{cases}\tag{13.6.15}$$

then C is said to be *adjustable*.

Lemma 13.6.4. *A non-negative solution of eq.* (13.6.11) *with the constraint* (13.6.12) *on* $\widetilde{\mathrm{Qui}}(\mu)$ *for a planar embedding* $\mu(G)$ *is optimal, or in other words minimum in eq. (13.6.13) is reached for the solution if, and only if, there is no adjustable circuit on* $\widetilde{\mathrm{Qui}}(\mu)$.

Proof. By contradiction for the necessity. Assume there is a circuit C which is adjustable for an optimal solution $x_e=\tau_e$, $e\in\tilde{E}$ on $\widetilde{\mathrm{Qui}}(\mu)$ of a planar embedding $\mu(G)$. Then we can find another non-negative solution of eq. (13.3.11) with constraint (13.6.12), $x_e=\hat{\tau}_e$, $e\in\tilde{E}$ as

$$\hat{\tau}_e=\begin{cases}\tau_e-\delta, & e\in\tilde{E}_+(C,\tau);\\ \tau_e+\delta, & e\in\tilde{E}(C)\setminus\tilde{E}_+(C,\tau);\\ \tau_e, & e\notin\tilde{E}(C),\end{cases}\tag{13.6.16}$$

where

$$0<\delta\le\min\{\tau_e\mid e\in\tilde{E}_+(C,\tau)\}.\tag{13.6.17}$$

However, for this solution, we have

$$\begin{aligned}\sum_{e\in\tilde{E}(F)}\hat{\tau}_e&=\sum_{e\in\tilde{E}}\mathrm{ch}(e)\hat{\tau}_e\\ &=\sum_{e\in\tilde{E}}\mathrm{ch}(e)\tau_e-\delta\Bigg(2\sum_{e\in\tilde{E}_+(C,\tau)}\mathrm{ch}(e)\\ &\qquad-\sum_{e\in\tilde{E}(C)}\mathrm{ch}(e)\Bigg)\\ &<\sum_{e\in\tilde{E}}\mathrm{ch}(e)\tau_e=\sum_{e\in\tilde{E}(F)}\tau_e.\end{aligned}$$

So a contradiction to the optimality of τ_e, $e\in\tilde{E}$, appears.

For the principle of linear programming, refered to read seen from Dantzig [66]. Because the problem is equivalent to a linear programming and the non-adjustment of circuits is just corresponding to the criterion of optimality in the linear programming, this is the sufficiency. □

In fact, we are allowed to find a non-negative solution of eqs. (13.6.11) and (13.6.12) by the above procedure from a tree $\tilde{T}$ on $\widetilde{\mathrm{Qui}}(\mu)$ chosen arbitrarily. It can be shown that this solution is just a basic feasible solution of the linear programming and that the solution obtained by the adjustment in the way of eq. (13.6.16) on an adjustable fundamental circuit associated with $\tilde{T}$ from a basic feasible solution is also a basic feasible solution. Because the minimum of the objective function in a linear programming can always be reached at a basic feasible solution, we soon find the following

Lemma 13.6.5. *The optimal non-negative solution of eq.* (13.6.11–13) *on* $\widetilde{\mathrm{Qui}}(\mu)$ *for a planar embedding* $\mu(G)$ *of a graph* G *of order* v *can always be found by* $O(v)$ *adjustments on the adjustable circuits from a non-negative solution obtained by the procedure along a spanning tree on* $\widetilde{\mathrm{Qui}}(\mu)$.

Proof. Because any circuit is a sum of some fundamental circuits in the cycle space as shown in Section 4.2 and for each edge in the common subset of the edge sets of two circuits C_1 and C_2, if $e \in \tilde{E}_+(C_1, \tau)$ then $e \in \tilde{E}_-(C_2, \tau)$ and vice versa for a solution $x_e = \tau_e, e \in \tilde{E}$. By a routine calculation, we may find that there is a circuit which is adjustable on $\widetilde{\mathrm{Qui}}(\mu)$ if, and only if, there is a fundamental circuit which is adjustable. Hence, the number of adjustments that have to be done for the optimal solution is at most the cyclomatic number of $\widetilde{\mathrm{Qui}}(\mu)$, which is $O(v)$ from eq. (13.6.1) and the planarity of G. □

Let $C^* = (e_1, e_2, \ldots, e_l)$ be a cocircuit on a rectilinear embedding $\gamma(G)$ without zigzag and $C = (f_1, f_2, \ldots, f_l)$ be the circuit corresponding to C^* in the planar dual of $\gamma(G)$. Of course, $f_1, f_2, \ldots, f_l$ can be seen as faces of $\gamma(G)$ with the relation

$$e_i \in E(f_i) \bigcap E(f_{i+1}) \tag{13.6.18}$$

for $i = 1, \ldots, l, f_{l+1} = f_1$. If an edge $e_i, 1 \le i \le l$, has an end at which the angle in f_i is not $\pi/2$ is said to be *secondary*; otherwise, *primary*. If an edge e_i $(1 \le i \le l)$ not necessary to be primary has a bend at which the angle in f_i is $3\pi/2$, then it is said to be *main oriented* for the cocircuit C^*. A cocircuit that has the number of main oriented edges greater than half the number of primary edges is called a *over load cocircuit*. It is easily checked that any minimum bend embedding has no overload cocircuit.

Theorem 13.6.1. *A rectilinear extension* $\gamma(G)$ *without zigzag of a planar embedding* $\mu(G)$ *without overload cocircuit is not a minimum bend one if, and only if, there exists a cocircuit* C^* *in which the number of all the edges with the same direction for bends is greater than half the number of edges in* C^* *of* $\gamma(G)$.

Proof. In fact, a translation of Lemma 13.6.4 is in terms of $\gamma(G)$ itself instead of $\widetilde{\text{Qui}}(\mu)$. □

From the corresponding result to Lemma 13.6.5, a more efficient algorithm can be designed to find a minimum bend extension of a planar embedding $\mu(G)$ based only on the adjustments of cocircuits in a rectilinear extension without zigzag of $\mu(G)$ itself instead of circuits in $\widetilde{\text{Qui}}(\mu)$.

13.7 Minimal area

The *area* occupied by a rectilinear embedding $\gamma(G)$ of a graph $G = (V, E)$ is here defined by the minimum of the areas of the rectangles which cover the embedding $\gamma(G)$ on the plane. This suggests us to find the horizontal and the vertical diameters of $\gamma(G)$ in order to determine its area.

For a rectilinear embedding $\gamma(G)$, let $V(\text{Hor})$ be the set of all maximal horizontal segments on $\gamma(G)$ and $E(\text{Hor})$ be the set of pairs (u, v) of elements in $V(\text{Hor})$ such that there is a face whose boundary has a point in common with each of u and v in $\gamma(G)$. Then the graph denoted by

$$\text{Hor}(\gamma) = \Big(V(\text{Hor}), E(\text{Hor})\Big) \tag{13.7.1}$$

is called the *horizontal graph* of $\gamma(G)$.

Lemma 13.7.1. *If the direction of an edge in* $\text{Hor}(\gamma)$ *is defined to be from the lower level to the upper level of the segments the two ends of the edge are in, then the directed graph* $\text{Hor}(\gamma)$ *is acyclic.*

Proof. By contradiction, suppose $(v_1, v_2, \ldots, v_s)$ is a dicircuit on $\text{Hor}(\gamma)$. Then, the y-coordinates of v_i, $i = 1, \ldots, s$, satisfy that

$$y(v_1) < y(v_2) < \ldots < y(v_s) < y(v_1).$$

This is a contradiction to the transitivity on the y-coordinates. □

If a new source s is introduced to join each source of $\text{Hor}(\gamma)$ and a new sink t is to each sink of $\text{Hor}(\gamma)$, then the resultant graph denoted by $\widetilde{\text{Hor}}(\gamma)$ is determined to be a lattice. Any longest path from s to t in $\widetilde{\text{Hor}}(\gamma)$ has its common part with $\text{Hor}(\gamma)$ being a longest path in $\text{Hor}(\gamma)$. Because a longest path in $\widetilde{\text{Hor}}(\gamma)$ can be found efficiently, a longest path in $\text{Hor}(\gamma)$ can be found efficiently as well. The length of a longest path in $\text{Hor}(\gamma)$ is defined to be the *vertical diameter* of $\gamma(G)$, denoted by $d_{\text{ver}}(\gamma G)$.

Similarly, for a rectilinear embedding $\gamma(G)$, we may also construct an associate graph denoted by

$$\text{Ver}(\gamma) = (V(\text{Ver}), E(\text{Ver}))\,, \tag{13.7.2}$$

where $V(\text{Ver})$ is the set of all maximal vertical segments in $\gamma(G)$ and $E(\text{Ver})$ is the set of all such pairs of elements in $V(\text{Ver})$ that there is a face in $\gamma(G)$ whose boundary has a point in common with each of them. Here, $\text{Ver}(\gamma)$ is called the *vertical graph* of $\gamma(G)$.

Moreover, the direction of an edge $(u, v) \in E(\text{Ver})$ is defined to be from the one of u and v with lower x-coordinate to the other.

Lemma 13.7.2. *The vertical graph* $\text{Ver}(\gamma)$ *of a rectilinear embedding* $\gamma(G)$ *is acyclic.*

Proof. The same as in the proof of Lemma 13.7.1 except for x-coordinate here instead of y-coordinate. □

Because $\text{Ver}(\gamma)$ is acyclic, we may introduce a new source s and a new sink t as well to obtain $\widetilde{\text{Ver}}(\gamma)$ which is in fact a lattice. From a longest path from s to t in $\widetilde{\text{Ver}}(\gamma)$, we may find a longest path in $\widetilde{\text{Ver}}(\gamma)$. The length of a longest path in $\text{Ver}(\gamma)$ is defined to be the *horizontal diameter* of $\gamma(G)$, denoted by $d_{\text{hor}}(\gamma G)$. The area of a rectilinear embedding $\gamma(G)$ is defined to be

$$\text{Ar}(\gamma G) = d_{\text{hor}}(\gamma G) d_{\text{ver}}(\gamma G). \tag{13.7.3}$$

In what follows, we discuss how to find a minimal area embedding from a rectilinear embedding $\gamma(G)$ in the way of adjusting a certain number of cocircuits on $\gamma(G)$.

Because we have known that a minimum bend extension of a planar embedding can be found efficiently from Section 13.6, the problem of finding a minimal area extension of a planar embedding can be only restricted to those are minimum bend extensions of the embedding.

The crucial step is to find a cocircuit from which a new extension whose area is smaller can be derived by doing a kind of operations on it.

Let $C^* = (e_1, e_2, \ldots, e_s)$ be a cocircuit in $\gamma(G)$ and

$$C = (f_1, f_2, \ldots, f_s)$$

be the circuit corresponding to C^* in the planar dual of $\gamma(G)$. Of course, all f_i, $1 \le i \le s$, correspond to faces in $\gamma(G)$. And, the edge sets $E(f_i)$ and $E(f_{i+1})$ on the boundaries of f_i and f_{i+1} satisfy that

$$e_i \in E(f_i) \bigcap E(f_{i+1}) \tag{13.7.4}$$

for $i = 1, \ldots, s$, where $f_{s+1} = f_1$. If the number of the main oriented edges is equal to half the sum of the number of primary edges and the number of main oriented edges which are not primary on C^* as mentioned in Section 13.6 or the maximum between the two numbers of bends in the two directions along C^* is half the number of all the

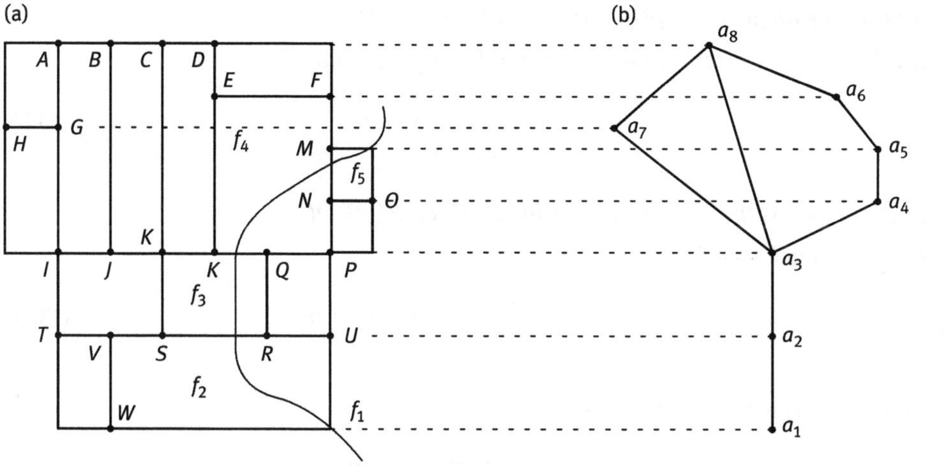

Figure 13.7.1: Rectilinear embedding and its horizontal graph.

edges in C^*, then C^* is said to be *bend balanced*. In Figure 13.7.1, (a) shows a rectilinear embedding $\gamma(G)$ in which the dashed line indicates the cocircuit

$$C^* = (UW, RS, QL, NM, OM).$$

Because UW, SR, QL and NM are all secondary edges, only OM is primary edge and UW is main oriented, we see that C^* is just bend balanced. And, (b) of Figure 13.7.1 shows the horizontal graph $\mathrm{Hor}(\gamma)$ in which the length of the longest path is 6 and hence the vertical diameter of $\gamma(G)$.

In order to describe a transformation on a cocircuit of a rectilinear embedding, we have to observe the relationship between rectilinear embedding and a system of angles associated with a planar embedding (straight line one of course).

Let $\mu(G)$ be a planar embedding on which some edges have additional 2-valent vertices. And, let $\mathscr{A}(\mu G)$ be the set of angles at ordinary vertices and $\mathscr{B}(\mu G)$ be the set of angles at additional vertices. Then, any straight-line embedding $\mu(G)$ of a graph G with additional 2-valent vertices on some edges determines a system of angles

$$\{\mathscr{A}(\mu G),\ \mathscr{B}(\mu G)\}.$$

However, the inverse case is not true in general.

Lemma 13.7.3. *A planar embedding $\mu(G)$ of a graph G with additional 2-valent vertices on some edges without articulate vertex determines a rectilinear embedding if, and only if, the system of angles*

$$\{\mathscr{A}(\mu G),\ \mathscr{B}(\mu G)\}$$

satisfies the following conditions:

Ans 1. *Any angle in* $\{\mathscr{A}(\mu G), \mathscr{B}(\mu G)\}$ *is* $\pi/2, \pi$ *or* $3\pi/2$.

Ans 2. *At each vertex v which is allowed to be additional,*

$$\sum_{A\in\tilde{\mathscr{A}}(\mu G)} A = 2\pi, \tag{13.7.5}$$

where $\tilde{\mathscr{A}}(\mu G)$ *is the set of all angles in* $\mathscr{A}(\mu G)$ *or* $\mathscr{B}(\mu G)$ *at v.*

Ans 3. *In each face f,*

$$\sum_{A\in\tilde{\mathscr{A}}(\mu G)} A = (\rho^*(f) - 2)\pi, \tag{13.7.6}$$

where $\tilde{\mathscr{A}}(\mu G)$ *is the set of angles in*

$$\mathscr{A}(\mu G) \cup \mathscr{B}(\mu G)$$

in the face f and $\rho^*(f)$ *includes the contribution of additional vertices on the boundary of f.*

Proof. The necessity is obvious by checking if Ans 1–3 are satisfied for a rectilinear embedding. The sufficiency can be derived from Theorem 12.3.1 because a rectilinear embedding of a graph with the bends as additional vertices is in fact a net embedding of a graph with additional 2-valent vertices. □

For a rectilinear embedding $\gamma(G)$, let $\tilde{\mathscr{A}}(\gamma G)$ be the set of all angles including those at bends, i.e.

$$\tilde{\mathscr{A}}(\gamma(G) = \mathscr{A}(\gamma G) + \mathscr{B}(\gamma G). \tag{13.7.7}$$

If $C^* = (e_1, e_2, \dots, e_s)$ be a cocircuit of $\gamma(G)$ and

$$C = (f_1, f_2, \dots, f_s)$$

be the circuit corresponding to C^* in the dual of $\gamma(G)$, then the operation on $\tilde{\mathscr{A}}(\gamma G)$ such that

A. The angle $A_{f_i}(e_i, S)$ which is greater than $\pi/2$ incident with e_i in face f_i becomes

$$A_{f_i}(e_i, S) - \pi/2$$

for e_i being a secondary edge which is not main oriented.

B. The bend on e_i is deleted for e_i being main oriented.

C. A bend is with the angle in f_{i+1} being $3\pi/2$ introduced on e_i for e_i being primary but not main oriented is called a *rotation with one side* for C^*.

Lemma 13.7.4. *If a system of angles $\tilde{\mathscr{A}}'$ is obtained by doing a rotation with one side for a cocircuit C^* on a rectilinear embedding $\gamma(G)$, then $\tilde{\mathscr{A}}'$ itself is also a system of angles of another rectilinear embedding $\gamma'(G)$ of G. And, the total number of bends of $\gamma'(G)$ is the same as that of $\gamma(G)$ if, and only if, C^* is bend balanced.*

Proof. The first statement is obtained by checking that $\tilde{\mathscr{A}}'$ satisfies the conditions: Ans1–3 in Lemma 13.7.3. The necessity of the last statement is obvious because if C^* is not bend balanced then $\gamma'(G)$ is not allowed to have the same number of bends as $\gamma(G)$ from the discussion in Section 13.6. From the bend balancedness of C^*, the number of bends in $\gamma'(G)$ introduced by the operation is the same as that deleted from $\gamma(G)$. This is the sufficiency of the last statement. □

In Figure 13.7.2, (a) shows the rectilinear embedding $\gamma'(G)$ obtained by doing a rotation with one side for the cocircuit identified by the dashed lines on $\gamma(G)$ as in Figure 13.7.1(a); and (b) shows the horizontal graph $\mathrm{Hor}(\gamma')$ because C^* is bend balanced, the total number of bends in $\gamma'(G)$ is the same as that in $\gamma(G)$. However, the vertical diameter of $\gamma'(G)$ is one less than that of $\gamma(G)$.

If an edge (u, v) in a rectilinear embedding $\gamma(G)$ has its two incident faces f_1 and f_2 with

$$A_{u,f_1} \geq \pi \text{ and } A_{v,f_2} \geq \pi$$

or vice versa, then e is said to be *turnable*. The operation on $\tilde{\mathscr{A}}(\gamma G)$ is such that

$$\begin{cases} A'_{v,f_2} &= A_{v,f_2} - \pi/2, \\ A'_{u,f_2} &= A_{u,f_2} + \pi/2, \\ A'_{v,f_1} &= A_{v,f_1} + \pi/2, \\ A'_{u,f_1} &= A_{u,f_1} - \pi/2, \end{cases} \tag{13.7.8}$$

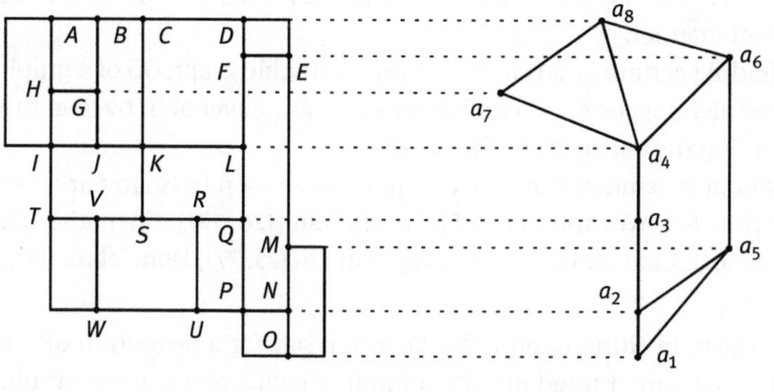

Figure 13.7.2: Rectilinear embedding and its horizontal graph.

and all other angles unchanged is called a *turnation* for an edge turnable. It is easily seen that a turnation transforms a rectilinear embedding into another rectilinear embedding without changing the total number of bends but with probably changing the area.

If a cocircuit C^* which is bend balanced in $\gamma(G)$ makes the embedding $\gamma'(G)$ obtained by the rotation with one side for C^* have the area less than that of $\gamma(G)$, then C^* is called *area adjustable*. Of course, we may find the characterization of C^* being area adjustable by the structure of $\gamma(G)$ itself via the vertical and the horizontal graphs. An edge which is turnable is also called *area adjustable* if the turnation for the edge makes the area of the resultant embedding lower than the original. This suggests us to employ the two kinds of operations for cocircuits and edges which are area adjustable to find a minimal area extension of a planar embedding with the restriction of total number of bends fixed.

Further, we are also allowed to find a minimal area extension with the bend number greater than that of the original but the area much less than that of a minimal area embedding for the restriction of bend number fixed by generalizing the area adjustable for bend unbalanced cocircuits.

13.8 Notes

13.8.1 Although the maximum non-orientable genus of a graph, except only for a tree, was determined to be the Betti number of the graph in 1979 from Liu [195], the maximum orientable genus of a graph has not yet be known, if there is a simple way, for its determination. One has only known that the determination can be done in a polynomial way in Furst et al. [94]. However, it is from Section 13.1 seen that the main result in Liu [196] directly leads to a polynomial way, for the determination, in two directions. One is by reductions $\pi(v'; x, y)$, for maintaining no occurrence of new cut vertex as possible. The other is by reductions $\pi_{x,y}$, via the transformation on the trees as shown in the proof of Theorem 13.1.2.

The well-known result, i.e. Corollary 13.1.1, for the maximum orientable genus of a graph in Xuong [408] has been shown to be directly deduced from the main result, but in Liu [196] in a different manner.

The technique of finding certain Eulerian circuit on the double graph GG of a graph G posed in Liu [195, 196] also suggests us to find a circuit double cover of G, by deleting circuit one by one avoiding the occurrence of a cut edge.

Other progresses about maximum genus of a graph can be seen in Huang and Liu [152, 153], Li and Liu [176–182], Wei and Liu [365], He and Liu [126, 127], Chen and Liu [45, 49], Wang and Liu [362], Chai and Liu [40], Dong and Liu [75, 77], Dong et al. [80], Lv and Liu [256, 257], etc.

13.8.2 The thread problem, in Hilbert and Cohn-Vossen [134], for a resolution of the Heawood map colour conjecture, turned out the minimum genus of a general graph, but not only the complete graph. Both the map colour problem and the four colour problem are famous not only in graph theory but also in topology, in the last two centuries, as shown in Heawood [128–132], Appel and Haken [6–8], Appel [10], Ore [274], Saaty [301, 302], Liu [194], etc. Because of extremal difficulty, only graphs with good symmetry have been considered, by a method involving a quotient embedding (particularly, current graphs and voltage graphs), as in Gustin [113], Ringel [287, 288], Youngs [418], Gross and Tucker [108], Liu [200], etc. Notably, a very general method is described, for graphs without such restriction, in Section 13.2.

13.8.3 The method of finding the optimal convex embedding, of a graph in Section 13.3, can be seen as a generalization of that used in Tutte [345], when the weights are restricted to the constant 1. However, Tutte did not discuss any kind of optimality, on the embeddings. His purpose at that time was to show the existence of a convex embedding.

13.8.4 The problem of finding a shortest triangulation, for vertices given on the plane, with the Euclidean distance in geometric case, or any positive weight in combinatorial case between two vertices as the length, is very difficult. No efficient algorithm has been found up to now, even in the geometric case. When vertices on the plane forms a convex polygon, it is polynomially solvable, both in the geometric and combinatorial cases. Dantzig et al. [67] provided an efficient algorithm, for this case, by linear programming. However, in Section 13.3, another efficient algorithm can also be found, and a more general problem is solved as well.

13.8.5 Similar problems to those mentioned in item 13.8.4 can also be proposed for surfaces, or higher dimensional spaces. However, no further results appear in literature yet.

13.8.6 On the bend minimization, or area minimization, in VLSI, many papers appear in recent year, with a variety of models. Among them, a minimum bend extension was transformed into a maximal flow of a network, in Tamassia [325]. However, what discussed in Section 13.6 can be realized as a development of the graphic method, used by Chinese in the 1950s, for solving the transportation problem, shown in Liu [213]. Some related problems can be seen in Basden and Nichols [17], Chen et al. [42], van Cleemput [52], Du and Zhang [81], Krentel [163], Larson and Li [169], Liu [211], de Rezende et al. [285] and so forth.

13.8.7 Although a method, similar to that in Section 13.6, can be used for finding minimal area extensions, it is not as efficient as that there. The minimality and the efficiencies are still needed to investigate further. Of course, the general problems of finding the minimum area (bend as well) embedding of a graph are more difficult, at least in computing complexity. Most papers are contributed to the estimation of the upper bound of the area, the minimum area embedding occupied, as in Bhatt and Leighton [30], Liu [212] by using separator theorem in Lipton and Tarjan [185, 186]. However, the least upper bounds of total number of bends, in the minimum bend embedding, for some kinds of graphs, have been found in a different way in Liu et al.

[239], Liu et al. [240] and Liu et al. [242, 243]. This method is also hopeful to estimate the upper bound of area occupied by the minimum area embedding for some kinds of graphs in Liu [212].

13.8.8 On the basis of Theorems 13.1.1, 13.1.2 and Corollary 13.1.1, the relationship between maximum genus with up-embeddability and other topological invariants has been sufficiently investigated as shown in Huang and Liu [146–152], Li and Liu [177–182], Chen and Liu [45], Chen and Liu [47, 49], He and Liu [126, 127], Lv and Liu [255–260], Cai et al. [38], Dong and Liu [75–77], Dong et al. [78, 79], Dong et al. [80], Wei and Liu [365], Chai and Liu [40], Wang and Liu [362], etc.

13.8.9 Since the minimum genus of a graph with high symmetry had been paid an attention on for a long period of time, motivated by the determination of the genus (minimum!), of the complete graph K_n of order n, via quotient graph, particularly, current graph or voltage graph, in this chapter, only graphs, not necessary to have symmetry, are much concentrated on, because of the occurrence of joint tree model, for representing embeddings of a graph on surfaces. Only a few results have been done worthily in this aspect such as Shao and Liu [312, 313, 315, 316], Shao [317], Wei et al. [371], Zeng and Liu [419], etc.

14 Matroidal Graphicness

14.1 Definitions

Many equivalent definitions used for matroids appear in literature. We here adopt that a *matroid* is a family $\mathscr{X}$ of subsets on a finite set E, which is called the *base set*, with the two axioms satisfied. The cardinality of E is called the *order* of M and that of $\mathscr{X}$, the *size* of M.

Axiom 1. No member of $\mathscr{X}$ is a proper subset of another.

Axiom 2. Let a and b be two distinct elements of E. Let X and Y be members of $\mathscr{X}$ such that $a \in X \cap Y$ and $b \in X \setminus Y$. Then there exists $Z \in \mathscr{X}$ such that $b \in Z \subseteq (X \cup Y) - a$.

A matroid is usually denoted by $M = (\mathscr{X}, E)$, or simply M when it is not necessary to identify the base set E. Members of $\mathscr{X}$ are said to be *circuits*. An element of E is called a *cell* of M. Two matroids $M_1 = (\mathscr{X}_1, E_1)$ and $M_2 = (\mathscr{X}_2, E_2)$ are said to be *isomorphic* if there is a bijection $\tau : E_1 \to E_2$ such that $\forall C_1 = \{e_1, e_2, \ldots, e_s\} \in \mathscr{X}_1$,

$$\tau(C_1) = \{\tau(e_1), \tau(e_2), \ldots, \tau(e_s)\} = C_2 \in \mathscr{X}_2. \tag{14.1.1}$$

Two isomorphic matroids M_1 and M_2 are represented by $M_1 \cong M_2$. Although only one element is allowed to be a circuit which is called a *loop*, we shall never consider a matroid with a loop because loops are unessential for our purpose here.

Let $\mathscr{E}$ be the space generated by $\{e \mid \forall e \in E\}$ and denoted by $\langle e \mid \forall e \in E\rangle$ over a field $\boldsymbol{F}$. For a vector $f \in \mathscr{E}$, let $E(f)$ be the subset of E, in which each element corresponds to a non-zero component of f. We also employ the notation f itself instead of $E(f)$ if without confusion and call $E(f)$ the *support* of f. An *integral vector* f that is one with all its component integers in the group $\mathscr{N}$ of vectors over the integer ring is called *elementary* if the support $E(f)$ is minimal for $\mathscr{N}$, i.e. no vector g in $\mathscr{N}$ has the property that $E(g)$ is a proper subset of $E(f)$.

If $\mathscr{E}$ is binary, i.e. over $\boldsymbol{F} = \mathrm{GF}(2)$, then it is easily checked that the family $\mathscr{X}(\mathscr{N}; 2)$ which consists of all the subsets corresponding to the elementary vectors in $\mathscr{N}$ forms a matroid which is denoted by $M(\mathscr{N}; 2)$. A matroid is isomorphic to $M(\mathscr{N}; 2)$ for a group $\mathscr{N}$, a subspace as well if $\mathscr{N} \subseteq \mathscr{E}$ over GF(2) is said to be *binary*. In general, we may take $\boldsymbol{F}$ to be the rational field or the real field. If an elementary vector for a group $\mathscr{N}$ of vectors over the integer ring has all its non-zero components be 1 or −1, then it is said to be *primitive*. It is also easily shown that the family of subsets which correspond to primitive vectors in $\mathscr{N}$ forms a matroid as well, which is denoted by $M(\mathscr{N})$. A matroid which is isomorphic to $M(\mathscr{N})$ for a group $\mathscr{N}$, called *regular*, i.e. to each elementary vector there corresponds a primitive one with the same support, is said to be *regular* as well. Because for each primitive vector there exists an elementary one with the same support, we soon see that any regular matroid is binary.

DOI 10.1515/9783110479492-014

14.2 Binary matroids

The characterization is investigated in the way by employing minors which reflect some local structure of a matroid.

Lemma 14.2.1. *If $M = (\mathscr{C}, E)$ is a binary matroid, then $\forall C_1, C_2 \in \mathscr{C}$,*

$$\exists C_3 \in \mathscr{C} : C_3 \subseteq C_1 \bigoplus C_2. \tag{14.2.1}$$

where $C_1 \bigoplus C_2 = (C_1 \cup C_2) \backslash (C_1 \cap C_2)$ is called the symmetric difference between C_1 and C_2.

Proof. Because M is binary, there exists a group $\mathscr{N}$ in $\mathscr{E}$ over GF(2) such that M is isomorphic to $M(\mathscr{N}; 2)$. Since the vector $f = f(C_1) + f(C_2) \in \mathscr{N}$ and $E(f) = C_1 \bigoplus C_2 \in \mathscr{C}(\mathscr{N}; 2) = \mathscr{C}$, we have an elementary vector f_0 in $\mathscr{N}$ such that $E(f_0) \subseteq E(f)$. Hence, $C_3 = E(f_0)$ satisfies eq. (14.2.1). □

If $M = (\mathscr{C}, E)$ is a binary matroid which is isomorphic to $M(\mathscr{N})$ for $\mathscr{N}$ as an Abelian group in $\mathscr{E}$ over GF(2), then the matroid which is isomorphic to $M(\mathscr{N}^{\perp})$ for the orthogonal subspace $\mathscr{N}^{\perp}$ of $\mathscr{N}$ in $\mathscr{E}$ is called the *dual matroid* of M and denoted by $M^* = (\mathscr{C}^*, E)$. Members of $\mathscr{C}^*$ are called *cocircuits*. Of course, $\mathscr{C}^*$ is a binary matroid as well.

Lemma 14.2.2. *If M is a binary matroid and M^*, its dual, then $\forall C \in \mathscr{C}, C^* \in \mathscr{C}^*$,*

$$| C \bigcap C^* | = 0 \pmod 2. \tag{14.2.2}$$

Proof. A direct result of the orthogonality between $\mathscr{C}$ and $\mathscr{C}^*$. □

A subset of E which does not contain a circuit is said to be *independent* and otherwise, *dependent*. Because $\emptyset$ is never a circuit, we see that $\emptyset$ is independent. Moreover, it is also easy to verify that any subset of an independent subset is independent.

Lemma 14.2.3. *For a binary matroid $M = (\mathscr{C}, E)$, let C_1 and C_2 be two distinct circuits and $A = C_1 \bigoplus C_2$. Then, $\exists C_1, \ldots, C_s \in \mathscr{C} \ni$*

$$A = \sum_{i=1}^{s} C_i, \tag{14.2.3}$$

where the summation represents the disjoint union of sets.

Proof. Suppose M is isomorphic to $M(\mathscr{N})$ for $\mathscr{N}$ in $\mathscr{E}$ over GF(2). We know that every non-zero vector of $\mathscr{N}$ corresponds to a dependent subset of E. From Lemma 14.2.1, A contains a circuit C_1. Because the vector $f(A_1)$ for $A_1 = A \bigoplus C_1$ is in $\mathscr{N}$, A_1 is dependent either. Let C_2 be a circuit in A_1. We consider $A_2 = A_1 \bigoplus C_2$ instead of A_1. From the finiteness of E, expression (14.2.3) can always be found. □

Because for two independent subsets X and Y, if $|X| = |Y| + 1$, then there exists $x \in X \backslash Y$ such that $Y \cup x$ is independent. It can be verified that all *maximal independent subsets*, i.e. each of which is not a proper subset of an independent subset, have the same cardinality which is called the *rank* of the matroid. A maximal independent subset is called a *base* denoted by $B(M)$, or simply B. We may also show that for $e \notin B$, there is exactly one circuit in $B \cup e$, which is called the *fundamental circuit* denoted by $C(B; e)$ of e on the matroid for B.

Lemma 14.2.4. *For a binary matroid $M = (\mathscr{L}, E)$, let B be a base of M. Then, for a circuit C, $C \backslash B = \{e_1, e_2, \dots, e_s\}$, we have*

$$C = \bigoplus_{1 \le i \le s} C(B; e_i). \tag{14.2.4}$$

Proof. From what discussed in Chapter 4, we see that all the fundamental circuits form a basis of the subspace generated by the group $\mathscr{N}$ for which M is isomorphic to $M(\mathscr{N}; 2)$. Because any circuit corresponds to a vector in $\mathscr{N}$, the expression of C as a linear combination of vectors in the basis is just corresponding to eq. (14.2.4). □

Let $M = (\mathscr{L}, E)$ be a matroid not necessary to be binary. For a subset S of E, let

$$\mathscr{L} = \{A \mid (A \subseteq S) \wedge (A \in \mathscr{L})\} = \mathscr{L} \cap S.$$

It is easy to check that $L = (\mathscr{L}, S)$ is a matroid. L is called the *reduction* of M to S and denoted by $M \cdot S$. For a subset $T \subseteq E$, let

$$\mathscr{P} = \{A \mid A = C \cap T \neq \emptyset, C \in \mathscr{L}\}.$$

It is also easy to check that $P = (\mathscr{P}, T)$ is a matroid. P is called the *contraction* of M to T and denoted by $M \times T$. A matroid that can be represented in the form $(M \cdot S) \times T$ is called a *minor* of M. It can be shown that a minor of a matroid M is a minor of M.

Lemma 14.2.5. *A minor of a binary matroid is binary.*

Proof. Because the corresponding operations such as reduction and contraction on a group in a space produce new groups in the space over GF(2), by the definition of binary matroid the lemma is found. □

By no means any matroid is binary. A simple example of a matroid not binary $U(4, 3)$ has the base set of four elements: a, b, c and d with circuits as all the three element subsets of the base set. Because of the two circuits $\{a, b, c\}$ and $\{a, b, d\}$ with their symmetric difference $\{c, d\}$ of no subset as a circuit, from Lemma 14.2.1, $U(4,3)$ is not binary. It can be seen that $U(4,3)$ is the only non-binary matroid with the least order and size.

Lemma 14.2.6. *For binary matroid M,*

$$\nexists S, T \subseteq E : \ (M \cdot S) \times T \cong U(4,3). \tag{14.2.5}$$

Proof. A direct result of Lemma 14.2.5. □

Furthermore, all the concepts on independence, cocircuits, base, duality and so on introduced above for binary matroids can be extended into those for general matroids. A subset of E is said to be *independent* for a matroid $M = (\mathscr{L}, E)$ not necessary to be binary if it does not contain a circuit in $\mathscr{L}$. A *base* of M is a maximal independent subset. Because all maximal independent subsets have the same cardinality, the cardinality of a base is called the *rank* of M. A subset of E is called a *cocircuit* for M if it is minimal for the property that it has non-null intersection with any base of M. It can be shown that the family $\mathscr{L}^*$ of all cocircuits for M determines a matroid on E which is called the *dual matroid* of M, denoted by $M^* = (\mathscr{L}^*, E)$. It is easy to see that $M^{**} = M$. The complement of a base of M on E is a base of M^* which is called a *cobase* of M. The cardinality of a cobase is called the *corank* of M. Because we also have that each element in a cobase $\bar{B}$ of M forms exactly one circuit with its corresponding base B, the circuit is called a *fundamental circuit* of M. Likewise, the cocircuit formed by an element in B with $\bar{B}$ is called a *fundamental cocircuit* of M.

Lemma 14.2.7. *For a matroid M not necessary to be binary, the conditions* (14.2.1)–(14.2.5) *are all equivalent to one another.*

Because of restricted space, we are not allowed to present a complete proof of the lemma, so the reader is referred to Tutte [340, 348], Welsh [381] and White [383–385].

Theorem 14.2.1. *A matroid $M = (\mathscr{L}, E)$ is binary if, and only if, one of the conditions* (14.2.1)–(14.2.5) *is satisfied.*

Proof. From Lemmas 14.2.1, 14.2.4 and Lemmas 14.2.6–14.2.7, it suffices to prove that one of the conditions (14.2.1)–(14.2.5) is sufficient for M being binary. We only take condition (14.2.4). If a subset S of E is represented by the vector $f(S)$ over GF(2) such that a component of f is non-zero, i.e. 1 if, and only if, its corresponding element is in the subset S. Then the symmetric difference on subsets is just the addition (mod 2) on vectors. Because the condition (14.2.4) provides the expression of a vector that corresponds to a circuit in M as a linear combination of vectors in a basis, this in turn correspond to the fundamental circuits. Thus, it is allowed to extend all the linear combinations of the vectors corresponding to fundamental circuits of M

into a subspace $\mathscr{N}$ which is an Abelian group as well in $\mathscr{E}$ over GF(2). From the axiom 2 of a matroid, all the elementary vectors have to be circuits in $\mathscr{L}$. Therefore, $M = (\mathscr{L}, E) \cong M.(\mathscr{N}; 2)$, and hence is binary. □

Lemma 14.2.8. *A matroid M is binary if, and only if, so is its dual matroid M^*.*

Proof. In fact, M and M^* are, respectively, produced by two subspaces: $\mathscr{N}$ and its orthogonal $\mathscr{N}^\perp$ in $\mathscr{E}$. The lemma is obtained. □

From Lemma 14.2.8 and Theorem 14.2.1, we may soon find

Theorem 14.2.2. *A matroid $M = (\mathscr{L}, E)$ is binary if, and only if, one of the following conditions is satisfied:*

(1) *For any two cocircuits C_1^* and C_2^* of M, there exists a cocircuit C_3^* such that $C_3^* \subseteq C_1^* \bigoplus C_2^*$.*
(2) *For any two distinct cocircuits C_1^* and C_2^*, the symmetric difference of them can be expressed as a disjoint union of cocircuits of M.*
(3) *For a cobase $\bar{B} = \{e_1, e_2, \dots e_s\}$, any cocircuit can be represented by the symmetric difference of the fundamental cocircuits.*
(4) *The dual matroid M^* does not have a minor isomorphic to $U(4, 3)$.*

In fact, one may see that any subset of two elements in $\{a, b, c, d\}$ is a base of the matroid

$$U(4, 3) = (\{abc, abd, acd, bcd\}, \{a, b, c, d\})$$

and hence a cobasis. That means $U(4, 3) = U^*(4, 3)$. A matroid whose dual matroid is isomorphic to itself is said to be *self-dual*. Thus, $U(4, 3)$ is a self-dual matroid.

14.3 Regularity

A matroid M on E is said to *representable over a field* $\mathbf{F}$ if there exists a bijection $\tau : E \to \mathscr{V}$ such that τ preserves the linear independence where $\mathscr{V}$ is a subset of a vector space over $\mathbf{F}$. For a matroid M, if there exists a field such that M is representable over the field then it is called *representable*.

Lemma 14.3.1. *Let $\mathscr{N}$ be a regular group on E over the integer ring. Then, for any $f \neq 0, f \in \mathscr{N}$, being an integral vector, there exist primitive vectors $f_1, f_2, \dots, f_s$ which are with the same support as f such that*

$$f = \sum_{i=1}^{s} f_i. \qquad (14.3.1)$$

Proof. Let

$$\mathrm{sum}(u) = \sum_{e \in E} | u(e) |$$

for a vector $u \in \mathcal{N}$. We may choose f to be one of those that does not satisfy the lemma with the *sum* minimum if the lemma fails. Suppose g is the primitive vector with the same support as f. We write $h = f - g$. Of course, if $h = 0$ then $f = g$. If $h \neq 0$, from the regularity, we have $\mathrm{sum}(h) < \mathrm{sum}(f)$ and $E(h) = E(f)$. Therefore, h is a sum of primitive vectors with the same support as f. In both the cases, we find f is such a sum as the lemma indicates. This is a contradiction to the choice of f. □

For an integer $p \geq 2$, if an integral vector f has all of its components with the absolute values less than p, then it is said to be *standard* for p.

Lemma 14.3.2. *Let $\mathcal{N}$ be a regular group on E. Then for each $q \geq 2$ and for each integral vector $g \in \mathcal{N}$, there is a standard vector f such that $f = g(\mathrm{mod}\ q)$.*

Proof. For an integral vector j, let $n_q(g)$ be the number of elements $e \in E(g) : | j(e) | \geq q$. We may choose an integral vector f:

$$n_q(f) = \min\{n_q(g) \mid j = g(\mathrm{mod}\ q)\}.$$

If $n_q(f) > 0$, suppose $| f(a) | \geq q, a \in E(f)$. From Lemma 14.3.1, let the primitive vector h be with $E(h) = E(f)$ and $h(a) \neq 0$. We write $f_1 = f - qh$. From $| f_1(a) | < | f(a) |$ and $f(a) < q \Rightarrow f_1(a) < q$, we have $n_q(f_1) \leq n_q(f)$ with the equality if, and only if, $f_1(a) \geq q$. If so, we may do the same procedure for f_1 instead of f. From the finiteness of $f(a)$, we may finally find f' such that $n_q(f') < n_q(f)$. This is a contradiction to the choice of f. □

Lemma 14.3.3. *Let $M = (\mathcal{X}, E)$ be a regular matroid. Then for any prime p, there exists a regular group $\mathcal{N}$ on E over* GF(*p*)*, the field of characteristic p, such that $M = M(\mathcal{N})$.*

Proof. Because M is regular, there exists a regular group $\mathcal{N}'$ on E over the rational field such that $M = M(\mathcal{N}')$. For each $f' \in \mathcal{N}'$, let f be the vector over GF(p) defined by

$$f(a) = f'(a)(\mathrm{mod}\ p) \ \forall a \in E. \tag{14.3.2}$$

It can be shown that the set $\mathcal{N}$ of all the standard vectors each of which satisfies eq. (14.3.2) for a vector f' in $\mathcal{N}'$ forms a group on E over GF(p). In what follows, we prove that $M(\mathcal{N}') = M(\mathcal{N})$.

Let $C' \in \mathcal{X}$. From $M = M(\mathcal{N}')$, there exists a primitive vector $h' \in \mathcal{N}'$ such that

$$E(h') = C'.$$

Consider the standard vector h that corresponds to h'. Then $E(h) = C'$. That implies C' is dependent on $M(\mathcal{N})$. Hence, there exists a circuit C of $M(\mathcal{N})$ such that $C \subseteq C$.

On the other hand, let C be a circuit of $M(\mathcal{N})$. There is a vector $f \in \mathcal{N}$ such that

$$E(f) = C.$$

But this means there exists $f' \in \mathcal{N}'$ such that $C \subseteq E(f')$. If $a \in E(f')\backslash C$, then $f'(a) = 0 \pmod{p}$. However, from Lemma 14.3.2, there exists a standard vector h' such that $E(h') = E(f) = C$, or in other words C is dependent on $M(\mathcal{N}')$. Hence, there exists a circuit C' in $M(\mathcal{N}')$ such that $C' \subseteq C'$.
Combining the two facts above, we have $M(\mathcal{N}) = M(\mathcal{N}')$. The lemma is obtained. □

Lemma 14.3.4. *If a matroid M is regular, then M is representable over every field.*

Proof. From the definition of a regular matroid, it is natural that M is representable over the rational field and hence over any field of characteristic zero. Further, from Lemma 14.3.3, we can see that M is representable over GF(p) and hence over any field of characteristic p. From the classification of fields according to the characteristic, the lemma is apparently obtained. □

Let $D(M)$ be the incident matrix of a matroid $M = (\mathscr{L}, E)$ with rows corresponding to circuits in $\mathscr{L}$ and columns, elements in E and be called the *circuit matrix* of M. Dually, the *cocircuit matrix* denoted by $D^*(M)$ of M is the incident matrix of cocircuits in $\mathscr{L}^*$ against elements in E where $M^* = (\mathscr{L}^*, E)$ is the dual matroid of M. In other words, the cocircuit matrix of M is the circuit matrix of its dual matroid M^*, i.e.

$$D^*(M) = D(M^*). \tag{14.3.3}$$

For a binary matroid M, if it is possible to assign the negative sign to some of the non-zero entries in $D = D(M)$ and $D^* = D^*(M)$ such that

$$DD^{*T} = 0 \tag{14.3.4}$$

over the integer ring, then M is said to be *orientable*.

Lemma 14.3.5. *If a matroid M is regular, then M is orientable.*

Proof. Because M on E is regular, there exists a group of integral vectors $\mathcal{N}$ such that $M = M(\mathcal{N})$. For a circuit C of M, we have a primitive vector $f \in \mathcal{N}$ such that $E(f) = C$. In the row of the circuit matrix D of M corresponding to C, we assign a positive or a negative sign to the entry according as the corresponding component is +1 or −1. And likewise, orient the cocircuit matrix D^* in agreement with the assignment of signs on the circuit matrix of the dual matroid $M^* = M(\mathcal{N}^*)$. Because of the orthogonality of the groups of vectors $\mathscr{N}$ and $\mathscr{N}^* = \mathscr{N}^{\perp}, D$ and D^* with the defined signs on entries satisfy eq. (14.3.4). This is the lemma. □

For a binary matroid M with a base B given, the incident matrix of fundamental circuits against elements is called the *fundamental matrix* of M. And, the *cofundamental matrix* of M is defined to be the fundamental circuit of its dual M^* for the cobase $\bar{B}$ corresponding to B. If a matrix over the rational field has all the determinants of its submatrices being 0, 1 or −1, then it is said to be *totally unimodular*.

Lemma 14.3.6. *If a matroid M is regular, then for any base its fundamental matrix is totally unimodular.*

Proof. From the regularity and Lemma 14.3.1, we are allowed only to discuss the totally unimodularity for the fundamental matrix of a regular space with the basis consisting of primitive vectors.

For convenience, let $J(S,T)$ for $S \subseteq \overline{B}, T \subseteq B, |S| = |T|$, be the submatrix of the fundamental matrix $J(B)$ for base B by the rows corresponding to the elements in S and the columns corresponding to the elements in T. If $\det J(S,T) \neq 0$, then it is seen that $J(B)$ can always be transformed into $J(B'), B' = (B \backslash T) \cup S$ by the operations: permuting the rows, adding to one row another row obtained by multiplying a row by −1 and multiplying a row by −1. This implies there is a non-singular matrix A with its entries 0, 1 and −1 such that

$$\det A \ \det J(S,T) = 1.$$

Because both $\det A$ and $\det J(S,T)$ are integers, it is only possible that $\det J(S,T)$ is 1 or −1. From the arbitrariness of the choices of B and (S,T), the lemma follows. □

A binary matroid is said to be *totally unimodular* if so is one of its fundamental matrices.

The simplest example of a binary matroid which is not regular is the *Fano matroid* which is on a set of seven elements $\{e_1, e_2, \dots, e_7\}$ with the circuits as

$$\{X_{126}, X_{135}, X_{247}, X_{234}, X_{257}, X_{367}, X_{456}\},$$

where $X_{ijk} = \{e_t \mid 1 \le t \le 7, t \neq i, j, \text{and } k\}$ for $1 \le i < j < k \le 7$.

The non-regularity of the Fano matroid can be shown in the following way. Because it is easily checked that $\{e_4, e_5, e_6, e_7\}$ is a base which corresponds to the three fundamental circuits X_{234}, X_{135} and X_{126}, the fundamental matrix is

$$J(\{e_4, e_5, e_6, e_7\}) = \begin{pmatrix} 1 & 0 & 0 & 0 & x_{12} & x_{13} & x_{14} \\ 0 & 1 & 0 & x_{21} & 0 & x_{23} & x_{24} \\ 0 & 0 & 1 & x_{31} & x_{32} & 0 & x_{34} \end{pmatrix},$$

where $x_{ij} = \pm 1, i \neq j, 1 \leq i, j \leq 4$. Suppose the matrix is totally unimodular. Because the determinants of the following four submatrices

$$\begin{pmatrix} x_{13} & x_{14} \\ x_{23} & x_{24} \end{pmatrix}, \begin{pmatrix} x_{21} & x_{24} \\ x_{31} & x_{34} \end{pmatrix}, \begin{pmatrix} x_{12} & x_{14} \\ x_{32} & x_{34} \end{pmatrix},$$
$$\begin{pmatrix} 0 & x_{12} & x_{13} \\ x_{21} & 0 & x_{23} \\ x_{31} & x_{32} & 0 \end{pmatrix}$$

have to be even, from the totally unimodularity we have the following equations with all unknowns being only allowed to be 1 or −1:

$$\begin{cases} x_{13}x_{24} = x_{14}x_{23}, \; x_{21}x_{34} = x_{24}x_{31}, \\ x_{12}x_{34} = x_{14}x_{32}, \; x_{21}x_{32}x_{13} + x_{31}x_{23}x_{12} = 0, \\ x_{13}x_{24}x_{14}x_{23} = x_{21}x_{34}x_{24}x_{31} \\ \qquad\qquad = x_{12}x_{34}x_{14}x_{32} = 1, \\ x_{21}x_{32}x_{13}x_{31}x_{23}x_{12} = -1. \end{cases}$$

However, these equations are contradictory for they imply

$$\begin{aligned} -1 &= (x_{21}x_{32}x_{13}x_{31}x_{23}x_{12})x_{14}^2x_{24}^2x_{34}^2 \\ &= (x_{13}x_{24}x_{14}x_{23})(x_{21}x_{34}x_{24}x_{31}) \\ &\quad \times(x_{12}x_{34}x_{14}x_{32}) = 1. \end{aligned}$$

From Lemma 14.3.6, the Fano matroid is not regular.

Lemma 14.3.7. *If a matroid is regular, then it does not contain Fano matroid or its dual as a minor.*

Proof. Because it can be verified that any minor of a regular matroid is also regular and that a matroid is regular if, and only if, so is its dual in the similar ways to the proofs of Lemmas 14.2.5 and 14.2.8 from the definition of a regular matroid, by the above description on the Fano matroid, the lemma is deduced. □

For a binary matroid M, if its fundamental matrix over GF(2) does not have a submatrix from which the following matrix

$$X = \begin{pmatrix} 0 & 1 & 1 & 1 \\ 1 & 0 & 1 & 1 \\ 1 & 1 & 0 & 1 \end{pmatrix}$$

or its transpose X^T can be obtained by permuting rows and columns, then M is said to be *X-free*.

Lemma 14.3.8. *If a binary matroid M is regular then its fundamental matrix over* GF(2) *is X-free.*

Proof. A direct result of Lemma 14.3.7 by the non-regularity of Fano matroid and its dual none of which is X-free. □

Lemma 14.3.9. *For a binary matroid M, the following statements are equivalent:*
(1) *M is representable over every field.*
(2) *M is orientable.*
(3) *M is totally unimodular.*
(4) *M has no minor which is isomorphic to Fano matroid or its dual.*
(5) *M is X-free.*

The proof is left to the reader to refer Tutte [340, 348], and/or Welsh [381].

Theorem 14.3.1. *A binary matroid M is regular if, and only if, M satisfies one of the statements* (1–5) *in Lemma* 14.3.9.

Proof. The necessity is obtained from Lemmas 14.3.4–14.3.8. It suffices to prove the regularity of M from the statement (1) for the sufficiency by using Lemma 14.3.9. However, this is obvious from the definition of a regular matroid. □

Of course, if M is replaced by its dual M^* in (1)–(5) of Lemma 14.3.9, then the theorem becomes its dual form.

14.4 Graphicness

For a graph $G = (V, E)$, let $\mathscr{R}(G)$ be the space generated by $\{e \mid \forall e \in E\}$ over the rational field. Let $\mathscr{C}$ and $\mathscr{C}^\perp$ be the groups of the integral vectors in $\mathscr{R}$, called the cycles and the cocycles of G as discussed in Section 3.1, respectively. Further, let $Z(G) = M(\mathscr{C})$ and

$$Z^\perp(G) = M(\mathscr{C}^\perp)$$

be the matroids called the *circuit matroid* and *cocircuit matroid* of G, respectively, because it can be verified that the families of subsets of E in $Z(G)$ and $Z^\perp(G)$ are all circuits and cocircuits on G.

Lemma 14.4.1. *Both $Z(G)$ and $Z^\perp(G)$ are binary.*

Proof. Because if $\mathscr{R}(G)$ is taken to be the space over GF(2), then $\mathscr{C}$ and $\mathscr{C}^\perp$ are the cycle space and the cocycle space as shown in Section 4.1. Therefore, from the definition of binary matroids in Section 14.2, the lemma naturally follows. □

Lemma 14.4.2. *Let $Z^*(G)$ be the dual matroid of the matroid $Z(G)$ for a graph $G = (V, E)$. Then,*

$$Z^*(G) = Z^{\perp}(G).$$

Proof. From the orthogonality of the cycle space and the cocycle space of a graph and the definition of the dual matroid of a binary matroid in Section 14.1, the lemma is obtained. □

Lemma 14.4.3. *Both $Z(G)$ and $Z^{\perp}(G)$ are regular.*

Proof. From Lemma 14.4.2 and that a matroid is regular if, and only if, so is its dual matroid, it suffices to prove the regularity of $Z(G)$ only. Because $Z(G)$ is binary from Lemma 14.4.1, all the members of subsets in $Z(G)$ can be represented by vectors over GF(2). We may arrange to each edge a direction on G and define the vector $f = f(C)$ for $C \in Z(G)$ as

$$f(e) = \begin{cases} 1, \text{ if } e \in C & \text{and } e \text{ is} \\ & \text{in the clockwise direction of } C; \\ -1, \text{ if } e \in C & \text{and } e \text{ is} \\ & \text{in the anticlockwise direction;} \\ 0, \text{ otherwise.} & \end{cases}$$

Because any cocircuit $C^* \in Z^{\perp}(G)$ has the form:

$$C^* = (X, Y) = \{(x, y) \mid \forall x \in X, y \in Y, (x, y) \in E\},$$

where $Y = V \backslash X$, the vector $f^* = f(C^*)$ for $C^* \in Z^{\perp}(G)$ is defined to be

$$f^*(e) = \begin{cases} 1, \text{ if } e = (u, v) \in C^* \text{and } u \in X, v \in Y; \\ -1, \text{ if } e = (u, v) \in C^* \text{and } u \in Y, v \in X; \\ 0, \text{ otherwise.} \end{cases}$$

Because for any $C \in Z(G)$ and $C^* \in Z^{\perp}(G)$, from the orthogonality we always have

$$\sum_{e \in E} f(e) f^*(e) = 0.$$

Such an orientation shows that $Z(G)$ is orientable. From Theorem 14.3.1, the lemma is derived. □

A matroid M is called a *graphic* or *cographic matroid* accordingly as there exists a graph G such that $M \cong Z(G)$, or $Z^*(G)$. For a matroid M, if there exist corank(M) + 1 circuits (in general, cycles) such that each element of E appears in exact two circuits among them, then M is said to be *over corank 2-coverable*. Dually, if there exist rank(M) + 1 cocircuits (in general, cocycles) such that each element of E appears in exact two cocircuits among them, then M is said to be *over rank 2-coverable*. Of course, from the duality M is over rank 2-coverable if, and only if, M^* is over corank 2-coverable.

Lemma 14.4.4. *If M is a graphic matroid, then M^* is over corank 2-coverable.*

Proof. Because M is graphic, there is a graph G such that $M \cong Z(G)$, the circuit matroid of $G = (V, E)$. Let $C^*(v)$ be the cocircuit (in general, cocycle) which consists of all the edges incident with v in G. Because each edge has two ends, it appears in exact two cocircuits among all $C^*(v), v \in V$. Therefore, M^* is over corank 2-coverable. □

For a binary matroid $M = (\mathscr{X}, E)$, let Cycl(M) be the space of vectors over GF(2) which is generated by all the vectors corresponding to subsets in $\mathscr{X}$. We call Cycl(M) the *cycle space* of M.

A family of subsets on E, each of which corresponds to a vector of the space Cycl(M) with the property that each element of E belongs to exactly two subsets in the family, is said to be a *double covering* of M. For a double covering $\mathscr{D}$ of M, let Boun$(M; \mathscr{D})$ be the space of vectors over GF(2), which is generated by all the vectors corresponding to the subsets in $\mathscr{D}$. We call the space Boun$(M; \mathscr{D})$, a *boundary space* of M for $\mathscr{D}$.

From the binariness, any binary matroid M has a double covering. In fact, the subset that consists of all the elements such that each of them appears exactly twice is itself a double covering of M since it corresponds to the zero vector. Because it is easily shown that Boun$(M; \mathscr{D})$ is always a subspace of Cycl(M) for any double covering $\mathscr{D}$ chosen, the space defined by

$$\mathscr{H}(M) = \text{Cycl}(M)/\text{Boun}(M; \mathscr{D}) \tag{14.4.1}$$

is called a *homology space* of M.

Lemma 14.4.5. *If M is a graphic matroid, then M^* has a homology space $\mathscr{H}(M^*) = 0$.*

Proof. Suppose M is isomorphic to the circuit matroid of a graph $G = (V, E)$. Let E_v be the set of all the edges incident with v in G. Because for any $e = (u, v) \in E$, we have $e = E_u \cap E_v$, the family of subsets $\mathscr{D} = \{E_v \mid \forall v \in V\}$ is a double covering of M^*, the dual of M. Since for any $u \in V$, $\{E_v \mid \forall v \in V, v \neq u\}$ forms a basis of Boun$(M^*; \mathscr{D})$ and $| \mathscr{D} \backslash \{E_u\} | = \text{rank}(M^*)$, we have

$$\text{Cycl}(M^*) = \text{Boun}(M^*; \mathscr{D})$$

and hence the lemma. □

By this lemma, we can show that the cocircuit matroids $M^*(K_{3,3})$ and $M^*(K_5)$ are not graphic because neither $M(K_{3,3})$ nor $M(K_5)$ has a homology space which is isomorphic to the trivial space in view of Theorem 4.2.5.

Lemma 14.4.6. *Any minor of a graphic matroid is graphic.*

Proof. Let M be a graphic matroid which is isomorphic to the circuit matroid of graph $G = (V, E)$. Because for $S \subseteq E$, the reduction of M is the circuit matroid of the graph obtained by deleting the edges not in S and the contraction of M is the circuit matroid of the graph obtained by contracting all the edges not in S, the lemma follows. □

Lemma 14.4.7. *Any graphic matroid does not have a minor which is isomorphic to the cocircuit matroid of K_5, or $K_{3,3}$.*

Proof. By the above discussion on the non-graphicness of the cocircuit matroid of K_5 and $K_{3,3}$ with Lemma 14.4.6, the lemma is obtained. □

For a matroid $M = (\mathcal{L}, E)$, if there is a base

$$B = \{e_1, e_2, \ldots, e_s\}$$

such that an order relation, denoted by $\preceq$ as usual, on B can be defined with the following conditions:

Cond.1. Only one element in B is the minimum.

Cond.2. On each fundamental circuit, all the elements in B induce a linear order,

then M is said to be *OD-orderable*.

For a matroid $M = (\mathcal{L}, E)$ which is OD-orderable, we introduce variables $y_e, e \in \bar{B}$, over GF(2) corresponding to the fundamental circuits $C(B; e), e \in \bar{B}$. Two elements a and b, $a, b \in \bar{B}$, are said to be *adjacent* if $C(B; a)$ and $C(B; b)$ are one of the three configurations: Confs.1–3 as shown in what follows.

Let $h(e)$ and $t(e)$, $e \in \bar{B}$, be the minimum and the maximum on $B \cap C(B; e)$. For $a, b \in \overline{B}$, $C(B; a) \cap (B; b) \neq \emptyset$ and $h(a) \neq h(b)$ we may assume $h(a) \prec h(b)$ without loss of generality.

Conf.1. $\exists c \in \bar{B}$,

$$\begin{cases} (i)\ h(c) \prec h(a) \prec h(b); \\ (ii)\ \langle t(a), \Lambda\rangle \prec \Lambda = \langle t(c), t(b)\rangle. \end{cases}$$

Conf.2. $\exists c, d \in \bar{B}$,

$$\begin{cases} (i)\ \max(h(c), h(d)) \prec h(a) = h(b); \\ (ii)\ \langle t(c), t(a)\rangle = \Lambda_1 \succ \langle \Lambda_1, \Lambda_2\rangle \Lambda_2 = \langle t(d), t(b)\rangle. \end{cases}$$

Conf.3. $\exists c \in \bar{B}$,

$$\begin{cases} (i)\ h(c) \prec h(a) \prec h(b); \\ (ii)\ \langle t(c), \Lambda\rangle \prec \Lambda = \langle t(a), t(b)\rangle. \end{cases}$$

In what is mentioned above, $\langle x, y\rangle = x \wedge y$, i.e. the g.l.b. (x, y) defined by the order.

By considering what is discussed in Section 8.4,one might soon find that Confs.1–3 are in fact Types A, B and C whenever a, b and c are replaced by α, β and γ, respectively. Of course, no much difficulty appears for justifying the OD-orderability and finding an OD-ordering on a matroid efficiently. A base with OD-ordering is called an *OD-base.*

Let $\mathscr{A}$ be the set of all adjacent pairs of elements not in B, then the equation system over GF(2)

$$y_a + y_b = \begin{cases} 1, \text{ if } a, b \text{ in Conf.1, or Conf.2;} \\ 0, \text{ otherwise, i.e., } a, b \text{ in Conf.3} \end{cases} \tag{14.4.2}$$

is said to be *critical equations* for the matroid.

A matroid $M = (\mathscr{X}, E)$, which satisfies that if it is OD-orderable, then the critical equations in eq. (14.4.2) for M has a solution which is said to be *consistent*; otherwise, *non-consistent*.

It is also easily seen that the cocircuit matroid of K_5 or $K_{3,3}$ is not graphic from Lemma 14.4.8 because its dual matroid can be verified to be non-consistent from the similar theory which appears in Section 5.4.

Lemma 14.4.8. *If $M = (\mathscr{X}, E)$ is a graphic matroid, then its dual M^* is consistent.*

Proof. By contradiction, suppose M^* is not consistent. Then from the OD-orderability, M^* can be seen as a circuit matroid of a graph by taking the base as an OD-tree. From what is similar to Lemma 5.4.8 in this case, we have seen that M^* has a minor that is isomorphic to the circuit matroid of K_5 or $K_{3,3}$. Hence, M has a minor which is isomorphic to the cocircuit of K_5 or $K_{3,3}$. From Lemma 14.4.7, M is not graphic. This contradicts to the given condition of the lemma. □

Further, we are also allowed to introduce auxiliary graphs for determining if a matroid is consistent in the way described in Section 8.4.

Lemma 14.4.9. *For a regular matroid M, the following statements are equivalent:*
(1) *M^* is over corank 2-coverable.*
(2) *M^* has a homology space, which is the trivial space.*
(3) *M has no minor which is isomorphic to the cocircuit matroid of K_5 or $K_{3,3}$.*
(4) *M^* is consistent.*

Proof.

(1) $\Rightarrow$ (2). Because of the over corank 2-coverability of M^*, the rank of the boundary space of M^* for $\mathscr{D}$, the set of the circuits that form an over corank 2-cover of M^*, is equal to that of the cycle space of M^*. This implies that the homology space is trivial and hence (2) is done.

(2) $\Rightarrow$ (3). By contradiction, suppose M has a minor isomorphic to $M^*(K_5)$ or $M^*(K_{3,3})$, then M^* has a minor isomorphic to $M(K_5)$ or $M(K_{3,3})$. However, it can be verified that neither $M(K_5)$ nor $M(K_{3,3})$ has a trivial homology space. From the heredity of having a trivial homology space, M^* has no trivial homology space. A contradiction to (2) appears.

(3) $\Rightarrow$ (4). By contradiction, suppose M^* is not consistent. From the OD-orderability, M^* is isomorphic to the circuit matroid of a graph G which is not planar. This implies M has a minor isomorphic to $M^*(K_5)$ or $M^*(K_{3,3})$, a contradiction to (3).

(4) $\Rightarrow$ (1). Left to the reader because of the lengthiness. □

Theorem 14.4.1. *A regular matroid* $M = (\mathscr{X}, E)$ *is graphic if, and only if, one of the statements* (1)–(4) *in Lemma* 14.4.9 *is satisfied.*

Proof. The necessity is obtained by Lemmas 14.4.4–14.4.8. From Lemma 14.4.9, it suffices to prove the sufficiency of the theorem for any one of the statements chosen. It looks to be the simplest one when we choose statement (1).

In fact, from the over corank 2-coverability of M^* we are allowed to define the vertices of the graph denoted by G as the circuits in the subfamily $\mathscr{D}$ of $\mathscr{X}^*$ which forms an over corank 2-cover of E, and the edges as the pairs of members in $\mathscr{D}$ having an element in common with a 1-to-1 correspondence to elements of E. Then from the orthogonality between M and M^*, we find $M \cong M(G)$, the circuit matroid of G. Therefore, M is graphic. □

14.5 Cographicness

This section is based on the duality of matroids for the usages in another direction.

Lemma 14.5.1. *A regular matroid M is graphic if, and only if, its dual M^* is cographic.*

Proof. From the definitions of graphic and cographic matroids, it is easily checked that the following diagram

$$\begin{array}{ccc} M & \xrightarrow{\quad\tau\quad} & M^* \\ {\scriptstyle\imath}\downarrow & & \downarrow{\scriptstyle\imath^*} \\ M(G) & \xrightarrow[\quad\tau'\quad]{} & M^*(G) \end{array}$$

is commutative, where τ, τ' determine the corresponding duality and $\imath$, $\imath^*$, the corresponding isomorphisms. Therefore, the conclusion of lemma is done. □

Theorem 14.5.1. *A regular matroid M is cographic if, and only if, one of the following four statements is satisfied:*
(1) *M is over corank 2-coverable.*
(2) *M has a homology space which is the trivial space.*
(3) *M has no minor, which is isomorphic to the circuit matroid of K_5 or $K_{3,3}$.*
(4) *M is consistent.*

Proof. A direct result of Lemma 14.5.1 and Theorem 14.4.1. □

From the two theorems above, we may soon find another criterion for the planarity of a graph.

Corollary 14.5.1. *A graph G is planar if, and only if, its circuit matroid is cographic or dually its cocircuit matroid is graphic.*

14.6 Notes

14.6.1 Although an axiomatic approach to linear and algebraic dependence in van der Waerden's *Moderne Algebra*, published in the last 1930s, can be seen as the idea of matroids, and the foundation of the theory is, as known, due to the Whitney's paper Whitney [395], in which the term "matroid" firstly appears. However, a very important and deep development to the matroid theory today comes from a series of papers such as Tutte [340, 341], published in the last 1950s. Classic papers on matroid theory are collected in Kung [164].

14.6.2 Regularity of a matroid comes along and out the totally unimodularity of matrices. The first theorem, which shows the equivalence between the regularity of matroid and the absence of Fano matroid with its dual as a minor, is due to Tutte from his homotopy theorem on matroids, in Tutte [340]. More and further development is made in Seymour [308, 309], from which the problem of justifying the totally unimodularity of a matrix is completely solved.

14.6.3 The graphic and cographic matroids, defined in Section 14.4, are in form different, from those originally defined in Tutte [341], with the interchangement commonly used, of the terms: graphic and cographic. However, the theory, with no difference, shows that any criterion for planarity can be used to form a theorem for the graphicness, or cographicness, in a proper way, from Tutte [342] and Weinberg [379]. The crucial step for doing this is to transform a criterion for planarity involving vertices into that without the involvement. In fact, the transformations in Section 14.4 for those appeared in Sections 4.2 and 8.4, or in Liu [192, 202], [205, 206], are presented.

This theory can also be applied to justifying geometric lattice in Sachs [303], and the *simplicity* of matroid in Crapo and Rota [59].

14.6.4 Recent progress of many aspects in matroid theory are reflected in the three volumes of books in White [383–385].

14.6.5 One might think of the generalization of the graph minor theorem in Robertson and Seymour [291] and the likes, for graphs to matroids.

14.6.6 Now, one might be awakened from Theorem 11 in Liu [238] to seek an order relation on the basis, or cobasis, of a matroid for establishing a criterion, to determine whether, or not, the matroid is graphic, or cographic, by only one forbidden configuration, not necessary to be a minor.

15 Knot Polynomials

15.1 Definitions

A *knot* (or a *link*), denoted by K (or L), is mathematically defined to be an image of (or more than one disjoint copies of) the circle $C_0 = \{(x, y, 0) \mid x^2 + y^2 = 1\}$ under a homeomorphism into the 3-dimensional space $\mathscr{R}^3$.

The two knots shown in Figure 15.1.1 are the simplest common knots: the one in (a) is called the *overhand knot*, or the *trefoil* and that in (b), the *figure eight knot*, or the *four knot*.

Two knots K_1 and K_2 are said to be *equivalent* if there exists a homeomorphism of $\mathscr{R}^3$ into itself which maps K_1 onto K_2. It is a triviality that the relation of knot equivalence is a true equivalence relation.

Equivalent knots are said to be of the same type, and each equivalent class of knots is called a *knot type*.

Those equivalent to the unknotted circle C_0 are called *trivial* and constitute the *trivial type*.

It will be seen that the trefoil and the four knot are in distinct knot types. A knot is said to be *polygonal* if it is the union of a finite number of closed straight segments called *edges*, whose ends are the *vertices* of the knot.

All the knots that are equivalent to a polygonal knot are said to be *tame*; otherwise *wild*. This distinction is of fundamental importance. In what follows, only tame knots are concerned with.

A knot K can be specified by the parallel projection denoted by $\mathrm{proj}_0\colon \mathscr{R}^3 \to \mathscr{R}^2$ such that

$$\mathrm{proj}_0(x, y, z) = (x, y, 0). \tag{15.1.1}$$

A point p of the image $\mathrm{proj}_0(K)$ is called a *multiple point* if the inverse $\mathrm{proj}_0^{-1}(p)$ contains more than one point of K. The *order* of $p \in \mathrm{proj}_0(K)$ is defined to be the *cardinality* of $\mathrm{proj}_0^{-1}(p) \cap K$.

A *double point* is a multiple point of order 2, and a *triple point* is that of order 3.

A knot is said to be in *regular position* if the projection has a finite number of multiple points all of which are of simply and doubly crossing as explained in Section 6.1.

Because it can be imagined that any knot (tame of course) can always be done in regular position at most within a small perturbation, what we concern with knots is always in regular position without loss of generality.

Each double point of the projected image of a tame knot in regular position is the image of two points of the knot. The one with the greater z-coordinate is called an *overcrossing*, and the other, the corresponding *undercrossing*.

DOI 10.1515/9783110479492-015

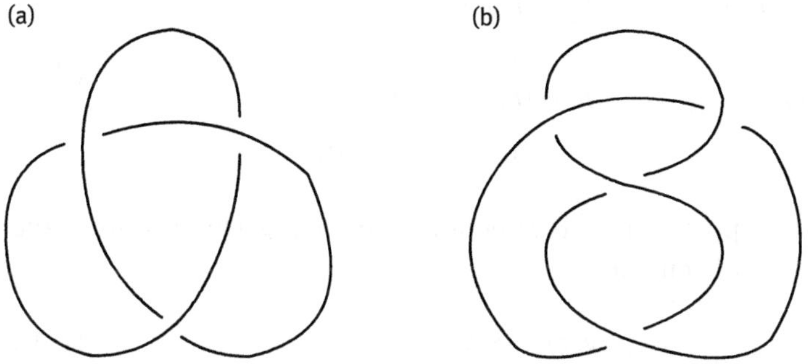

Figure 15.1.1: (a) Trefoil knot and (b) Four knot.

An *isotopic deformation* of a topological space $\mathscr{X}$ is defined to be a family of homeomorphism h_t, $0 \le t \le 1$, of $\mathscr{X}$ onto itself such that h_0 is the identity, i.e. $h_0(p) = p$, $\forall p \in \mathscr{X}$ and the function $H(t,p) = h_t(p)$ is simultaneously continuous in t and p.

Two knots K_1 and K_2 are said to belong to the same *isotopy type* if there exists an isotopic deformation $\{h_t \mid 0 \le t \le 1\}$ of $\mathscr{R}^3$ such that $h_1(K_1) = K_2$.

The parameter t is intentionally chosen to suggest time. Thus, for a fixed point $p \in K_1 \subset R^3$, the point $h_t(p)$ traces out, so to say, the path as one moves originally at p during the motion from the initial position at K_1 to K_2.

It is obvious that if two knots K_1 and K_2 are of the same isotopy type, then they are equivalent. But, the converse is false in general.

The following discussion on orientation serves to illustrate the difference between the two definitions.

A homeomorphism h of $\mathscr{R}^3$ onto itself is said to be *orientation preserving* if the image of every point with clockwise (anticlockwise) rotation is again with clockwise (anticlockwise) rotation; otherwise, *orientation reversing*.

From the continuity of h, the neighbourhood of every point with a fixed rotation is preserved by h is also a neighbourhood of the image with the same rotation (without considering the directions: clockwise and anticlockwise).

Since h is a homeomorphism, the image of every point p of $\mathscr{R}^3$ belongs to one of these two disjoint sets: a set with the rotation of the same direction; and a set with the rotation of the different direction compared with that of p. And, since $\mathscr{R}^3$ is connected, it follows that one of the two sets must be empty.

This implies that a homeomorphism h of $\mathscr{R}^3$ is orientation reversing if, and only if, the image of every point with clockwise (anticlockwise) rotation is with anticlockwise (clockwise) rotation.

Let H_+ and H_- be, respectively, the sets of orientation-preserving and orientation-reversing homeomorphisms of $\mathscr{R}^3$ onto itself. Then we soon find that the composition h_1h_2 of two homeomorphisms h_1 and h_2 satisfies the following rules:

$$\begin{cases} h_1h_2 \in H_+ \Longleftrightarrow (h_1 \in H_+ \wedge h_2 \in H_+) \\ \qquad\qquad\qquad \vee (h_1 \in H_- \wedge h_2 \in H_-); \\ h_1h_2 \in H_- \Longleftrightarrow (h_1 \in H_+ \wedge h_2 \in H_-) \\ \qquad\qquad\qquad \vee (h_1 \in H_- \wedge h_2 \in H_+). \end{cases} \tag{15.1.2}$$

Of course, the identity mapping on $\mathscr{R}^3$ is orientation preserving. On the other hand, the reflection ref : $\mathscr{R}^3 \longrightarrow \mathscr{R}^3$ defined by

$$\mathrm{ref}(x, y, z) = (x, y, -z) \tag{15.1.3}$$

is orientation reversing.

The justification of a homeomorphism h to be orientation preserving or orientation reversing is not hard. If h is a linear transformation, then it is orientation preserving or orientation reversing as its determinant is positive or negative. Otherwise, whenever h is differentiable at a point, the orientation preserving or orientation reversing of h is determined by the Jacobian at the point to be positive or negative, respectively.

Lemma 15.1.1. *Two knots K_1 and K_2 are of the same isotopy type if, and only if, there exists an isotopic deformation $\{h_t\}, 0 \le t \le 1$, such that*

$$h_1 \in H_+,\ h_1(K_1) = K_2. \tag{15.1.4}$$

Proof. The sufficiency is obvious according to the definition of isotopy types. Conversely, since h_0 is the identity which is orientation preserving, from the continuity of $H(t, p) = h_t(p)$, $\{h_t\}$ is all orientation preserving for $0 \le t \le 1$ and hence eq. (15.1.4). □

A knot K is said to be *amphicheiral* if there exists an orientation-reversing homeomorphism h of $\mathscr{R}^3$ onto itself such that $h(K) = K$. For the reflection defined by eq. (15.1.3), the image ref(K) of a knot K is called the *mirror image* of K.

Lemma 15.1.2. *A knot K is amphicheiral if, and only if,*

$$\exists h' \in H_+ : h'(K) = \mathrm{ref}(K). \tag{15.1.5}$$

Proof. If K is amphicheiral, then there exists $h \in H_-, h(K) = K$. From eq. (15.1.2), we have $h' = \mathrm{ref} \bullet h \in H_+$ and hence $h'(K) = \mathrm{ref}(K)$. This is the necessity. Conversely, since $h'(K) = \mathrm{ref}(K)$, from eq. (15.1.2) $\mathrm{ref} \bullet h' \in H_-$. Moreover, $\mathrm{ref} \bullet h'(K) = \mathrm{ref} \bullet \mathrm{ref}(K) = K$. Hence, K is amphicheiral. This is the sufficiency. □

It can be shown from Figure 15.1.2 that the figure-eight knot in Figure 15.1.1(b) is amphicheiral. However, the overhand knot is not amphicheiral. It will be seen by using further results that appear in Section 15.3 that the overhand knot and its mirror

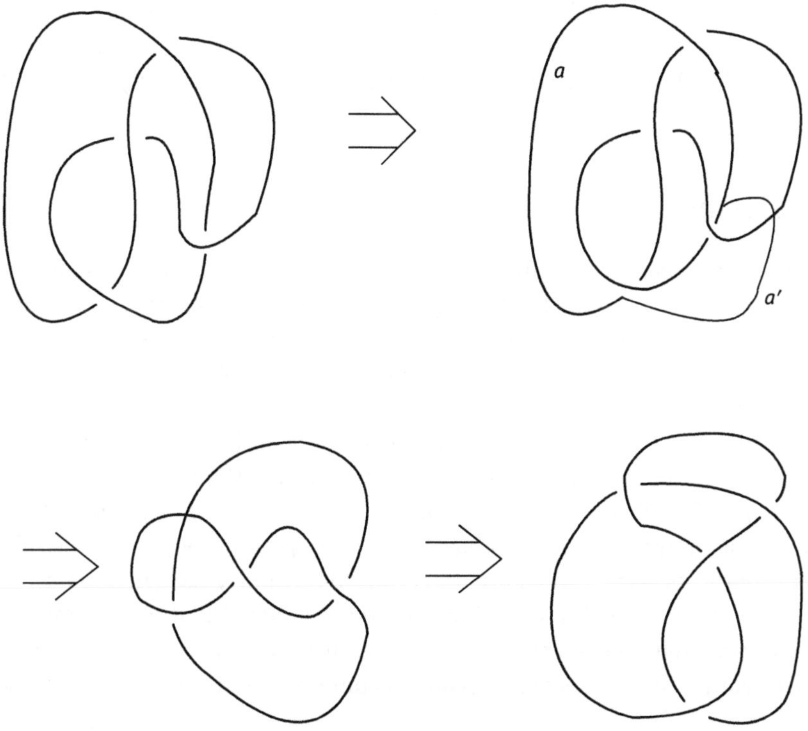

Figure 15.1.2: Four knot is amphicheiral.

image, or the *left-hand trefoil* and the *right-hand trefoil* are equivalent but not of the same isotopy type.

Now, one might ask whether or not every orientation-preserving homeomorphism h of $\mathscr{R}^3$ onto itself is realizable by an isotopic deformation, i.e., for the given h does there exist $\{h_t\}, 0 \leq t \leq 1$, such that $H(t,p) = h_t(p)$ is simultaneously continuous in t and p, $h_0(p) = p$, and $h_1(p) = h(p), p \in \mathscr{R}^3$? If the answer were no, we would have a third kind of knot type. However, we have the following improvement of Lemma 15.1.1.

Theorem 15.1.1. *Two knots K_1 and K_2 are of the same isotopy type if, and only if,*

$$\exists h \in H_+ : h(K_1) = K_2. \tag{15.1.6}$$

Proof. The necessity is a direct result of Lemma 15.1.1. The sufficiency, which is not easy, can be found as shown in Fisher [89]. □

Just as every homeomorphism of $\mathscr{R}^3$ onto itself is either orientation preserving or orientation reversing, so is every homeomorphism of K onto itself. We are allowed to define a knot K to be *invertible* if there exists an orientation-preserving homeomorphism h of $\mathscr{R}^3$ onto itself such that the restriction $h|_K$ is an orientation-reversing homeomorphism of K onto itself. It can be shown that both the trefoil and the four knot are invertible. Of course, any knot which is of the same isotopy type as its mirror image is invertible. However, there is still an infinite set of non-invertible knots in Trotter [330].

15.2 Knot diagram

For a link L in regular position described in Section 15.1, the projection $\mathrm{proj}_0(L)$ under the mapping proj_0 defined by eq. (15.1.1) is called a *universe* denoted by U of L. If for each double point on U, the two corresponding points on L are distinguished to be the overcrossing and the undercrossing, then it is called a *diagram* of L. In fact, in Figure 15.1.1(a) shows a diagram of the trefoil and (b) that of the four knot. If all the double points in U are treated as vertices with the segments without inner double point as edges between two points which correspond to vertices, then the resultant graph is called a *base graph* of L. Of course, the unknotted circle, or in other words the trivial knot without double point, is defined to have its base graph as a loop. A link is said to be *preconnected* if its base graphs are all connected. A link mentioned below is always meant preconnected without specific interpretation. If a link is equivalent to a disjoint union of two links whose universes have no point in common, then it is called *disconnected*. Of course, all connected links are preconnected. But, the converse is not true.

Lemma 15.2.1. *A graph is a base graph of a link if, and only if, it is a 4-regular planar graph except only for the loop which is the base graph of the circle, the trivial knot.*

Proof. The necessity is obvious from the regularity of the position a link is on as described in Section 15.1. Conversely, since the graph is Eulerian, by tracing edges from a vertex and an edge incident to the vertex chosen to be the starting point with the rule: whenever a vertex is met the foregoing edge is not successive with the incoming edge on the rotation at the vertex to find a tour. If the tour is Eulerian, then it is a base graph of a knot and hence of a specific link; otherwise, by doing the procedure again from a vertex that has not yet met on the graph induced by all the edges not traced. From the finiteness of the graph, an edge disjoint union of tours which forms the graph is found. Hence, the graph is a base graph of a general link. □

Because each double point in a diagram of a knot is in one of two possible cases as shown in Figure 15.2.1, it is allowed to assign a weight: 0 or 1 to a vertex according to the case (a) or (b) in Figure 15.2.1 appears on a base graph. Such a base graph with vertices weighted above is called a *signed graph* of the corresponding link.

From the proof of Lemma 15.2.1, we have seen that any 4-regular planar graph has a *tour partition*, a partition of the edge set to those induce tours such that on each tour any pair of consecutive edges is not successive on the rotation at the common end.

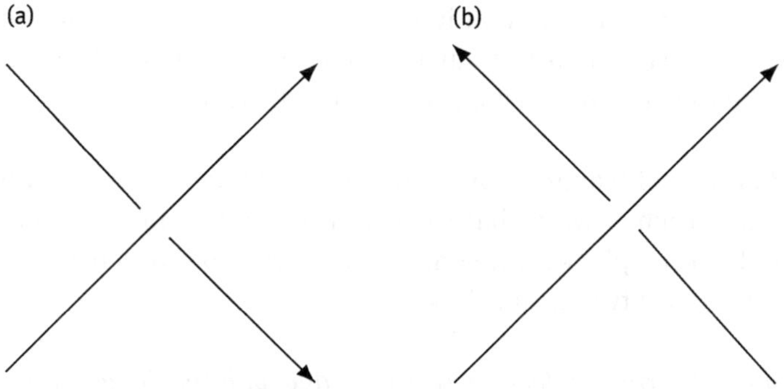

Figure 15.2.1: Two types of double point.

Such an tour is called a *non-successive tour*. And, such a tour partition is said to be *non-successive* as well. A 4-regular planar graph with a non- successive tour partition and with vertices weighted by 0 or 1 is called a *NoTPaPW4-graph* (Non-successive Tour-Partitioned Planar-Weighted 4-regular graph).

Lemma 15.2.2. *There is a* 1*-to-*1 *correspondence between NoTPa- PW4-graphs and diagrams of links except the one whose projection is equivalent to the circle on the plane.*

Proof. Because whenever the direction of each tour in a NoTPa- PW4-graph is chosen, according to the weight the overcrossing point is determined at each vertex and hence a diagram is uniquely defined. From Lemma 15.2.1, the lemma follows. □

Because any 4-regular planar graph has its face set partitioned into two classes such that in each of the two classes, any two faces are with at most one vertex in common on their boundaries, each class is allowed to form a graph (planar of course) with vertices as faces and edges corresponding to common vertices. It is easy to show that such two graphs are a pair of those planar dual one to another. They are both called a *face graph* of the original one as discussed in Section 7.1. A face graph of a signed graph of a link with its edges weighted as the same as the vertices of the signed graph is called an *edge-signed face graph*, or in short *ESF graph* of the link.

Lemma 15.2.3. *A graph with edges weighted by* 0 *and* 1 *is an ESF graph of a link (or knot) if, and only if, it is planar (or planar without bicycle).*

Proof. The necessity for links is obvious from the definition of an ESF graph. For knots, it is obtained by Lemma 4.4.3.

Conversely, by the similar procedure to that in the proof of Lemma 15.2.2 for the arrangement of overcrossing at each crossing point on what is obtained by T.T. rule in Section 4.4 according to the weights, from Lemma 4.4.4, the sufficiency is derived. □

In general, for a given dual pair of planar graphs with bicycles there are at least two 4-regular planar graphs, and each of which has the pair as its face graphs. This implies that a dual pair of planar graphs with weights 0 and 1 are allowed to have the corresponding diagrams of at least two distinct links.

Lemma 15.2.4. *There is a 1-to-1 correspondence between the dual pairs of planar graphs without bicycle with binary weights on edges and the diagrams of knots.*

Proof. By following the similar procedure used in the proof of Lemma 15.2.2 to determining the overcrossings according to the weights on what is uniquely obtained from the T.T. rule in Section 4.4 along all the edges of one in the pair, the unique diagram of a knot is found. However, from the duality for planar graphs without bicycle, the diagram for another in the pair is the same as that for its dual and hence the lemma is true. □

According to what was mentioned in Chapter 6, we are also allowed to represent the diagrams of links by sequences. Suppose edges on an ESF graph are numbered by distinct letters that are considered to have their values binary, i.e. 0 and 1. Then, for a planar graph without bicycle G, a sequence of letters is obtained by recording the letters corresponding to edges appearing on the way of using T.T. rule on the graph. If G is an ESF graph of a knot K, or in general a link L if G has a bicycle, then the sequence $\mathrm{Seq}(G)$ is called a *sequence of* the knot K, or the link L, denoted by $\mathrm{Seq}(K)$, or $\mathrm{Seq}(L)$ as well. A sequence which is a sequence of a knot or a link is called a *knot* or *link sequence*.

Lemma 15.2.5. *A sequence of letters that have binary values is a knot sequence if, and only if, it is a crossing sequence.*

Proof. A direct result of Lemma 15.2.3. □

Now, we have seen that any one of signed graphs, ESF graphs 1 and link sequences can be applied to links instead of diagrams. Here, we only restrict ourselves to ESF graphs.

Let G be a planar graph with binary weights on edges. For any articulate vertex v, let

$$\alpha_1(G; v) = G - v, \tag{15.2.1}$$

then α_1 and its inverse α^{-1} are called the *first elementary operator* denoted by π_1.

For a 2-valent vertex v with $e_1 = (w, v)$ and $e_2 = (u, v)$ as the two incident edges, if the weights of them are different, then let

$$\alpha_2(G; E_v) = (G \bullet e_1) \bullet e_2 \tag{15.2.2}$$

be the resultant graph of deleting the 2-valent vertex v and then identifying u with w. Of course, the inverse α_2^{-1} of α_2 is adding a new 2-valent vertex v with different weights of the two incident edge (u, v) and (v, w) on what is obtained by splitting a vertex into u and w. We call any of α_2 and α_2^{-1} a *second elementary operator* denoted by π_2.

Suppose $St(u; i, j, k) = (\{u, v_i, v_j, v_k\}, \{(u, v_t)|t = i, j, k\})$ be a 3-star in G with weight $w(u, v_t)$ on edge (u, v_t), $t = i, j, k$. If the travel obtained by following the T.T. rule passes through the double points corresponding to (u, v_i) and (u, v_j) being consecutively both undercrossing or consecutively both overcrossing according to the procedure suggested in the proof of Lemma 15.2.2, then let

$$\alpha_3(G; St) = (G - u) + \{(v_i, v_j), (v_j, v_k), (v_k, v_i)\}, \tag{15.2.3}$$

with $w(v_i, v_j) = w(u, v_k)$, $w(v_j, v_k) = w(u, v_i)$ and $w(v_k, v_i) = w(u, v_j)$. Here, α_3 and its inverse are called the *third elementary operator* denoted by π_3. Any of π_1, π_2 and π_3 is called an *elementary operator*.

Two planar graphs G_1 and G_2 with binary weights on edges are said to be *elementary equivalent* if G_1 can be transformed into G_2 by doing a series of elementary operators. If G_1 and G_2 are elementary equivalent only under π_2 and π_3, then they are called *regular equivalent*; otherwise, *ambient equivalent*, a synonym of elementary equivalent.

Problem 1. To identify the regular or ambient equivalent classes for planar graphs with binary weights on edges.

Now, we turn back to topology. In the twenties, Reidemeister simplified the study of knots by introducing three kinds of operations denoted by $\gamma_1 = R1, \gamma_2 = R2$ and $\gamma_3 = R3$ as shown in Figure 15.2.2 such that they do not change the knot and any two diagrams of the same knot can be converted into each other by doing a series of them. Any of γ_1, γ_2 and γ_3 is called a *Reidemeister operation*.

Two diagrams of knots, or in general, links are said to be in the same *ambient isotopy type* if one of them can be converted into the other by a series of γ_1, γ_2 and γ_3. If only γ_2 and γ_3 are restricted for the conversion, then they are called in the same *regular isotopy type*.

Problem 2. To identify the knots, or in general, the links in the same ambient isotopy type or in the same regular isotopy type.

In order to find the relationship between the two problems, we have to observe the relationship between elementary operators for graphs and Reidemeister operations for diagrams of links.

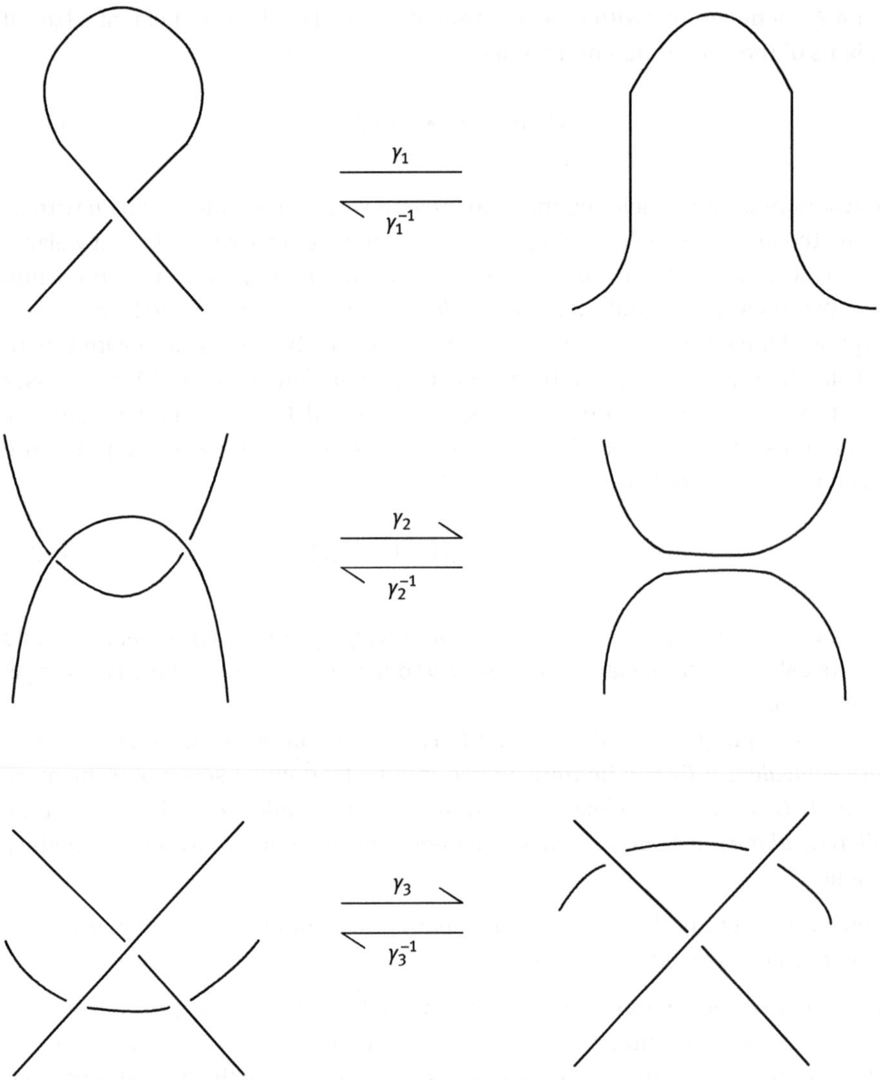

Figure 15.2.2: Three Reidemeister operations.

Lemma 15.2.6. *There is a 1-to-1 correspondence between γ_i on diagrams of links and π_i on planar graphs with binary weights of edges, for $i = 1, 2, 3$.*

Proof. For $i = 1$, or 2, it is easily checked from the correspondence between planar graphs with binary weights of edges and diagrams of links shown in Lemma 15.2.4. For $i = 3$, by considering that the condition for which π_3 is allowed to use on graphs is corresponding to that for γ_3 on diagrams and that the weights of the new edges in the transformation by π_3 just reflect the change on diagrams by γ_3, the fact is also true. □

This lemma suggests us that it is not necessary to distinguish the two problems above.

Theorem 15.2.1. *Two diagrams of links are in the same ambient isotopy type (or in the same regular isotopy type) if, and only if, their corresponding ESF graphs are in the same ambient equivalent class (or in the same regular equivalent class).*

Proof. A direct result of Lemma 15.2.4 and Lemma 15.2.6. □

Of course, there is no difficulty for describing the equivalent problems above in the version of sequence according to Lemma 15.2.5. The crucial step is to establish the corresponding operations on a sequence. It is left to the reader.

15.3 Tutte polynomial

By the *dichromatic polynomial* of a diagram M, denoted by $Q(M)$, we shall mean such a polynomial of M determined by the following recursion:

$$Q(M) = \begin{cases} Q(M-e) + Q(M \bullet e), & \text{if } e \text{ is neither a loop nor an isthmus;} \\ (1+\nu)Q(M-e), & \text{if } e \text{ is a loop;} \\ (1+\mu)Q(M \bullet e), & \text{if } e \text{ is an isthmus} \end{cases} \tag{15.3.1}$$

in which the conditions below are satisfied.

Q1 When M is the vertex map, then

$$Q(M) = \mu. \tag{15.3.2}$$

Q2 When M is the union of two maps M_1 and M_2 which have only one vertex in common, then

$$Q(M) = \mu^{-1}Q(M_1)Q(M_2). \tag{15.3.3}$$

Theorem 15.3.1. *For any diagram M its dichromatic polynomial has the following form:*

$$Q(M) = \sum_{S} \mu^{p_0(S)}\nu^{p_1(S)} \tag{15.3.4}$$

where S is over all spanning subgraphs of G(M); $p_0(S)$ and $p_1(S)$ are the component number and the cyclomatic number of S, respectively.

Proof. Because the vertex diagram only has one spanning graph which is itself from eq. (15.3.4) the condition Q1 is satisfied.

For a diagram M which is the union of M_1 and M_2 with only one vertex in common, any spanning subgraph S of M is also a union of S_1 and S_2 which are spanning subgraphs of M_1 and M_2, respectively.

Since it is easily checked that

$$p_0(S) = p_0(S_1) + p_0(S_2) - 1$$

and

$$p_1(S) = p_1(S_1) + p_1(S_2),$$

the condition Q2 is also satisfied.

We then prove that $Q(M)$ expressed by eq. (15.3.4) satisfies eq. (15.3.1).

When e is neither a loop nor an isthmus. All the spanning subgraphs can be divided into two classes: those that contain e and those do not contain e. Both of them consist of all the spanning subgraphs of $M - e$ and $M \bullet e$, respectively. From

$$\begin{cases} p_0(S) = p_0(S; M - e), \\ \qquad p_1(S) = p_1(S; M - e), \\ \qquad\qquad \text{if } e \notin S; \\ p_0(S) = p_0(S; M \bullet e), \\ \qquad p_1(S) = p_1(S; M \bullet e), \\ \qquad\qquad \text{if } e \in S, \end{cases}$$

the form (15.3.4) leads to eq. (15.3.1).

When e is a loop, for any spanning subgraph S of M, we have

$$\begin{cases} p_0(S) = p_0(S; M - e), \\ \qquad p_1(S) = p_1(S; M - e), \\ \qquad\qquad \text{if } e \notin S; \\ p_0(S) = p_0(S; M - e), \\ \qquad p_1(S) = p_1(S; M - e) + 1, \\ \qquad\qquad \text{if } e \in S, \end{cases}$$

that is, the form (15.3.4) still leads to eq. (15.3.1).

When e is an isthmus, for any S as mentioned earlier, we have

$$\begin{cases} p_0(S) = p_0(S; M \bullet e) + 1, \\ \qquad p_1(S) = p_1(S; M \bullet e), \\ \qquad\qquad \text{if } e \notin S; \\ p_0(S) = p_0(S; M \bullet e), \\ \qquad p_1(S) = p_1(S; M \bullet e), \\ \qquad\qquad \text{if } e \in S, \end{cases}$$

that is, the form (15.3.4) leads to eq. (15.3.1) as well.

In consequence, from the well definedness suggested by Theorem 1.2.4 of Liu [228], the theorem is obtained. □

From eq. (15.3.3) the two undeterminates in the dichromatic polynomial do not have symmetry. However, whenever we introduce another polynomial

$$\chi(M; x, y) = \mu^{-1} Q(M; \mu, \nu), \tag{15.3.5}$$

it is easily verified that the two undeterminates $x = \mu + 1$ and $y = \nu + 1$ in χ have the symmetry.

Attention 15.3.1. Both polynomials of $\chi(M; x, y)$ and $Q(M; \mu, \nu)$ are valid for M as a general graph not necessary to be a diagram which is planar here.

In fact, from eqs. (15.3.1) to (15.3.3), we soon see that χ satisfies the following recursion:

$$\begin{cases} \chi(M) = \begin{cases} \chi(M-e) + \chi(M \bullet e), \\ \qquad \text{if } e \text{ is neither a loop} \\ \qquad\qquad \text{nor an isthmus;} \\ x\chi(M \bullet e), \ \text{if } e \text{ is an isthmus;} \\ y\chi(M-e), \ \text{if } e \text{ is a loop;} \end{cases} \\ \chi(M) = 1, \ \text{if } M \text{ is the vertex map;} \\ \chi(M) = \chi(M_1)\chi(M_2), \ \text{if } M = M_1 \cup M_2 \\ \qquad \text{provided } M_1 \cap M_2 \in V, \ \text{a vertex.} \end{cases} \tag{15.3.6}$$

The polynomial χ is called the *dichromate* in Tutte's terminology, or *Tutte polynomial* in literature.

Suppose the edges in a diagram M are numbered $e_1, e_2, \ldots, e_\epsilon$. Let T be a spanning tree on M. If a tree edge e_i has the property that no cotree edge in the fundamental cocircuit formed in $\bar{T} + e_i$ has its suffix exceeding i, then e_i is said to be *internally active*. On the other hand, if a cotree edge e_j has the property that no tree edge in the fundamental circuit formed in $T + e_j$ has its suffix exceeding j, then e_j is said to be *externally active*. It is shown that the number of spanning trees T of M with given numbers $\text{int}(T)$ and $\text{ext}(T)$ of internally or externally active edges, respectively, is independent of the way of numbering edges.

Theorem 15.3.2. [343], *The dichromate $\chi(M)$ of a diagram M has the following spanning tree expansion:*

$$\chi(M) = \sum_{\substack{0 \le i \le \epsilon \\ 0 \le j \le \epsilon}} \tau(M; i, j) x^i y^j, \tag{15.3.7}$$

where $\tau(M; i, j)$ is the number of spanning trees of M with $\text{int}(T) = i$ and $\text{ext}(T) = j$, $0 \le i \le \epsilon, 0 \le j \le \epsilon$.

Proof. The crucial step is to show the case when e is neither a loop nor an isthmus because other cases can be easily obtained in a similar way. Let $T = T(M;i,j)$ be a spanning tree on M with $\text{int}(T) = i$ and $\text{ext}(T) = j$. For e internally active in T, $M - e$ has exactly one

$$T_1 = T(M - e; i, j) = T - e + e_l$$

corresponding to T where $l = \max\{s | e_s \in C_e^*,\ C_e^*$ is the fundamental cocircuit formed by $\bar{T} + e\}$. Moreover, $T \bullet e$ is a tree on $M \bullet e$ but not a $T(M \bullet e; i, j)$. If e is externally active for T, then $T - e$ is a tree on $M - e$ but not a $T(M - e; i, j)$, and $T \bullet e - e_t$ is the unique $T(M \bullet e; i, j)$ corresponding to T when $t = \max\{s | e_s \in C_e - e,\ C_e$ is the fundamental circuit formed by $T + e\}$. Therefore,

$$\tau(M;i,j) = \tau(M - e; i, j) + \tau(M \bullet e; i, j)$$

is satisfied. By virtue of eq. (15.3.6), the theorem is soon obtained. □

Attention 15.3.2. The expansion of polynomial $\chi(M)$ in eq. (15.3.7) is determined by the inner structure from particular trees of M. It is also valid for M as a general graph.

15.4 Pan-polynomial

By an *invariant* on knots (or links), we shall mean such a function f of knots (or links) that all the knots (or links) in the same type has the same value of f. The purpose of this section is to find the invariants that are all polynomials in form. From what we obtained in Section 15.2, we are allowed to restrict ourselves only to consider graphs (planar with binary weights on edges of course).

Let G be a planar graph with binary weights on edges. For an edge e, its weight $w(e) = 0$, or 1. Of course, e is allowed to be either a loop or an isthmus in our case here. Let $k_i = k_i(G)$ be the number of isthmuses of weight i, $i = 0, 1$; and $l_i = l_i(G)$ be the number of loops of weight i, $i = 0, 1$.

Now, we define a function $\Phi(G)$, $G = (V, E)$, on the set of all planar graphs with binary weights of edges by following the recursion:

$$\Phi(G) = \begin{cases} A(e)\Phi(G - e) + B(e)\Phi(G \bullet e), \\ \qquad \text{if } e \in E \text{ is neither a loop} \\ \qquad \text{nor an isthmus;} \\ (X + Yz)^{\bar{w}(e)}(Xz + Y)^{w(e)}\Phi(G - e), \\ \qquad \text{if } e \text{ is a loop;} \\ (X + Yz)^{w(e)}(Xz + Y)^{\bar{w}(e)}\Phi(G \bullet e), \\ \qquad \text{if } e \text{ is an isthmus,} \end{cases} \tag{15.4.1}$$

where

$$A(e) = \bar{w}(e)X + w(e)Y; \ B(e) = w(e)X + \bar{w}(e)Y$$

and $G - e$, $G \bullet e$ are the usual operation called deletion, contraction of e on G with the conditions:

Cond.1 If G is the trivial graph of a single vertex without edge, then

$$\Phi(G) = 1. \tag{15.4.2}$$

Cond.2 If G is disconnected, i.e., $G = G_1 + G_2$, then

$$\Phi(G) = z\Phi(G_1)\Phi(G_2). \tag{15.4.3}$$

From the recursion, it is easily shown that if G is a connected graph with each edge as either a loop or an isthmus, then

$$\Phi(G) = (X + Yz)^{k_1+l_0}(Xz + Y)^{k_0+l_1}.$$

Further, we may also see that if G^* is the planar dual of G, then $\Phi(G^*)$ has the same form as $\Phi(G)$ within the interchangement between X and Y.

Attention 15.4.1. In eqs. (15.4.1)–(15.4.3), G has to be planar. Of course, G can be chosen as a diagram of knots. On the basis of Note 4.5.9 in Section 4.5, it looks interesting for extending G as a general graph not necessary to be planar. However, what mentioned in Note 4.5.9 has been done.

Lemma 15.4.1. *For any graphs G considered, $\Phi(G)$ is a polynomial of X, Y and z.*

Proof. By induction. From condition (15.4.2), it is allowed to assume that for the graphs considered of size less than that of G, the function Φ is a polynomial of X, Y and z. If it is disconnected, condition (15.4.3) allows us to use the hypothesis on each component for showing that $\Phi(G)$ is a polynomial. Otherwise, by using the recursive relation (15.4.1) to derive $\Phi(G)$ is a polynomial from the hypothesis on $G - e$ and $G \bullet e$ for an edge e. □

In what follows, we investigate what happens to Φ under the elementary operators.

Let v be an articulate vertex of $G = (V, E)$ considered. And, let $e = (u, v)$ of weight 1 be the edge incident with v. According to recursion (15.4.1), we have

$$\Phi(G) = \Big(A(e)z + B(e)\Big)\Phi(\pi_1 G), \tag{15.4.4}$$

in which eq. (15.4.3) is used for the resultant graph of deleting e with that for the trivial graph $\Phi = 1$ according to eq. (15.4.2).

Lemma 15.4.2. *Φ is invariant for π_1 if, and only if, $Xz + Y = X + zY = 1$.*

Proof. From eq. (15.4.4), Φ is invariant for π_1 if, and only if,

$$A(e)z + B(e) = \bar{w}(e)(Xz + Y) + w(e)(X + Yz) = 1.$$

By the symmetry between $\bar{w}(e)$ and $w(e)$, the lemma is obtained. □

For π_2, let v be the 2-valent vertex having the incident edges $e_1 = (u, v)$ and $e_2 = (v, w)$ with the condition $w(e_1) \neq w(e_2)$. According to eq. (15.4.1), we have

$$\begin{aligned}\Phi(G) = \Big(&zA(e_1)A(e_2) + A(e_1)B(e_2)\\ &+A(e_2)B(e_1)\Big)\Phi(G - v)\\ &+B(e_1)B(e_2)\Phi(\pi_2 G).\end{aligned} \tag{15.4.5}$$

Lemma 15.4.3. *Φ is invariant under π_2 if, and only if,*

$$\begin{aligned}&XY = 1;\\ &XYz + X^2 + Y^2 = 0.\end{aligned} \tag{15.4.6}$$

Proof. By the symmetry between e_1 and e_2, we may assume $w(e_1) = 1$ and $w(e_2) = 0$. From eq. (15.4.5), Φ is invariant if, and only if, the coefficient of $\Phi(G - v)$ is 0 and that of $\Phi(\pi_2 G)$ is 1. Because of

$$\begin{aligned}&A(e_1) = Y, A(e_2) = X,\\ &B(e_1) = X, B(e_2) = Y,\end{aligned}$$

the lemma is soon obtained. □

For π_3, let St be the star of centre u and articulate vertices v_i, $i = 1, 2, 3$ with the condition required. Then, $\pi_3 G$ has triangle

$$Tr = (v_1, v_2, v_3).$$

If $e_i = (u, v_i)$, $i = 1, 2, 3$, are edges in G, then we have

$$\begin{aligned}w(v_1, v_2) = w(e_3),\\ w(v_2, v_3) = w(e_1)\end{aligned}$$

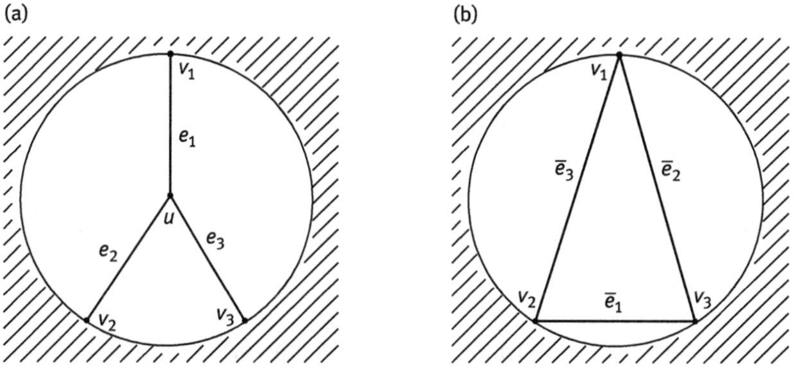

Figure 15.4.1: (a) St in G and (b) Tr in $\pi_3 G$.

and

$$w(v_1, v_3) = w(e_2).$$

Here it is allowed to denote the edges (v_1, v_2), (v_2, v_3) and (v_1, v_3) by $\bar{e}_3$, $\bar{e}_1$ and $\bar{e}_2$ in $\pi_3 G$, respectively, as shown in Figure 15.4.1.

For convenience, let us write

$$\begin{cases} G_0 = G - u; \\ G_1 = ((G - e_1) \bullet e_2) \bullet e_3; \\ G_2 = ((G \bullet e_1) - e_2) \bullet e_3; \\ G_3 = ((G \bullet e_1) \bullet e_2) - e_3; \\ G_4 = ((G \bullet e_1) \bullet e_2) \bullet e_3. \end{cases} \tag{15.4.7}$$

Then, it is easily verified that

$$\begin{cases} ((G - e_1) - e_2) - e_3 = \emptyset_1 + G_0; \\ ((G - e_1) - e_2) \bullet e_3 \\ \quad = ((G - e_1) \bullet e_2) - e_3 \\ \quad = ((G \bullet e_1) - e_2) - e_3 \\ \quad = G_0, \end{cases} \tag{15.4.8}$$

where $\emptyset_1$ is the trivial graph.

From eqs. (15.4.1) and (15.4.7)–(15.4.8), by routine calculation we find

$$\begin{aligned} \Phi(G) = {} & \Lambda_0 \Phi(G_0) + \Lambda_1 \Phi(G_1) + \Lambda_2 \Phi(G_2) \\ & + \Lambda_3 \Phi(G_3) + \Lambda_4 \Phi(G_4), \end{aligned} \tag{15.4.9}$$

where

$$\begin{aligned}
\Lambda_0 &= z\prod_{i=1}^{3} A(e_i) + A(e_1)A(e_2)B(e_3),\\
&\quad + A(e_1)B(e_2)A(e_3) + B(e_1)A(e_2)A(e_3),\\
\Lambda_1 &= A(e_1)B(e_2)B(e_3),\\
\Lambda_2 &= B(e_1)A(e_2)B(e_3),\\
\Lambda_3 &= B(e_1)B(e_2)A(e_3),\\
\Lambda_4 &= \prod_{i=1}^{3} B(e_i).
\end{aligned}$$

Similarly, since $\bar{e}_i$ is the edge in the triangle of $\pi_3 G$, which corresponds to e_i in the 3-star of G, $i = 1, 2, 3$, then by considering that

$$\begin{aligned}
((\pi_3 G - \bar{e}_1) - \bar{e}_2) - \bar{e}_3 &= G_0;\\
((\pi_3 G - \bar{e}_1) \bullet \bar{e}_2) - \bar{e}_3 &= G_2;\\
((\pi_3 G \bullet \bar{e}_1) - \bar{e}_2) - \bar{e}_3 &= G_1;\\
((\pi_3 G - \bar{e}_1) - \bar{e}_2) \bullet \bar{e}_3 &= G_3;
\end{aligned} \tag{15.4.10}$$

and the relations

$$\begin{aligned}
((\pi_3 G - \bar{e}_1)\bullet\bar{e}_2) \bullet \bar{e}_3 &= ((\pi_3 G \bullet \bar{e}_1) - \bar{e}_2) \bullet \bar{e}_3\\
&= ((\pi_3 G \bullet \bar{e}_1) \bullet \bar{e}_2) - \bar{e}_3\\
&= ((\pi_3 G \bullet \bar{e}_1) \bullet \bar{e}_2) \bullet \bar{e}_3\\
&= G_4,
\end{aligned} \tag{15.4.11}$$

we can find

$$\begin{aligned}
\Phi(\pi_3 G) &= \Lambda_0(\pi_3)\Phi(G_0) + \Lambda_1(\pi_3)\Phi(G_1)\\
&\quad + \Lambda_2(\pi_3)\Phi(G_2) + \Lambda_3(\pi_3)\Phi(G_3)\\
&\quad + \Lambda_4(\pi_3)\Phi(G_4),
\end{aligned} \tag{15.4.12}$$

where

$$\begin{aligned}
\Lambda_0(\pi_3) &= \prod_{i=1}^{3} B(\bar{e}_i),\\
\Lambda_1(\pi_3) &= A(\bar{e}_1)B(\bar{e}_2)B(\bar{e}_3),\\
\Lambda_2(\pi_3) &= B(\bar{e}_1)A(\bar{e}_2)B(\bar{e}_3),\\
\Lambda_3(\pi_3) &= B(\bar{e}_1)B(\bar{e}_2)A(\bar{e}_3),\\
\Lambda_4(\pi_3) &= B(\bar{e}_1)A(\bar{e}_2)A(\bar{e}_3)\\
&\quad + A(\bar{e}_1)B(\bar{e}_2)A(\bar{e}_3) + A(\bar{e}_1)A(\bar{e}_2)\\
&\quad \times (X + Yz)^{w(\bar{e}_3)}(Xz + Y)^{\bar{w}(\bar{e}_3)}.
\end{aligned}$$

Lemma 15.4.4. *Φ is invariant for π_3 if, and only if,*

$$X^2 + Y^2 + XYz = 0. \tag{15.4.13}$$

Proof. From eqs. (15.4.9) and (15.4.12) with the relations $A(\bar{e}_i) = B(e_i)$, $B(\bar{e}_i) = A(e_i)$, $i = 1, 2, 3$, which are determined by the duality, we have

$$\Phi(G) - \Phi(\pi_3 G) = \Delta_0\Phi(G_0) + \Delta_1\Phi(G_1) + \Lambda_2\Phi(G_2) \\ + \Lambda_3\Phi(G_3) + \Lambda_4\Phi(G_4),$$

where

$$\begin{aligned}\Delta_0 &= z\prod_{i=1}^{3} A(e_i) - \prod_{i=1}^{3} B(e_i) + A(e_1)A(e_2)B(e_3) \\ &\quad + A(e_1)B(e_2)A(e_3) + B(e_1)A(e_2)A(e_3), \\ \Delta_1 &= A(e_1)B(e_2)B(e_3) - A(e_1)B(e_2)B(e_3), \\ \Delta_2 &= B(e_1)A(e_2)B(e_3) - B(e_1)A(e_2)B(e_3), \\ \Delta_3 &= B(e_1)B(e_2)A(e_3) - B(e_1)B(e_2)A(e_3), \\ \Delta_4 &= B(e_1)B(e_2)B(e_3) - B(e_1)A(e_2)A(e_3) \\ &\quad - A(e_1)B(e_2)A(e_3) - A(e_1)A(e_2)(X + Yz)^{w(e_3)} \\ &\quad \times (Xz + Y)^{\bar{w}(e_3)}.\end{aligned}$$

Because of $\Delta_1 = 0$, $\Delta_2 = 0$ and $\Delta_3 = 0$, we have

$$\Phi(G) - \Phi(\pi_3 G) = \Delta_0\Phi(G_0) + \Delta_4\Phi(G_4). \tag{15.4.14}$$

From the definition of π_3, by symmetry, only two cases should be considered.

Case 1. $w(e_1) = 0$, $w(e_2) = 1$, and $w(e_3) = 0$. Because of the relations:

$$\begin{cases} A(e_1) = X,\ A(e_2) = Y,\ A(e_3) = X; \\ B(e_1) = Y,\ B(e_2) = X,\ B(e_3) = Y, \end{cases}$$

we have

$$\Delta_0 = (X^2Yz - XY^2 + XY^2 + X^3 + XY^2);$$

$$\Delta_4 = -XY^2 - X^3 + XY^2 - XY(Xz + Y);$$

and hence

$$\Phi(G) - \Phi(\pi_3 G) = (X^2Yz + XY^2 + X^3)\Big(\Phi(G_0) - \Phi(G_4)\Big).$$

Case 2. $w(e_1) = 0$, $w(e_2) = 0$, $w(e_3) = 1$. Because of the relations:

$$\begin{cases} A(e_1) = X,\ A(e_2) = Y,\ A(e_3) = Y; \\ B(e_1) = Y,\ B(e_2) = X,\ B(e_3) = X, \end{cases}$$

we have

$$\Delta_0 = XY^2z - X^2Y + X^2Y + X^2Y + Y^3;$$

$$\Delta_4 = -Y^3 - X^2Y + X^2Y - XY(X + Yz);$$

and hence

$$\Phi(G) - \Phi(\pi_3 G) = (XY^2z + X^2Y + Y^3)\Big(\Phi(G_0) - \Phi(G_4)\Big).$$

By combining the two cases, we find

$$\begin{aligned}\Phi(G) - \Phi(\pi_3 G) &= \delta(X, Y)(XYz + X^2 + Y^2)\\ &\quad \times\Big(\Phi(G_0) - \Phi(G_4)\Big),\end{aligned} \tag{15.4.15}$$

where

$$\delta(X, Y) = \begin{cases} X, & \text{if the case 1 appears;} \\ Y, & \text{otherwise,} \end{cases}$$

and hence the lemma. □

For a planar graph G with binary weights on edges, let

$$\Phi(x; G) = \Phi(G)|_{X=x, Y=1/x, z=-(x^2+1/x^2)}, \tag{15.4.16}$$

then from Lemma 15.4.1, $\Phi(x; G)$ is a polynomial of the undeterminate x.

Theorem 15.4.1. *Polynomial $\Phi(x; G)$ for planar graphs with binary weights on edges is an invariant for the regular equivalence defined by the two elementary operators π_2 and π_3.*

Proof. A direct result of Lemma 15.4.3–Lemma 15.4.4. □

According to Lemma 15.2.3, we have that the operations $sp1$, $sp2$ and their inverses at a crossing point on a universe of a link as shown in Figure 15.4.2 correspond to the deletion and contraction for an edge on the ESF graph and that at each crossing point in the diagram of a link, there are two states denoted by X and Y as shown in Figure 15.4.3. So we find that the polynomial $\Phi(L)$ of X, Y and z correspond to that for G from the discussion in Section 15.2 on the relationship between the elementary operators for graphs and the Reidemeister operations for diagrams of links.

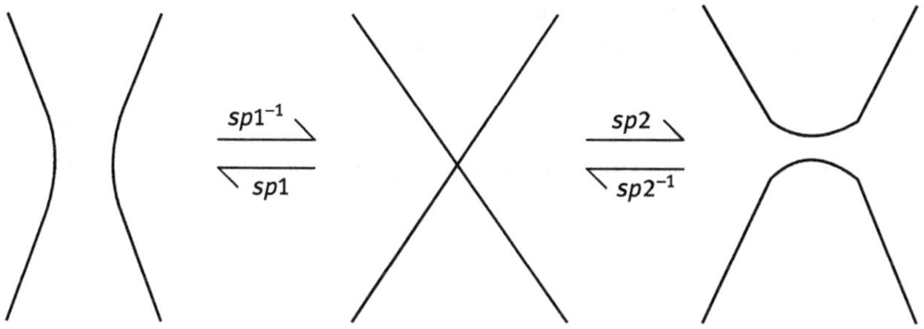

Figure 15.4.2: Two operations.

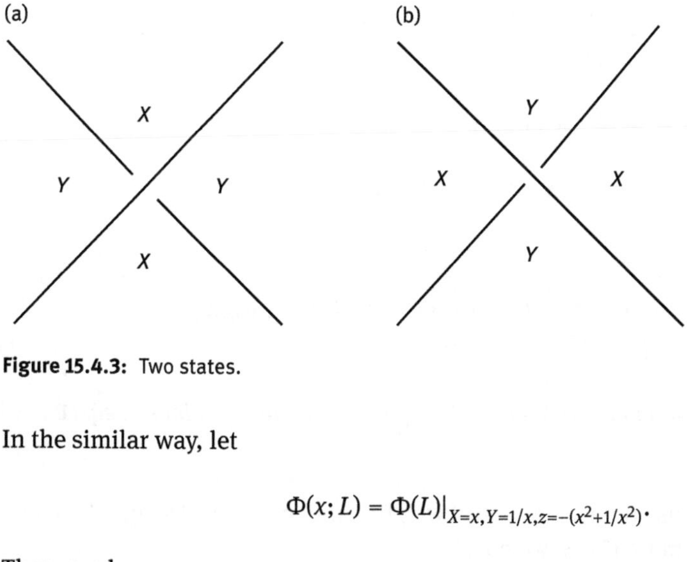

Figure 15.4.3: Two states.

In the similar way, let

$$\Phi(x; L) = \Phi(L)|_{X=x, Y=1/x, z=-(x^2+1/x^2)}. \tag{15.4.17}$$

Then, we have

Corollary 15.4.1. *The polynomial $\Phi(x; K)$ for knots is an invariant for the regular isotopy type.*

15.5 Jones Polynomial

From Lemma 15.4.2, $\Phi(x; G)$ is apparently not an invariant for the ambient equivalence on planar graphs with binary weights on edges.

However, if we introduce

$$\Psi(G) = (-Y)^{c_0-2c_1}(-X)^{c_1-2c_0}\Phi(G), \tag{15.5.1}$$

where c_0 and c_1 are the numbers of edges whose weights are 0 and 1, respectively, then Ψ would lead to an invariant for ambient equivalence.

Lemma 15.5.1. *The polynomial $\Psi(G)$ for planar graphs with the known conditions is invariant for π_1 when the relations in* (15.4.6) *are satisfied.*

Proof. Only discuss the case that edges on which π_1 is allowed to use are of weight 1 because the other case of weight 0 can be done in the similar way. Since the edge is of weight 1, we have that

$$c_0(\pi_1 G) = c_0(G), \ c_1(\pi_1 G) = c_1(G) - 1.$$

From eq. (15.4.4), we see that

$$\begin{aligned}\Psi(G) &= (-Y)^{c_0(G)-2c_1(G)}(-X)^{c_1(G)-2c_0(G)}(X + Yz)\Phi(\pi_1 G)\\ &= (-Y)^{c_0(\pi_1 G)-2c_1(\pi_1 G)-2}(-X)^{c_1(\pi_1 G)+1-2c_0(\pi_1 G)}\\ &\quad \times(X + Yz)\Phi(\pi_1 G).\end{aligned}$$

By the conditions in eq. (15.4.6), we obtain

$$X + Yz = -Y^2 X^{-1}.$$

Thus, we find

$$\begin{aligned}\Psi(G) &= (-Y)^{c_0(\pi_1 G)-2c_1(\pi_1 G)}(-X)^{c_1(\pi_1 G)-2c_0(\pi_1 G)}\Phi(\pi_1 G)\\ &= \Psi(\pi_1 G).\end{aligned}$$

□

Lemma 15.5.2. *The polynomial $\Psi(G)$ is invariant for π_2 when the relations in eq. (15.4.6) are satisfied.*

Proof. Because in any case G has one edge of weight 1 and one edge of weight 0 greater than $\pi_2 G$ does, from Lemma 15.4.3, we have

$$\begin{aligned}\Psi(G) &= (-Y)^{c_0(G)-2c_1(G)}(-X)^{c_1(G)-2c_0(G)}\Phi(G)\\ &= (-Y)^{+1-2}(-X)^{+1-2}\Psi(\pi_2 G),\end{aligned}$$

by the first relation in eq. (15.4.6),

$$= \Psi(\pi_2 G).$$

□

Lemma 15.5.3. *The polynomial $\Psi(G)$ is invariant for π_3 when eq. (15.4.6) is satisfied.*

Proof. Because the numbers of edges of weights 0 and 1 are unchanged, respectively, from Lemma 15.4.4 the lemma follows. □

For planar graph G with binary weights on edges, let

$$\Psi(x; G) = \Psi(G)\big|_{X=x,Y=1/x,z=-(x^2+1/x^2)}. \tag{15.5.2}$$

Theorem 15.5.1. *The polynomial $\Psi(x; G)$ for planar graphs of binary weights on edges is an invariant for ambient equivalence.*

Proof. A direct result of Lemmas 15.5.1–15.5.3. □

Similarly to eq. (15.4.17), let

$$\Psi(x; L) = \Psi(L)|_{X=x, Y=1/x, z=-(x^2+1/x^2)}, \tag{15.5.3}$$

then we have

Corollary 15.5.1. *The polynomial $\Psi(x; K)$ for knots is an invariant for the ambient isotopy type.*

The polynomial $\Psi(x; K)$ by eq. (15.5.3) can be shown to be the Jones polynomial $V_K(t)$ for knots whenever setting $t = Y^{-2}X^2$. Thus, Corollary 15.5.1 shows again that the Jones polynomial for knots is an invariant for the ambient isotopy type in Jones [158], Kauffman [160], and Liu [215].

15.6 Notes

15.6.1 The polynomial $\Phi(x; K)$ defined by eq. (15.4.17) is in fact the bracket polynomial first introduced in Kauffman[160]. From Corollary 15.4.1, we see again that the bracket polynomial for knots is an invariant of the regular isotopy type in Liu [215].

15.6.2 The first attention to the relation between the Tutte polynomial discovered by Tutte in the fifties in graph theory and the Jones polynomial by Jones in the eighties in topology is due to Thistlethwaite [326]. In fact, both of the Jones polynomial and the bracket polynomial as mentioned in Sections 15.4 and 15.5 satisfy the similar type of recursion that the Tutte polynomial does.

Further, they all have the corresponding spanning tree expansion that Tutte exploited in Tutte [343] of the last 1950s. Fundamental theory, in the study of knots through diagrams that enable us to apply graph theory, is due to Reidemeister, who found the three operations, for identifying two knots of the same type, in Reidemeister [282].

15.6.3 It looks that the oldest and most successful techniques of knot theory bypass the diagrams and the Reidemeister operations, at least in theory, and employ topological transformation. These techniques start by the knot complement in the 3-dimensional space.

The natural question is that whether knots are determined by their complements. This was conjectured to be true. However, it is not proved until 1989 in Gordon and Luecke [106]. Those about knot groups can be seen in Crowell and Fox [62] (1963), an introductory textbook.

15.6.4 Applications of knot theory to statistical mechanics and quantum field theory can be seen in Baxter [19] and Witten [400].

15.6.5 The classification of knots is still an open problem. Although the Jones polynomial and the bracket polynomial are considerably important progress in the knot theory, none of them is shown to be the *complete invariant*, which can be used to identify different classes of knots in ambient isotopy type or regular isotopy type. In correspondence, there is also the classification problem for planar graphs.

15.6.6 More polynomials on planar graphs can be seen in Liu [215].

15.6.7 Group approach to knot invariants is via observing the complement of a knot on the basis of the fact that two knots are topological equivalent if, and only if, so are their complements in Gordon and Luecke [106].

15.6.8 On the basis of Section 15.3, one might be motivated to find a criterion for determining whether, or not, two knots, or links, have the same pan-polynomial but not topological equivalent, or to characterize two knots, or links, in the same class for the pan-polynomial equivalence via their inner constructions.

Bibliography

[1] Abrams, L. and D.C. Slilaty, An algebraic characterization of projective-planar graphs, Manuscript, 2002; Also in *J. Graph Theory*, **42** (2003), 320–331.

[2] Agoston, M.K., *Algebraic Topology*: A First Course, Dekker, New York, 1976.

[3] Aho, A.V., J.E. Hopcroft and J.D. Ullman, *The Design and Analysis of Computer Algorithms*, Addison-Wesley, 1974.

[4] Alekseev, V.B. and V.S. Gonchakov, Thickness of arbitrary complete graphs, *Mat. Sbornik*, **101** (1976), 212–230.

[5] Alexandroff, P.S., *Combinatorial Topology* I and II, Greylock, 1957.

[6] Appel, K. and W. Haken, The existence of unavoidable sets of geographically good configurations, *Illinois J. Math.*, **20** (1976), 218–297.

[7] Appel, K. and W. Haken, Every planar map is four colorable I: discharging, *Illinois J. Math.*, **21** (1977), 429–490.

[8] Appel, K. and W. Haken, An unavoidable set of configurations in planar triangulations, *J. Comb. Theory*, **B26** (1979), 1–21.

[9] Appel, K., W. Haken and J. Kock, Every planar map is four colorable II: reducibility, *Illinois J. Math.*, **21** (1977), 491–567.

[10] Appel, K., W. Haken and J. Mayer, Triangulations a v_5 separes dans le probleme des quatre couleurs, *J. Comb. Theory*, **B27** (1979), 135–150.

[11] Archdeacon, D., A Kuratowski theorem for the projective plane, *J. Comb. Theory*, **5** (1981), 243–246.

[12] Archdeacon, D. and P. Huneke, A Kuratowski theorem for non-orientable surfaces, *J. Comb. Theory*, **B46** (1989), 173–231.

[13] Archdeacon, D. and P. Huneke, On cubic graphs which are irreducible for non-orientable surfaces, *J. Comb. Theory*, **B39** (1985), 233–264.

[14] Archdeacon, D. and J. Siran, Characterizing planarity using theta graphs, *J. Graph Theory*, **27** (1998), 17–20.

[15] Auslander, L., T. Brown, and J.W.T. Youngs, The imbedding of graphs in manifolds, *J. Math. Mech.*, **12**:4 (1963), 229–234.

[16] Auslander, L. and S.V. Parter, On imbedding graphs in sphere, *J. Math. Mech.*, **10** (1961), 517–523.

[17] Basden, A. and K.G. Nichols, New topological method for layout printed circuits, *Proc. IEEE*, **120**:3 (1973), 325–328.

[18] Batini, C., M. Talamo and R. Tamassia, Computer aided layout of entity-relationship diagrams, *IEEE J. Syst. Software*, **4** (1984), 163–173.

[19] Baxter, R.J., *Exactly Solved Models in Statistical Mechanics*, Academic. Press, 1982.

[20] Beineke, L. and F. Harary, On the thickness of the complete graph, *Bull. Am. Math. Soc.*, **70** (1964), 618–620.

[21] Beineke, L. and F. Harary, Inequalities involving the genus of a graph and its thickness, *Proc. Glasgow Math. Assoc.* **7** (1965), 19–21.

[22] Beineke, L. and F. Harary, The thickness of the complete graph, *Canad. J. Math.*, **17** (1965), 850–859.

[23] Beineke, L., F. Harary and J.W. Moon, On the thickness of the complete bipartite graph, *Proc. Camb. Phil. Soc.*, **60** (1964), 1–5.

[24] Bellman, R., K.L. Cook and J.A. Lockett, *Algorithms, Graphs, and Computers*, Academic Press, 1970.

[25] Berge, C., *Graphs and Hypergraphs*, North-Holland, 1973.

[26] Berman, K., Three-coloring of planar 4-valent maps, *J. Comb. Theory*, **B30** (1981), 82–88.

[27] Berman, K., Bicycles and spanning trees, *SIAM J. Algeb. Discrete Method*, **7** (1987), 1–12.

[28] Berman, K. and Y.P. Liu, Bicycles, matchings and trees, *Proc. 100 Ann. Jiaotong University*, 1996, 1–10; also in *J. Northern Jiaotong University*, **20** (1996), 694–703.

[29] Berman, K. and Y.P. Liu, Generalized bicycles, *Discrete Appl Math.*, **78** (1997), 27–40.

[30] Bhatt, S.N. and F.T. Leighton, A framework for solving VLSI graph layout problems, *J. Comput. Syst. Sci.*, **28** (1984), 300–343.

[31] Biggs, N., *Algebraic Graph Theory*, Cambridge University Press, 1974.

[32] Biggs, N.L. and G.H.J. Meredith, A theorem on planar partitions, Proc. 5-th British Comb. Conf., Aberdeen, 1975, 73–78.

[33] Birkhoff, G.D., The reducibility of maps, *Am., J. Math.*, **35** (1913), 115–128.

[34] Birkhoff, G.D. and D.C. Lewis, Chromatic polynomials, *Trans. Am. Math. Soc.*, **60** (1946), 355–455.

[35] Bondy, J.A. and U.S.R. Murty, *Graph Theory with Applications*, Macmillan, 1976.

[36] Braun, D. et al, Techniques for multi-layer channel routing, *IEEE Trans. CAD*, **7** (1988), 698–712.

[37] Brooks, R.L., C.A.B. Smith, A.H. Stone and W.T. Tutte, The dissection of rectangles into squares, *Duke Math. J.*, **7** (1940), 312–340.

[38] Cai, J.L., G.H. Dong and Y.P. Liu, A sufficient condition on upper embeddability of graphs, *Sci. China, Math.*, **53** (2010), 5: 1377–1384.

[39] Capobianco, M. and J.C. Molluzzo, *Examples and Counterexamples*, North-Holland, 1978.

[40] Chai, Z. and Y.P. Liu, The up-embeddability of graphs with maximum face degree 6 (in Chinese), *J. Northern Jiaotong University*, **26** (2005), 94–97.

[41] Chan, S.P., *Introductory Topological Analysis of Electrical Networks*, Holt, 1969.

[42] Chen, R.W., Y. Kajitani and S.P. Chan, A graph-theoretic via minimization algorithm for two- layer printed circuit boards, *IEEE Trans. Circuit Syst. Case*-**30** (1983), 284–299.

[43] Chen, W.K., *Applied Graph Theory*, North-Holland, 1971.

[44] Chen, W.K., On vector spaces associated with a graph, *SIAM J. Appl. Math.*, **20** (1971), 526–529.

[45] Chen, Y.C. and Y.P. Liu, Maximum genus, girth and maximum non-adjacent edge set, *Ars Comb.*, **79** (2006), 145–159.

[46] Chen, Y.C. and Y.P. Liu, On a conjecture of S. Stahl, *Can. J. Math.*, **62**:5 (2010), 1058–1059.

[47] Chen, Y.C. and Y.P. Liu, On the average crosscap number II: bounds for a graph, *Sci China*, **A50** (2007), 292–304.

[48] Chen, Y.C. and Y.P. Liu, Up-embeddability of a graph by order and girth, *Graphs Comb.*, **23** (2007), 521–527.

[49] Chen, Y.C. and Y.P. Liu, A note on lower bounds for maximum genus, *Util. Math.*, **73** (2007), 23–31.

[50] Chen, Y.C., Y.P. Liu and R.X. Hao, Structure for a graph with average genus, *J. Appl. Math. Comput.*, **21** (2006), 45–55.

[51] Chen, Y.C., Y.P. Liu and T. Wang, The total embedding distributions of cacti and necklaces, *Acta Math. Sin.* (Engl. Ser.), **22** (2006), 1583–1590.

[52] van Cleemput, W.M., Mathematical models for the circuit layout problem, *IEEE Trans. Circuit Syst.* Case-**23**: 12 (1976), 759–767.

[53] van Cleemput, W.M. and J.G. Linders, An improved graph theoretical model for the circuit layout problem, Proc. 11-th Design Automation Workshop, Denver, 1974.

[54] Cook, S.A., The complexity of theorem proving procedures, Proc. 3-rd ACM Symp. Comput., 1971, 151–158.

[55] Cori, R., Graphes planaires et systemes de parentheses, Doct. Thesis, Paris, 1969.

[56] Cori, R. and A. Machi, Note: construction of maps with prescribed automorphism group, *Theor. Comput. Sci.*, **21** (1982), 91–98.

[57] Cori, R. and A. Machi, Flows on hypermaps, *Glasgow Math. J.*, **30** (1988), 17–29.

[58] Coxeter, H.S.M., *Regular Polytopes*, 3rd Edition, Dover, 1973.

[59] Crapo, H.H. and G.C. Rota, *On the Foundation of Combinatorial Theory: Combinatorial Geometry*, MT. Press, 1970.

[60] Crapo, H.H. and G.C. Rota, Simplicial geometries, *Am. Math. Soc. Symp.*, **XIX** (1971), 71–75.

[61] Cribb, D.W., R.D. Ringeisen and D.R. Shier, On cycle basis of a graph, *Cong. Numer.*, **32** (1981), 221–129.

[62] Crowell, R.H., R.H. Fox, *Introduction to Knot Theory*, Springer, 1963.

[63] Cui, X.F. and Y.P. Liu, Efficient algorithms for the rectilinear extension of a standard graph (in Chinese), Research Report, Inst. Appl. Math., Academia Sinica, 1992.

[64] Dai, W.W-M, B. Eschermann, E.S. Kuh, and M. Pedram, Hierarchical placement and floorplanning in BEAR, *IEEE Trans. CAD*, **8** (1989), 1335–1349.

[65] Dambit, Ja., Embedding of a graph into the plane, *Latvian Math.*, Yearbook **2** (1966), 79–93.

[66] Dantzig, G.B., *Linear Programming and Extensions*, Princeton, 1963.

[67] Dantzig, G.B., A.J. Hoffman and T.C. Hu, Triangulations (tiling) and certain block triangular matrices, *Math. Programming*, **31** (1985), 1–14.

[68] Dehn, M., Uber Kombinatoriche Topologie, *Acta Math.*, **67** (1936), 123–168.

[69] Demoucron, G., Y. Malgrange and R. Pertuiset, Graphes planaires, reconnaissance et construction de representations planaires topologiques, *Rev. francaise Recherche Operation-nelle*, **8** (1964), 33–47.

[70] Deo, N., Note on Hopcroft and Tarjan's planarity algorithm, *J. ACM*, **33** (1976), 74–75.

[71] Deo, N., Minimum length fundamental cycle set, *IEEE Trans. Circuit Syst.* **26** (1979), 894–895.

[72] Deo, N., G.M. Prabhu and M.S. Krishnamoorthy, Algorithms for generating fundamental cycles in a graph, *ACM Trans. Math. Software* , **8** (1982), 26–42.

[73] Dixon, J.D. and B. Mortimer, *Permutation Groups*, Springer, 1996.

[74] Dixon, E.T. and S.E. Goodman, An algorithm for the longest cycle problem, *Networks*, **6** (1976), 139–149.

[75] Dong, G.H. and Y.P. Liu, Results of the maximum genus of graphs, *Sci China*, **A50** (2007), 1563–1570.

[76] Dong, G.H. and Y.P. Liu, Up-embeddability via girth and the degree-sum of adjacent vertices, *Sci China*, **A52** (2009), 3: 597–604.

[77] Dong, G.H. and Y.P. Liu, Lower bound on the maximum genus of triangle-free graphs, *Adv. Math.* (China) **38**:6 (2009), 692–698.

[78] Dong, G.H., Y.P. Liu and N. Wang, Up-embeddable graphs via the degree-sum of nonadjacent vertices: non-simple graphs, *Ars Combin.*, **95** (2010), 129–141.

[79] Dong, G.H., Y.P. Liu and N. Wang, Up-embeddable graphs via the degree-sum of nonadjacent vertices, *Ars Combin.*, **95** (2010), 305–319.

[80] Dong, G.H., N. Wang, Y.Q. Huang and Y.P. Liu, Joint-tree model and the maximum genus of graphs, *Int J. Math. Combin.*, **4** (2012), 80–91.

[81] Du, D. and Y. Zhang, On heuristics for minimum length rectilinear partitions, *Algorithmica* **5** (1990), 111–128.

[82] Duke, R.A., The genus, regional number and Betti number, *Can. J. Math.*, **13** (1966), 817–822.

[83] Edmonds, J.R., A combinatorial representation for polyhedral surfaces, *Notices AMS*, **7** (1960), 646.

[84] Fary, I., On straight line representation of planar graph, *Acta Univ. Szeged. Sect. Sci. Math.*, **11** (1948), 229–233.

[85] Fiorini, S., On the chromatic index of outerplanar graphs, *J. Comb. Theory*, **B18** (1975), 35–38.

[86] Fiorini, S. and R.J. Wilson, *Edge- Colorings of Graphs*, Pitman, 1977.

[87] Fisher, G.J., Computer identification and extraction of planar graphs, Eng. Sc. D. Dissertation, Columbia Univ., New York, 1965.

[88] Fisher, G.J. and O. Wing, Computer recognition and extraction of planar graphs from the incidence matrix, *IEEE Trans. Circuit Theory* CT-**23**:2 (1966), 254–163.

[89] Fisher, G.M., On the group of all homeomorphisms of a manifold, *Trans. AMS*, **97** (1960), 193–212.

[90] Francis, G.K., Null genus realizability criterion for abstract intersection sequences, *J. Comb. Theory*, **7** (1969), 331–341.

[91] Fraysseix, H., J. Pach and R. Pollack, Small sets supporting Fary embeddings of planar graphs, IMA Preprint Series No. 387, 1988.

[92] Fraysseix, H. and P. Rosenstiehl, A depth – first search characterization of planarity, *Ann. Discrete Math.*, **13** (1982), 75–80.

[93] Fraysseix, H. and P. Rosenstiehl, A characterization of planar graphs by Tremaux order, *Combinatorica*, **5** (1985), 127–155.

[94] Furst, M.L., J.L. Gross and L.A. McGeoch, Finding a maximum-genus graph embedding, *J. ACM*, **35** (1988), 523–534.

[95] Garey, M.R. and D.S. Johnson, *Computers and Intractability*, A Guide to the Theory of NP – Completeness, W.H. Freeman, 1979.

[96] Gauss, C.F., *Werke*, (Teubner, Leipzig, 1900) **272**, 282–286.

[97] Giblin, P.J., *Graphs, Surfaces and Homology* (An Introduction to Algebraic Topology), Chapman and Hall, 1977.

[98] Gilbert, W.J., *Modern Algebra with Applications*, Wiley, 1976.

[99] Glover, H. and J. Huneke, Graphs with bounded valence that do not embed in projective plane, *Discrete Math.*, **18** (1977),155–165.

[100] Glover, H. and J. Huneke, The set of irreducible graphs for the projective plane is finite, *Discrete Math.*, **22** (1978), 243–256.

[101] Glover, H., J. Huneke and C.S. Wang, 103 graphs that are irreducible for the projective plane, *J. Comb. Theory*, **B27** (1979), 232–370.

[102] Goldstein, A.J., An efficient and constructive algorithm for testing whether a graph can be imbedded in a plane, Graph Comb. Conf., Princeton University, 1963.

[103] Goldstein, A.J. and D.G. Schweikert, A proper model for testing the planarity of electrical circuits, *Bell Syst. Tech. J.*, **52** (1973), 135–142.

[104] Golumbic, M.C., Comparabality graphs and a new matroid, *J. Comb. Theory*, **B22** (1977), 68–90.

[105] Golumbic, M.C., *Algorithmic Graph Theory and Perfect Graphs*, Academic Press, 1980.

[106] Gordon, C.McA. and J. Luecke, Knots are determined by their complements, *J. Am. Math. Soc.*, **2** (1989), 371–415.

[107] Greenberg, M.J., *Lectures on Algebraic Topology*, Benjamin, 1967.

[108] Gross, J.L. and T.W. Tucker, *Topological Graph Theory*, John Wiley & Sons, 1987.

[109] Grunbaum, B., Arrangements and Spreads, Reg. Conf. Ser. in Math. No. 10, AMS, 1972.

[110] Grunbaum, B., Shortness exponents of families of graphs, *J. Comb. Theory*, **14** (1973), 346–385.

[111] Grunbaum, B. and G.C. Shephard, The theorem of Euler and Eberhard for tilings of the plane, *Resultate Math.*, **5** (1982), 19–44.

[112] Grunbaum, B. and J. Zaks, The existence of certain planar maps, *Discrete Math.*, **10** (1974), 93–115.

[113] Gustin, W., Orientable embedding of Cayley graphs, *Bull. Am. Math. Soc.*, **69** (1963), 272–275.

[114] Haken, W., An existence theorem for planar maps, *J. Comb. Theory*, **14** (1973), 180–184.

[115] Hammer, P.L. and Y.P. Liu, 0-1 Programming and order relations (in Chinese with English abstract) (I), *J. Math. Res. Expos.*, **8**:2 (1988), 315–325; (II), ibid. **8**:3 (1988), 481–489.

[116] Hammer, P.L., Y.P. Liu and B. Simeone, Boolean approaches to combinatorial optimization (in Chinese with English abstract) (I), *J. Math. Res. Expos.*, **10**:2 (1990), 300–312; (II), ibid. **10**:3 (1990), 455–468; (III), ibid. **10**:4 (1990), 619–628.

[117] Hao, R.X. and Y.P. Liu, Evaluation of crossing numbers by applying auxiliary graphs, *J. Henan Normal Univ.* (NS), **30** (2002), 2: 7–13.
[118] Hao, R.X. and Y.P. Liu, The genus polynomials of cross ladder digraphs in orientable surfaces, *Scin. China*, **A50** (2008), 5: 889–896.
[119] Hao, R.X. and Y.P. Liu, The distribution of directed antiladders in orientable surfaces, *Appl. Math. Lett.*, **21** (2008), 161–164.
[120] Hao, R.X. and Y.P. Liu, Auxilliary graphs of projective planar signed graphs (in Chinese), *J. Syst. Sci. Math. Sci.*, **30** (2010), 9:1251–1258.
[121] Hao, R.X., Y.P. Liu, T.Y. Zhang and L.S. Xu, The genus distributions of 4-regular digraphs, *Australas. J. Combin.*, **43** (2009), 79–90.
[122] Harary, F., Recent progress in topological graph theory, *Acta Math. Acad. Sci. Hung.*, **15** (1964), 405–412.
[123] Harary, F., R. Karp and W.T. Tutte, A criterion for the planarity of the square of a graph, *J. Comb. Theory*, **2** (1967), 395–405.
[124] Harary, F. and W.T. Tutte, A dual form of Kuratowski's theorem, *Can. J. Bull.*, **8** (1965), 17–20.
[125] Harary, F. and W.T. Tutte, On the order of the group of a planar map, *J. Comb. Theory*, **1** (1966), 194–195.
[126] He, W.L. and Y.P. Liu, For a neighbor condition on up-embeddability of graphs (in Chinese), *OR Trans.*, **7** (2003), 92–96.
[127] He, W.L. and Y.P. Liu, A note on the maximum genus of graph G^3, *J. Syst. Sci. Inf.*, **1** (2003), 615–519.
[128] Heawood, P.J., Map colour theorem, *Quart. J. Math.*, **24** (1890), 332–338.
[129] Heawood, P.J., On extended congruences connected with the four color theorem, Proc. London Math. Soc. 2nd Series **33** (1932), 253–266.
[130] Heawood, P.J., Failures in congruences with the four color theorem, ibid. **40** (1936), 189–202.
[131] Heawood, P.J., Note on a correction in a paper on map – congruences, *J. London Math. Soc.*, **18** (1943), 160–167.
[132] Heawood, P.J., Note on a correction in a paper on map – congruences, ibid. **19** (1944), 18–22.
[133] Heffter, L., Ueber das Problem der Nachbargebiete, *Math. Ann.*, **38** (1891), 615–620.
[134] Hilbert, D. and S. Cohn-Vossen, *Anschauliche Geometrie*, Julius Springer, 1932.
[135] Hopcroft, J.E, An $n \log n$ algorithm for isomorphism of planar triply connected graphs, in *Theory of Machines and Computation*, ed. by Z. Kohavi et al., Academic Press, 1971, 189–196.
[136] Hopcroft, J.E. and R.E. Tarjan, Planarity testing in $V \log V$ steps: extended abstract, Proc. IFIP Cong., 1971, 85–90.
[137] Hopcroft, J.E. and R.E. Tarjan, A V^2 algorithm for determining isomorphism of planar graphs, *Inf. Proc. Lett.*, **1** (1971), 32–34.
[138] Hopcroft, J.E. and R.E. Tarjan, Isomorphism of planar graphs, in *Complexity of Computer Computations*, ed. by R. Miller et al., Plenum, 1972, 131–152.
[139] Hopcroft, J.E. and R.E. Tarjan, Dividing a graph into triconnected components, *SIAM J. Comput.*, **2** (1973), 135–158.
[140] Hopcroft, J.E. and R.E. Tarjan, Efficient planarity testing, *J. ACM*, **21** (1974), 549–568.
[141] Hopcroft, J.E. and J.K. Wong, Linear time algorithm for isomorphism of planar graphs (extended abstract), 6-th Ann. ACM Symp. Comput., Seattle, 1974.
[142] Hotz, G., The embedding of graphs in the 2- sphere, *Z. Angew. Math. Mech.*, **45** (1965).
[143] Hotz, G., Embedding of graphs in the plane, *Math. Ann.*, **167** (1967), 214–223.
[144] Hu, T.C., Multicommodity network flows, *Oper. Res.*, **11** (1973), 344–360.
[145] Hu, T.C. and E.S. Kuh, Theory and concepts of circuit layout, in *VLSI Circuit Layout: Theory and Design*, IEEE Press, 1985, 3–18.
[146] Huang Y.Q. and Y.P. Liu, Maximum genus and maximum nonseparating independent set of a 3-regular graph, *Discrete Math.*, **176** (1997), 149–158.

[147] Huang Y.Q. and Y.P. Liu, Some classes of upper embeddable graph, *Acta Math. Sci.*, **17** (Supp.) (1997), 154–161.

[148] Huang Y.Q. and Y.P. Liu, Maximum genus and 2-factors of graphs, *Chin J. Contemp. Math.*, **18** (1997), 425–434.

[149] Huang Y.Q. and Y.P. Liu, Extensions on 2-edge connected 3-regular up-embeddable graphs, *Acta Math. Appl. Sinica*, **14** (1998), 337–346.

[150] Huang Y.Q. and Y.P. Liu, Upper embeddability of graphs, *Sci. China*, **41A** (1998), 493–504.

[151] Huang Y.Q. and Y.P. Liu, The maximum genus of graphs with diameter three, *Discrete Math.*, **194** (1999), 139–149.

[152] Huang Y.Q. and Y.P. Liu, Face size and the maximum genus of graph 1. Simple graphs, *J. Comb. Theory*, **80B** (2000), 356–370.

[153] Huang Y.Q. and Y.P. Liu, On the average genus of 3-regular graphs, *Adv. Math.*, **31** (2002), 56–64.

[154] Hubicka, E. and M.M. Syslo, Minimal basis of cycles of a graph, in *Recent Advances in Graph Theory*, ed. by M. Fiedler, Academic press Praha, 1975, 283–293.

[155] Huneke, J., On the genus of a graph, *Proc. Symp. Pure Math.*, **24** (1979), 357–364.

[156] Iri, M., *Network Flow, Transportation and Scheduling*, Academic Press, 1969.

[157] Jacobson, N., *Basic Algebra* I, Freeman, San Francisco, 1974.

[158] Jones, V.F.R., A polynomial invariant for links via Von Neumann algebras, *Bull. Am. Math. Soc.*, **12** (1985), 103–112.

[159] Karp, R.M., Computational complexity of combinatorial problems, *Networks*, **5** (1975), 45–68.

[160] Kauffman, L.H., New invariants in knot theory, *Am. Math. Mon.*, **95**:3 (1988), 195–243.

[161] Kauffman, L.H., A Tutte polynomial for signed graphs, *Discrete Math.*, **25** (1989), 105–127.

[162] Kaufman, A., *Graphs, Dynamic Programming and Finite Games*, Academic Press, 1967.

[163] Krentel, M.W., The complexity of optimization problems, *J. Comput. Syst. Sci.*, **36** (1988), 190–509.

[164] Kung, J.P.S., *A Source Book in Matroid Theory*, Birkhauser, 1986.

[165] Kuo, C.L., *Linear Network and System*, McGraw-Hill, 1967.

[166] Kuratowski, K., Sur le Problem des Courbes Gauches en Topologie, *Fund. Math.*, **15** (1930), 271–283.

[167] Lai, Y.-T. *et al.*, Algorithmsfor floorplan design via rectangular dualization, *IEEE Trans. CAD* , **7** (1988), 1278–1289.

[168] Larson, J.A., Some graphs with chromatic number three, *J. Comb. Theory*, **B26** (1979), 317–322.

[169] Larson, R.C. and V.O. Li, Finding minimum rectilinear distance paths in the presence of barriers, *Networks*, **11** (1981), 285–304.

[170] Lee, D.T. and F.P. Preparata, Location of a point in a planar subdivision and its applications, 8th Ann. ACM Symp. Comput., 1976, 231–235.

[171] Lefschetz, S., Planar graphs and related topics, *Proc. Nat. Acad. Sci.*, **54** (1965), 1763–1765.

[172] Lefschetz, S., *Applications of Algebraic Topology*: *Graphs and Networks*, Springer, 1975.

[173] Lefschetz, S., *Algebraic Topology*, AMS, 1942.

[174] Lerda, F. and E. Majoranic, An algorithm for connecting n points with a minimum number of crossings, *Calcolo* , **1** (1964), 257–365.

[175] Levow, R.B., On Tutte's algebraic approach to the theory of crossing numbers, Proc. 3-rd S-E Conf. Comb. Graph Theory Comput., 1972, 315–324.

[176] Li, D.M. and Y.P. Liu, The number of 4-regular Hamiltonian planar maps, *J. Northern Jiaotong Univ.*, **21** (1997), 548–553.

[177] Li, D.M. and Y.P. Liu, The maximum genus of a 3-regular simplicial graph, *Appl. Math. -JCU*, **14B** (1999), 203–214.

[178] Li, D.M. and Y.P. Liu, A tight lower bound on the maximum genus of 3-edge connected loopless graphs, *Acta Math. Appl. Sinica*, **15** (1999), 361–367.

[179] Li, D.M. and Y.P. Liu, A polynomial algorithm for finding the minimum feedback vertex set of a 3-regular simple graph, *Acta Math. Sci.*, **10B** (1999), 375–381.

[180] Li, D.M. and Y.P. Liu, Maximum genus, girth and connectivity, *Eur. J. Comb.*, **21** (2000), 651–657.

[181] Li, D.M. and Y.P. Liu, A tight lower bound on the maximum genus of a 3-connected loopless multigraph, *Appl. Math.-JCU*, **15B** (2000), 369–376.

[182] Li, D.M. and Y.P. Liu, A relative maximum genus graph embedding and its local maximum genus, *Acta Math. Appl. Sinica*, **16.** (2000), 366–372.

[183] Li, L.F. and Y.P. Liu, Genus distribution of embeddings for a type of 3-regular graphs(in Chinese with English abstract), *J. Beijing Jiaotong Univ.*, **28** (2004), 39–42.

[184] Lin, P.M., On methods of deleting planar graphs, Proc. 8-th Midwest Symp. Cir. Theory, Colorado State Univ., Boulder, 1965, 11–19.

[185] Lipton, R.J. and R.E. Tarjan, A separator theorem for planar graphs, *SIAM J. Appl. Math.*, **36** (1979), 177–189.

[186] Lipton, R.J. and R.E. Tarjan, Applications of a planar separator theorem, *SIAM J. Comput.*, **9** (1980), 615–627.

[187] Lipton, R.J., D.J. Rose and R.E. Tarjan, Generalized nested dissection, *SIAM J. Numer. Anal.*, **16** (1979), 177–189.

[188] Liu, C.L., *Introduction to Combinatorial Mathematics*, McGraw- Hill, 1968.

[189] Liu, T.Y. and Y.P. Liu, Near-triangulations of small circuit double cover with at most $n - 2$, *J. Northern Jiaotong Univ.*, **24** (2000), 65–67, 72.

[190] Liu, T.Y. and Y.P. Liu, Strong embeddings of planar graphs on higher surfaces, *Acta Math. Sci.*, **22B** (2002), 542–548.

[191] Liu, T.Y. and Y.P. Liu, On the crossing number of circular graphs, *OR Trans.*, **2** (1998), 32–38.

[192] Liu, Y.P., Modulo 2 programming and planar embeddings (in Chinese), *Acta Math. Appl. Sinica*, **1** (1978), 395–406.

[193] Liu, Y.P., Planarity testing and planar embeddings of graphs (in Chinese), *Acta Math. Appl. Sinica*, **2** (1979), 350–365.

[194] Liu, Y.P., Planar graph theory and the four color problem (in Chinese with English abstract) I, *J. Math. Res. Expos.*, **3**:3 (1983), 122–136; II, ibid. **4**:1 (1984), 121–136; III, ibid. **5**:1 (1985), 125–144; IV, ibid. **5**:3 (1985), 123–136; V, ibid. **6**:2 (1986), 175–188.

[195] Liu, Y.P., The maximum non-orientable genus of a graph (in Chinese with English abstract), *Scientia Sinica* Special Issue on Math. **I** (1979), 191–201.

[196] Liu, Y.P., The maximum orientable genus of a graph, *Scientia Sinica* Special Issue on Math. **II** (1979), 41–55.

[197] Liu, Y.P., The maximum orientable genus of some kinds of graphs (in Chinese), *Acta Math. Sinica*, **24** (1981), 817–832.

[198] Liu, Y.P., A note on Edmonds' surface duality theorem of graphs (in Chinese), *Chin. J. Oper. Res.*, **2**:1 (1983), 62–63.

[199] Liu, Y.P., The orientable genus of $K_n - K_3$, *KEXUE TONGBAO* (Chinese Ed.) **25** (1980), 959; Or see *KEXUE TONGBAO* (English Ed.) **26** (1981), 188.

[200] Liu, Y.P., Map color theorem and surface embeddings of graphs(in Chinese) I, *Math. Theory Prac.* No.1 (1981), 65–78; II, ibid. No.2 (1981), 59–69; III, ibid. No.3 (1981), 33–44; IV, ibid. No.4 (1981), 33–41; V, ibid, No.1 (1982), 34–44.

[201] Liu, Y.P., Four color problem, *Chin. Encycl., Math.* (1988), 623–624.

[202] Liu, Y.P., On the linearity of testing planarity of a graph, Comb. Optim. CORR 84-5, University of Waterloo, 1984; Also in *Chin. Ann. Math.*, **7B** (1986), 425–434.

[203] Liu, Y.P., Extending Kotzig's theorem into the non-orientable cases, Comb. Optim. CORR82-36, University of Waterloo, 1982.

[204] Liu, Y.P., K-valent maps on the surfaces, Comb. Optim. CORR82- 35, University of Waterloo, 1982; Also in *Acta Math. Appl. Sinica* Eng. Series **1** (1984), 57–62.

[205] Liu, Y.P., A new approach to the linearity of testing planarity of graphs, Report, Rutgers University, 1984; Also in *Acta Math. Appl. Sinica* Eng. Series **4** (1988), 257–265.

[206] Liu, Y.P., Boolean planarity characterization of graphs, RUTCOR Research Report RRR38-87, Rutgers University, 1987; Also in *Acta Math. Sinica*, New Series, **4** (1988), 316–329.

[207] Liu, Y.P., Boolean approach to planar embeddings of a graph, RUTCOR Research Report RRR 39-87, Rutgers University, 1987; Also in *Acta Math. Sinica*, New Series, **5** (1989), 64–79.

[208] Liu, Y.P., Boolean characterizations of planarity and planar embeddings of graphs, *Ann. Oper. Res.*, **24** (1990), 165–174.

[209] Liu, Y.P., On the net-embeddability of graphs, RUTCOR Research Report RRR8-90, Rutgers University, 1990. Also in *Acta Math. Sinica*, New Series, **8** (1992), 413–423.

[210] Liu, Y.P., The efficient recognition on net- embeddability of graphs, Research Report, Inst. Appl. Math., Academia Sinica, 1991. Also in *Chin. Sci. Bull.*, **38** (1993), 533–536.

[211] Liu, Y.P., Some combinatorial optimization problems arising from VLSI circuit design, Research Report Series A, No.6, Department of Statistics, University "La Sapienza" of Rome, 1992. Also in *Applied Math.* (A Chinese J. Univ.) **8B** (1993), 218–235.

[212] Liu, Y.P., *Rectilinear Embeddings: Theory and Methods* (in Chinese), Science Press, 1994.

[213] Liu, Y.P., Transportations, layouts, orientations, partitions and others (in Chinese), Proc. the 3-rd National Conf. O.R., Chengdu, 1992, 500–507.

[214] Liu, Y.P., A polyhedral theory on graphs, *Acta Math. Sinica*, New Series, **10** (1994), 136–142.

[215] Liu, Y.P., Combinatorial invariants on planar graphs, *Acta Math. Sinica*, New Series, **11** (1995), 211–220.

[216] Liu, Y.P., *Theory of Graphic Embeddability* (in Chinese), Science Press, Beijing, 1994.

[217] Liu, Y.P., *Embeddability in Graphs*, Kluwer, Dordrecht, 1995.

[218] Liu, Y.P., *Advances in Combinatorial Maps* (in Chinese), Northern Jiaotong University Press, 2003.

[219] Liu, Y.P., *Algebraic Principles of Maps* (in Chinese), Higher Education Press, 2006.

[220] Liu, Y.P., Asymmetrization of combinatorial maps (in Chinese with English abstract), *J. Shenyang Normal Univ.* (NS), **23** (2005), 97–103.

[221] Liu, Y.P., Surface embedding of graphs(I) (in Chinese with English abstract), *J. Tianjin Inst. Tech.*, **19** (2003), 2:1–5.

[222] Liu, Y.P., Surface embedding of graphs(II) (in Chinese with English abstract), *J. Tianjin Inst. Tech.*, **19** (2003), 3: 1–5.

[223] Liu, Y.P., *Theory of Rectilinear Layouts* (in Chinese), China Railway Publishing House, 1997.

[224] Liu, Y.P., On polyhedra-surfaces-embeddings-maps (in Chinese with English abstract), *J. Jishou Univ.* (NS), **28** (2007), 1–6.

[225] Liu, Y.P., *Graph Theory and Algorithms* (in Chinese), Graduate School, Academia Sinica, 1979.

[226] Liu, Y.P., Boolean approaches to graph embeddings related to VLSI, *Science in China* (Series A), **44** (2001), 1034–1041.

[227] Liu, Y.P., *Theory of Polyhedra*, Science Press, 2008.

[228] Liu, Y.P., *Enumerative Theory of Maps*, Kluwer, 1999.

[229] Liu, Y.P., Surface embeddability of graphs via reduction, *Int. J. Math. Combin.*, **3** (2011), 61–67.

[230] Liu, Y.P., *Theory of Graphic Embeddability* 2nd Edition., (in Chinese), Science Press, 2010.

[231] Liu, Y.P., Surface embeddability of graphs via homology, *Sci. Chin. Math.*, **53** (2010), 2889–2892.

[232] Liu, Y.P., Surface embeddability of graphs via tree travels, *Int. J. Math. Combin.*, **1** (2011), 73–81.

[233] Liu, Y.P., Algebraic principles on polyhegons, *Orient. J. Math.*, **2**:2(2010), 83–93.

[234] Liu, Y.P., *Introductory Map Theory*, Kappa & Omega, 2010.

[235] Liu, Y.P., Homology and cohomology on graphs, *J. Kunming Univ. Sci. & Tech.* (NS), **36**:6(2011), 67–71, 78.

[236] Liu, Y.P., Surface embeddability of graphs via joint trees, *Int. J. Math. Combin.*, **2** (2011), 15–20.
[237] Liu, Y.P., *Theory of Combinatorial Functional Equations* (in Chinese), USTC Press, 2015.
[238] Liu, Y.P., Advanced graph theory as I have initially realized (I): On surface of genus 0 (in Chinese), *J. Kunming Univ. Sci. Tech.* (NS), **41**:3(2016), 142–146.
[239] Liu, Y.P., P. Marchioro and R. Petreschi, At most single- bend embeddings of cubic graphs, Research Report SI-92/01, Dept. Comput. Sci, Univ. "La Sapienza" of Rome, 1992. Also in *Applied Math.* (A J. Chinese Univ.), **9B** (1994), 127–142.
[240] Liu, Y.P., P. Marchioro, R. Petreschi and B. Simeone, Theoretical results on at most 1-bend embeddability of graphs, Research Report Series A No.3, Department of Statistics, University of Rome "La Sapienza", 1990; Also in *Acta Math. Appl. Sinica*, Eng. Series, **8** (1992), 188–192.
[241] Liu, Y.P., P. Marchioro, R. Petreschi and B. Simeone, On theoretical results of at most 1- embeddability of graphs, *Chin. Sci. Bull.*, **36** (1991), 1054–1055.
[242] Liu, Y.P., A. Morgana and B. Simeone, On the general theoretical results for rectilinear embeddabilityof graphs, *KEXUE TONGBAO*(Chinese Ed.) **35** (1990), 1513–1514. Or see *Chin. Sci. Bull.* (KEXUE TONGBAO, English Ed.) **36** (1991), 1490.
[243] Liu, Y.P., A. Morgana and B. Simeone, General theoretical results on rectilinear embeddability of graphs, Research Report Series A No.2, Department of Statistics, University of Rome " La Sapienza", 1990; Also in *Acta Math. Appl. Sinica*, Eng. Series **7** (1991), 187–192.
[244] Liu, Y.P., A. Morgana and B. Simeone, A linear time algorithm for finding 3-embeddings of a graph, Research Report, No.7, Inst. Appl. Math., Academia Sinica, 1993.
[245] Liu, Y.P., A. Morgana and B. Simeone, A graph partition problem, Research Report, No.27, Inst. Appl. Math., Academia Sinica, 1992.
[246] Liu, Y.P., A. Morgana and B.Simeone, Another linear time algorithm for finding 3-embeddings of a graph, Research Report, No.1, Inst. Appl. Math., Academia Sinica, 1993.
[247] Liu, Y.P., A. Morgana and B.Simeone, Characterizations of a kind of orientations of a graph, Research Report, No.2, Inst. Appl. Math., Academia Sinica, 1993.
[248] Liu, Yi. and Y.P. Liu, A characterization of the embeddability of graphs on the surface of given genus, *Chin. Ann. Math.*, **17B** (1996), 457–462.
[249] Liu, Yi. and Y.P. Liu, Embeddability of graphs on surfaces, *J. Northern Jiaotong Univ.*, **21**:2 (1997), 127–136.
[250] Liu, Yi. and Y.P. Liu, Embeddings of one kind of graphs, *Appl. Math. -JCU*, **12B** (1997), 253–258.
[251] Liu, Yi. and Y.P. Liu, Lower bound number of irreducible graphs on surfaces, *Acta Math. Scientia*, **18** (1998), 333–339.
[252] Lovasz, L., *Combinatorial Problems and Exercises*, North-Holland, 1979.
[253] Lovasz, L. and M.L. Marx, A forbidden subgraphs characterization of Gauss codes, *Bull. Am. Math. Soc.*, **82** (1976), 121–122.
[254] Lu, J.J., H. Ren and D.J. Ma, The crossing number of $C(m, 3)$ (in Chinese with English abstract), *J. Syst. Sci. Math. Sci.*, **26**:4 (2006), 504–512.
[255] Lv, S.X. and Y.P. Liu, Up-embeddability of graphs with small order, *Appl. Math. Lett.*, **23**:3 (2010), 267–271.
[256] Lv, S.X. and Y.P. Liu, A lower bound on maximum genus of graphs with girth and minimal degree, *Adv. Math.*, **39**:6 (2010), 729–735.
[257] Lv, S.X. and Y.P. Liu, A new bound on maximum genus of simple graphs, *Ars Combin.*, **98** (2011), 7–14.
[258] Lv, S.X. and Y.P. Liu, Up-embeddability and independent number of simple graphs, *Ars Combin.*, **102** (2011), 237–243.
[259] Lv, S.X. and Y.P. Liu, Up-embeddability of graphs with new degree-sum of independent vertices, *J. Math. Res. Appl.*, **32**:4 (2012), 399–406.
[260] Lv, S.X. and Y.P. Liu, Extensions on 2-edge connected 3-regular none up-embeddable graphs, *Adv. Math.* (China), **44**:1 (2015), 55–60.

[261] MacLane, S., A structural charaterisation of planar combinatorial graphs, *Duke Math. J.*, **3** (1937), 460–472.
[262] MacLane, S., A combinatorial condition for planar graphs, *Fund. Math.*, **28** (1937), 22–32.
[263] MacLane, S. and G. Birkhoff, *A Survey of Modern Algebra*, th Edition Macmillan, New York, 1977.
[264] Mao, L.F. and Y.P. Liu, New automorphism groups identity of trees, *Adv. Math.*, **32** (2003), 113–117.
[265] Mao, L.F. and Y.P. Liu, Group action for enumerating maps on surfaces, *J. Appl. Math. Comp.*, **13** (2003), 201–215.
[266] Mao, L.F. and Y.P. Liu, A new approach for enumerating maps on orientable surfaces, *Aust. J. Combin.*, **30** (2004), 247–259.
[267] Mao, L.F., Y.P. Liu and E.L. Wei, The semi-arc automorphism group of a graph with application to map enumeration, *Graphs Combin.*, **22** (2006), 83–101.
[268] Marx, M.L., The Gauss realizability problem, *Trans. AMS*, **134** (1972), 612–613.
[269] Massey, W.S., *Algebraic Topology*: An Introduction, Harcourt, New York, 1967.
[270] Mayer, J., Ineqalites nouvelles le probleme des quatre couleurs, *J. Comb. Theory*, **B19** (1975), 119–149.
[271] Mondshein, L., Combinatorial ordering and embedding of graphs, Tech. Note 1971- 35, Lincohn Lab., M.I.T., 1971.
[272] Mullin, R.C. and P.J. Schellenberg, The enumeration of c-nets via quadrangulations, *J. Comb. Theory*, **4** (1968), 259–276.
[273] Ore, O., *Theory of Graphs*, AMS, 1962.
[274] Ore, O., *The Four Color Problem*, Academic Press, 1967.
[275] Ore, O. and J. Stemple, Numerical calculations on the four-color problem, *J. Comb. Theory*, **8** (1970), 65–78.
[276] Owens, A.B., Graphs and printed circuit embeddings, A survey and bibliography, Naval Res. Lab., Washington DC., Bibl. No. 33, 1969.
[277] Pralts, J.L., *Computer Data Structure*, McGraw-Hill, 1977.
[278] Read, R.C., Graph theory algorithms, in *Graph Theory and Its Applications*, ed. by B. Harris, Academic Press, 1970, 51–78.
[279] Read, R.C., A note on tiling rectangles with dominoes, *Fibonacci Quart.*, **18**:1 (1980), 24–27.
[280] Read, R.C., The dimer problem for narrow rectangular arrays: a unified method of solution, and some extensions, *Aequationes Math.*, **24** (1982), 47–65.
[281] Read, R.C. and P. Rosenstiehl, On the Gauss crossing problem, in *Combinatorics*, eds. by Hajnal et al., North-Holland, 1978, 843–876.
[282] Reidemeister, K., *Knot Theory* (English Edition), BCS Associate, 1883.
[283] Ren, H., Y.P. Liu, D.J. Ma and J.J. Lu, Generating cycle spaces for graphs on surfaces with small genera, *Eur. J. Combin.*, **25** (2004), 1087–1105.
[284] Ren, H., Y.P. Liu, D.J. Ma and J.J. Lu, Cycle spaces of graphs on the sphere and the projective plane, *Acta Math. Sci.*, **25B** (2005), 41–49.
[285] de Rezende, P.J., D.T. Lee and Y.F. Wu, Rectilinear shortest paths in the presence of rectangular barriers, *Discrete Comput. Geom.*, **4** (1989), 41–53.
[286] Ringel, G., The toroidal thickness of complete graphs (in German), *Math. Z.*, **87** (1985), 14–26.
[287] Ringel, G., *Farbungsprobleme auf Flachen und Graphen*, VEB Deutscher Verlag der Wissenschafteng, 1959.
[288] Ringel, G., *Map Color Theorem*, Springer, 1974.
[289] Robertson, N. and P. Seymour, Graph minors VIII: Kuratowski theorem for general surfaces, Preprint.
[290] Robertson, N. and P. Seymour, Generalizing Kuratowski's theorem, *Cong. Numer.*, **45** (1984), 129–138.

[291] Robertson, N. and P. Seymour, Graph minors–A survey, in *Survey in Combinatorics*, 153–171, London Math. Soc. Lect. Notes Ser., Vol. 105, Cambridge Univ Press, 1975.

[292] Robinson, D.J.S., *A Course in the Theory of Groups*, Springer, 1982.

[293] Roman, S., *Advanced Linear Algebra*, Springer, 1992.

[294] Rosenstiehl, P., Bicycles et diagonales des graphes planaires, Colloq. sur Theorie des Graphes, Paris, 1974; Also in *Cahier Centre Etudes Rech. Oper.*, **17** (1975), 365–383.

[295] Rosenstiehl, P., Caracterisation des graphes planaires par une diagonale absreacte, *Cong. Numer.*, **15** (1976), 521–527.

[296] Rosenstiehl, P., Preuve algebrique du critere de planarite de Wu- Liu, *Ann. Discrete Math.*, **9** (1980), 67–78.

[297] Rosenstiehl, P., Solution algebrique du probleme de Gauss sur la permutation des points d'intersection d'une ou plusieurs courbes fermees du plan, *Compt. Rend. Acad. Sci. Paris* **283** (1976), 8: 551–553.

[298] Rosenstiehl, P. and R.C. Read, On the principal edge tripartition of a graph, *Ann. Discrete Math.*, **3** (1978), 195–226.

[299] Rosenstiehl, P. and R. Tarjan, Rectilinear planar layouts and bipolar presentations of planar graphs, *Discrete Comput. Geom.*, **1** (1986), 343–353.

[300] Rosenstiehl, P. and R. Tarjan, Gauss codes, planar Hamiltonian graphs and stack- sortable permutations, *J. Algorithms*, **5** (1984), 375–390.

[301] Saaty, T.L., Remarks on the four-color problem: The Kempe catastrophe, *Math. Mag.* **40** (1967), 31–36.

[302] Saaty, T.L., Thirteen colorful variations on Guthries four-color conjecture, *Am. Math. Mon.*, **79** (1972), 2–43.

[303] Sachs, D., Graphs, matroids and geometric lattices, *J. Comb. Theory*, **9** (1970), 192–199.

[304] Sarrafzadeh, M. and D.T. Lee, A new approach to topological via minimization, *IEEE Trans. CAD*, **8** (1989), 890–900.

[305] Seymour, P.D., On odd cuts and plane multicommodity flows, *Proc. London Math. Soc.*, **42**:3 (1981), 178–192.

[306] Seymour, P.D., Nowhere- zero 6- flows, *J. Comb. Theory*, **B30** (1981), 130–135.

[307] Seymour, P.D., On Tutte's extension of the four color problem, *J. Comb. Theory*, **B31** (1981), 82–94.

[308] Seymour, P.D., Matroid representation over *GF*(3), *J. Comb. Theory*, **B26** (1979), 159–173.

[309] Seymour, P.D., Decomposition of regular matroids, *J. Comb. Theory*, **B28** (1980), 305–359.

[310] Shank, H., The theory of left- right paths, *Combinatorical Mathematics* III, Lect. Notes in Math. **452**, Springer, 1975, 42–54.

[311] Shao, Z.L. and Y.P. Liu, Genus distributions of orientable embeddings for two types of graphs, *Aust. J. Combin.*, **41** (2008), 205–217.

[312] Shao, Z.L. and Y.P. Liu, Genus embeddings of a type of graphs, *J. Appl. Math. Comput.*, **28** (2008), 69–77.

[313] Shao, Z.L. and Y.P. Liu, The genus of a type of graph, *Science in China*, **A53**:2 (2010), 457–464.

[314] Shao, Z.L. and Y.P. Liu, Genus polynomials for three types of graphs, *Ars Combin.*, **96** (2010), 65–73.

[315] Shao, Z.L. and Y.P. Liu, Minimum genera of certain types of graphs, *Utilitas Math.* **81** (2010), 85–98.

[316] Shao, Z.L. and Y.P. Liu, On the number of genus embeddings of complete tripartite graph $K_{n,n,l}$, *Ars Combin.*, **104** (2012), 129–142.

[317] Shao, Z.L., Y.P. Liu and Z.G. Li, On the number of genus embeddings of complete bipartite graphs, *Graphs Combin.*, **29** (2013), 1909–1919.

[318] Stahl, S., Generalized embedding schemes, *J. Graph Theory*, **2** (1978), 41–52.

[319] Stahl, S., A combinatorial analog of the Jordan curve theorem, *J. Comb. Theory*, **B35** (1983), 1–21.

[320] Stahl, S. and L.W. Beineke, Blocks and non- orientable genus of graphs, *J. Graph Theory*, **1** (1977), 75–78.

[321] Stillwell, J., *Classical Topology and Combinatorial Group Theory*, Springer, New York, 1980.

[322] Storer, J.A., The node cost measure for embedding graphs in the planar grid, Proc. 12th ACM Symp. Comput., 1980, 201–210.

[323] Sun, X.R., On the complexity of testing the planarity of graphs by using Wu-Liu theorem (in Chinese with English abstract), *Chin. J. Comput.*, **12**:1 (1989), 33–37.

[324] Sun, X.R., Wu-Liu theorem and planarity testing of graphs, MS Thesis, Inst. Appl. Math., Academia Sinica, 1987.

[325] Tamassia, R., On embedding a graph in the grid with the minimum number of bends, *SIAM J. Comput.*, **16** (1987), 421–444.

[326] Thistlethwaite, M.B., A spanning tree expansion of the Jones polynomial, *Topology*, **26** (1987), 197–309.

[327] Thomassen, C., Planar and infinite hypohamiltonian and hypotraceable graphs, *Discrete Math.*, **14** (1976), 377–289.

[328] Titus, C.J., A theory of normal curves and some applications, *Pacific J. Math.*, **10** (1960), 1083–1096.

[329] Treybig, L.B., A characterization of the double point structure of the projection of a polygonal knot in regular position, *Trans. AMS*, **130** (1968), 223–247.

[330] Trotter, H.F., Noninvertible knots, *Topology*, **2** (1964), 275–280.

[331] Trotter, W.T. and J.I. Moore Jr., The dimension of planar posets, *J. Comb. Theory*, **B22** (1977), 54–67.

[332] Tutte, W.T., On the non-biplanar character of the complete 9-graph, *Canad. Math. Bull.*, **6** (1963), 319–330.

[333] Tutte, W.T., The thickness of a graph , *Indag. Math.*, **25** (1963), 567–577.

[334] Tutte, W.T., Bridges and Hamiltonian circuits in planar graphs, *Aequa. Math.*, **15** (1977), 1–33.

[335] Tutte, W.T., A class of Abelian groups, *Can. J. Math.*, **8** (1956), 13–28.

[336] Tutte, W.T., The subgraph problem, *Ann. Discrete Math.*, **3** (1978), 289–295.

[337] Tutte, W.T., On the four color conjecture, *Proc. London Math. Soc.*, **50** (1948), 137–149.

[338] Tutte, W.T., A theorem on planar graphs, *Trans. AMS*, **82** (1956), 99–116.

[339] Tutte, W.T., How to draw a graph, *Proc. London Math. Soc.* Ser.3, **13** (1963), 743–768.

[340] Tutte, W.T., A homotopy theorem for matroids I, *Trans. AMS*, **88** (1958), 144–160; II, ibid. **88** (1958), 161–174.

[341] Tutte, W.T., Matroids and graphs, *Trans. AMS*, **90** (1959), 527–552.

[342] Tutte, W.T., An algorithm for determining whether a given binary matroid is graphic, *Proc. AMS*, **11** (1960), 905–917.

[343] Tutte, W.T., A contribution to the theory of chromatic polynomials, *Can. J. Math.*, **6** (1954), 80–91.

[344] Tutte, W.T., On the problem of decomposing a graph into n connected factors, *J. London Math. Soc.*, **36** (1961), 221–230.

[345] Tutte, W.T., Convex representations of graphs, *Proc. London Math. Soc.*, **10** (1960), 304–320.

[346] Tutte, W.T., Toward a theory of crossing numbers, *J. Comb. Theory*, **8** (1970), 45–53.

[347] Tutte, W.T., *Connectivity in Graphs*, Oxford University Press, 1966.

[348] Tutte, W.T., *Introduction to the Theory of Matroids*, Amer. Elsevier, 1971.

[349] Tutte, W.T., *Graph Theory*, Encycl. Math. **21**, Addison- Wesley, 1984.

[350] Tutte, W.T., *Graph Theory as I Have Known It*, Oxford, 1998.

[351] Tutte, W.T. and H. Whitney, Kempe chains and four color problem, *Utilitas Math.*, **2** (1972), 241–281.
[352] Valiant, L.G., Universality consideration in VLSI circuits, *IEEE Trans. Comput.* C-**30** (1981), 135–140.
[353] Vanlier, M.C. and R.H. Otten, On the mathematical formulation of the wiring problem, *Int. J. Circuit Theory Appl.*, **1** (1973), 137–147.
[354] Vince, A. and C.H.C. Little, Discrete Jordan curve theorem, *J. Comb. Theory*, **B47** (1989), 251–261.
[355] Wan, L.X. and Y.P. Liu, The orientable genus distribution of a new type of graphs (in Chinese with English abstract), *J. Beijing Jiaotong Univ.*, **26** (2005), 65-68.
[356] Wan, L.X. and Y.P. Liu, Orientable embedding distributions by genus for certain type of non-planar graphs, *Ars Combin.*, **79** (2006), 97–105.
[357] Wan, L.X. and Y.P. Liu, Orientable embedding distributions of certain types of graphs, *J. Comb. Theory*, **B98** (2008), 19–32.
[358] Wan, L.X. and Y.P. Liu, On the embedding distribution of ladders and crosses, *Appl. Math. Letters*, **21** (2009), 738–742.
[359] Wan, L.X. and Y.P. Liu, Orientable embedding distributions by genus for certain type of non-planar graphs, *Ars Combin.*, **94** (2010), 201–210.
[360] Wan, L.X., K.Q. Feng, Y.P. Liu and D.J. Wang, Genus distribution of ladder type and cross type graphs, *Sci. China*, **A52** (2009), 8: 1760–1768.
[361] Wan, L.X., H.J. Lai and Y.P. Liu, Genus distribution for a graph, *Int. J. Math. Combin.*, **4** (2009), 99–109.
[362] Wang, H.Y. and Y.P. Liu, The maximum genus of a type of 2-connected graphs with diameter 3 (in Chinese), *J. Xinyang Normal College* (NS), **17** (2004), 381–384.
[363] Wang, T. and Y.P. Liu, Implements of some new algorithms for combinatorial maps, *OR Trans.*, **12**:2 (2008), 58–66.
[364] Wei, E.L., Equivalence of auxiliary graph and signed triple graph, *J. Graduate School CAS* (in Chines with English abstract), **21** (2004), 451–453.
[365] Wei, E.L. and Y.P. Liu, Strong maximum genus of a type of graphs, *J. Northern Jiaotong Univ.*, **26** (2002), 19–21.
[366] Wei, E.L. and Y.P. Liu, Weak embedding of planar graphs, *J. Appl. Math. Comput.*, **21** (2006), 175–187.
[367] Wei, E.L. and Y.P. Liu, A note of interpolation theorem, Proc. 9th Bellam Continuum: International Workshop on Uncertain System and Soft Computing, Beijing, China, July 24–27, 2002, 318–322.
[368] Wei, E.L. and Y.P. Liu, Fewer face strong embeddings of complete bipartite graphs (in Chinese with English abstract), *J. Qufu Normal Univ.*, **29** (2003), 1–5.
[369] Wei, E.L. and Y.P. Liu, Planarity of 3,4-jump graphs, *Northeastern Math. J.*, **20** (2004), 383–395.
[370] Wei, E.L. and Y.P. Liu, Planarity of iterated jump graphs, *Northeastern Math. J.*, **21** (2005), 9–17.
[371] Wei, E.L., Y.P. Liu and Z.X. Li, The minimal genus of circular graph $C(n, m)$, *OR Trans.*, **14** (2010), 3: 11–18.
[372] Weinberg, L., On the maximum order of the automorphism group of a planar triply connected graphs, *J. SIAM Appl. Math.* **14** (1966), 729–738.
[373] Weinberg, L., A simple and efficient algorithm for determining isomorphism of planar triply connected graphs, *IEEE Trans. Circuit Theory*, CT-**13** (1966), 142–148.
[374] Weinberg, L., Plane representations and codes for planar graphs, Proc. 3rd Ann. Allerton Conf. Circuit Syst., 1965, 733–744.
[375] Weinberg, L., Algorithms for determining automorphism groups for planar graphs, Proc. 3rd Ann. Allerton Conf. Circuit Syst., 1965, 913–929.
[376] Weinberg, L., Additional simple codes for planar graphs, Proc. 4-th Allerton Conf. Circuit Syst., 1966.

[377] Weinberg, L., Alternative algorithms for determining isomorphism, automorphisms and automorphism groups of planar graphs, Proc. 4-th Allerton Conf. Cir. Syst., Univ. IL., 1966.

[378] Weinberg, L., Two new characterizations of planar graphs, Proc. 5-th Allerton Conf. Cir. Syst.,1967, Univ. IL., 1967.

[379] Weinberg, L., Planar graphs and matroids, Lect. Notes in Math. **303**, Springer, 1972, 313–329.

[380] Weinberg, L. and P. Slepian, Series- parallel networks, *IRE Trans. PGCT* CT-**4**:3 (1957).

[381] Welsh, D.J.A., *Matroid Theory*, Academic Press, 1976.

[382] White, A.T., *Graphs, Groups and Surfaces*, North-Holland, 1973.

[383] White, N.L., *Theory of Matroids* (ed.), Cambridge University Press, 1986.

[384] White, N.L., *Combinatorial Geometries* (ed.), Cambridge University Press, 1987.

[385] White, N.L., *Matroid Applications* (ed.), Cambridge University Press, 1992.

[386] Whitney, H., A theorem on graphs, *Ann. Math.*, **32** (1931), 378–390.

[387] Whitney, H., Congruent graphs and the connectivity of graphs, *Am. J. Math.*, **54** (1932), 150–168.

[388] Whitney, H., A set of topological invariants for graphs, *Am. J. Math.*, **55** (1933), 231–135.

[389] Whitney, H., On the classification of graphs, *Am. J. Math.*, **55** (1933), 236–244.

[390] Whitney, H., The coloring of graphs, *Ann. Math.*, **33** (1932), 688–718.

[391] Whitney, H., Non- separable and planar graphs, *Trans. AMS*, **34** (1932), 339–162.

[392] Whitney, H., Planar graphs, *Fund. Math.*, **21** (1933), 73–84.

[393] Whitney, H., 2- isomorphic graphs, *Am. J. Math.*, **55** (1933), 245–254.

[394] Whitney, H., On regular closed curves in the plane, *Compositio Math.*, **4** (1937), 276–284.

[395] Whitney, H., On the abstract properties of linear dependence, *Am. J. Math.*, **57** (1935), 509–533.

[396] Wilde, P.J., Matroids with given restrictions and contractions, *J. Comb. Theory*, **B22** (1977), 122–130.

[397] Willamson, S.G., Embedding graphs in the plane-algorithm aspects, *Ann. Discrete Math.*, **6** (1980), 349–384.

[398] Williams, N.H., *Combinatorial Set Theory*, North-Holland, 1977.

[399] Wing, O., On drawing a planar graph, *IEEE Trans. Circuit Theory*, **13** (1966), 112–114.

[400] Witten, E., Quantum field theory and the Jones polynomial, *Comm. Math. Phys.*, **121** (1989), 351–399.

[401] Wong, D.F., Floorplan design of VLSI circuits, *Algorithmica*, **4** (1989), 263–291.

[402] Wu, W.T., The realization problem of complexes in the Euclidean space (in Chinese), *Acta Math. Sinica*, **5** (1955), 505–552.

[403] Wu, W.T., Planar embedding of linear graphs (in Chinese), *KEXUE TONGBAO*, **19**:2 (1974), 226–228.

[404] Wu, W.T., *Rational Homotopy Type*, Lect. Notes in Math. 1246, Springer, 1987.

[405] Wu, W.T., Mathematical problems in the design of integrate circuits (in Chinese), *Math. Theory Prac.* No.1 (1973), 20–40.

[406] Wu, W.T., On the planar embedding of linear graphs I, *J. Syst. Sci. Math.*, **5** (1985), 290–320; II, ibid. **6** (1986), 23–35.

[407] Xu, W.X., An efficient algorithm for planarity testing based on Wu-Liu's criterion, Proc. 1-st China-USA Conf. Graph Theory and its Applications, *Ann. N. Y. Acad. Sci.*, **576** (1989), 641–652.

[408] Xuong, N.H., How to determine the maximum genus of a graph, *J. Comb. Theory*, **B26** (1979), 217–225.

[409] Yang, Y. and Y.P. Liu, On three types of embedding of pseudowheels on the projective plane, *Utilitas Math.*, **78** (2009), 267–278.

[410] Yang, Y. and Y.P. Liu, Classification of (p, q, n) dipoles on nonorientable surfaces, *The Electron. J. Combin.*, **17** (2010), 1:#N12, 1–6.

[411] Yang, Y. and Y.P. Liu, Number of embeddings of circular and Mobius ladders on surfaces, *Sci. China, Math.*, **53** (2010), 5: 1393–1405.

[412] Yang, Y. and Y.P. Liu, Flexibility of circular graphs $C(2n, 2)$ on the projective plane, *Ars Combin.*, **101** (2011), 75–83.

[413] Yang, Y. and Y.P. Liu, Number of embeddings of wheel graphs on surfaces of small genus, *Ars Combin.*, **101** (2011), 225–249.

[414] Yang, Y. and Y.P. Liu, Embedding of circulant graphs and generalized Petersen graphs on projective plane, *Front Math. China*, **10** (2015), 1: 209–220.

[415] Yannakakis, M., Embedding planar graphs in four pages, *J. Comput. Syst. Sci.*, **38** (1989), 36–67.

[416] Yao, X., M. Yamada and C.L. Liu, A new approach to the pin assignment problem, *IEEE Trans. CAD*, **8** (1989), 999–1006.

[417] Yoshida, K. and T. Nakagawa, Topological layout design of monolithic IC in CAD, in Digest ISSCC Philadelphia, PA, 1969, 136–137.

[418] Youngs, J.W.T., Minimal embeddings and the genus of graphs, *J. Math. Mech.*, **12** (1963), 353–315.

[419] Zeng, J.C. and Y.P. Liu, A genus inequality of the union graphs, *Ars Combin.*, **92** (2009), 445–452.

[420] Zeng, J.C. and Y.P. Liu, Genus distributions for double pearl-ladder graphs, *Ars Combin.*, **98** (2011), 511–520.

[421] Zeng, J.C. and Y.P. Liu, Genus polynomial for similar benzene structure graphs, *Adv. Appl. Discrete Math.*, **11** (2013), 121–133.

[422] Zeng, J.C., Y.P. Liu and R.X. Hao, Counting orientable embeddings by genus for a type of 3-regular graphs, *Graphs and Combin.*, **28**:1 (2012), 133–142. Published online on (February 18, 2011), 10 pages.

[423] Zhao, X.M. and Y.P. Liu, Genus polynomial for tree-like graphs (in Chinese with English abstract), *J. Northern Jiaotong Univ.*, **28** (2004), 7–11.

[424] Zhao, X.M. and Y.P. Liu, Unimodality of genus distribution for tree-graphs (in Chinese with English abstract), *J. Shanxi Univ.* (NS), **29** (2006), 242–244.

[425] Zhao, X.M. and Y.P. Liu, Genus distribution of circle-like graphs (in Chinese), *Acta Math. Sci.*, **C28** (2008), 4:768–778.

[426] Zhu, Z.L. and Y.P. Liu, Genus distribution of two types of graphs, *J. Shenyang Normal Univ.* (NS), **24** (2006), 1–5.

[427] Zykov, A.A., *Foundations of Graph Theory* (Russian), Nauka, 1987.

Subject Index

Author Index

Bei Fragen zur Produktsicherheit wenden Sie sich bitte an:
If you have any questions regarding product safety,
please contact:

Walter de Gruyter GmbH
Genthiner Straße 13
10785 Berlin
productsafety@degruyterbrill.com